PHYSICS OF COAL AND MINING PROCESSES

Anatoly D. Alexeev

CRC Press
Taylor & Francis Group
Boca Raton   London   New York

CRC Press is an imprint of the
Taylor & Francis Group, an **informa** business

CRC Press
Taylor & Francis Group
6000 Broken Sound Parkway NW, Suite 300
Boca Raton, FL 33487-2742

First issued in paperback 2017

ISBN 13: 978-1-4398-7634-3 (hbk)
ISBN 13: 978-1-138-07494-1 (pbk)

Library of Congress Cataloging-in-Publication Data

Alexeev, Anatoly D.
　　Physics of coal and mining processes / Anatoly D. Alexeev.
　　　　p. cm.
　　Includes bibliographical references and index.
　　ISBN 978-1-4398-7634-3 (hardcover : alk. paper)
　　1. Coal mines and mining--Ukraine. 2. Mining engineering--Ukraine. I. Title.

TN808.U4A44 2012
622'.33409477--dc23   2011038367

Visit the Taylor & Francis Web site at
http://www.taylorandfrancis.com

and the CRC Press Web site at
http://www.crcpress.com

Contents

Foreword

Coal represents the wealth of Ukraine and ensures the nation's energy independence to a considerable extent. Therefore, increasing coal production is a very important strategic objective for our future. Unfortunately, the coal industry is often connected with loss of human lives. The National Academy of Sciences of Ukraine implements regular basic studies in the physics of coal and physics of mining processes to enhance our understanding of the nature of methane emissions in mines and provide the safest mining conditions.

This book is an exhaustive monograph summarizing physical approaches to coal beds (including deep beds) and coals as complex physical media based on fundamental experimental and theoretical studies conducted at the Institute for Physics of Mining Processes of the National Academy of Sciences of Ukraine (Donetsk) headed by Anatoly D. Alexeev, the author and a corresponding member of the Academy. These studies focused on analyses of coal structures using radiophysical methods such as nuclear magnetic and electronic paramagnetic resonances; discovery of phase states of methane in coals and methane solubility in less matured coals; observations of methane mass transfer in coal–methane systems; and many other relevant topics. Results of these studies obtained using modern equipment made it possible to implement novel methods of coal bed behavior prediction and develop measures to prevent highly explosive coal and gas outbursts, with the primary goal of increasing worker safety in Ukraine mines.

I am sure that the publication of *Physics of Coal and Mining Processes* will be an important event for all specialists interested in coal mining and will introduce them to the achievements of Ukrainian scientists in this important field.

B. E. Paton
President of National Academy of Sciences of Ukraine

Preface

The physics of mining processes is a relatively new branch of science that evolved to solve problems arising during mineral products recovery (this volume concentrates on coal), particularly safety issues such as prevention of rock failures, coal and gas outbursts, and methane explosions. Around the globe, on average, four coal miners die for each million tons of coal recovered. Substantially increasing coal production and improving the safety of mining work are impossible without further development of the physics of mining processes. The main mission of the Institute for Physics of Mining Processes of the National Academy of Sciences of Ukraine established in 2003 is to serve the needs of the Ukrainian major coal mining industry.

This volume discusses research work carried out at the institute and information gathered from international observations. It includes chapters relating to massif mechanics. I found it reasonable to include a chapter addressing the most important safety issues such as control of coal and gas outbursts and estimates for accumulation times of explosive methane concentrations in different phase states (adsorbates on surfaces, solid solutions in coal, etc.). Solutions to the specified problems may be found in research discoveries in coal physics and the physics of mining processes.

Most studies on the physics of mining processes are dedicated to specific critical problems. This study is the first to present a unifying methodology for addressing various problems arising in the mining industry. Concentrating on the key issue of each particular problem, I discuss the adopted assumptions and approximations and apply the fundamental laws of physics. The reader will easily recognize the unity of the approaches used. In each case, the validity criterion of the adopted model is the solution of a particular industrial problem.

Because of the vital importance of mineral products recovery for modern industry, the physics of coal and mining processes has been rapidly developing around the world for decades. I attempt to cover the main discoveries and achievements to the fullest extent possible because I have not found a general study of the subject in the research literature. The reader will find many original results along with discussions and development of various prior research studies. I found it reasonable to include certain material from my earlier publications. For example, Chapter 5 is devoted to solutions of particular industrial problems and implementation of scientific results.

Anatoly D. Alexeev
Donetsk, Ukraine

Acknowledgments

I gratefully acknowledge all the contributors who participated in writing this book. They, like me, endeavored to contribute their best to the theory and applications of the physics of coal and mining processes. Most of them are researchers at the Institute for Physics of Mining Processes of the National Academy of Sciences of Ukraine. They are Professor A. Y. Granovskii (Introduction, Sections 1.1 and 1.2); Candidate V. V. Slusarev (Section 1.3); Dr. E. V. Ulyanova (Section 1.4 and Chapter 5); Candidate A. K. Kirillov (Section 1.5); Professor E. P. Feldman and Candidate V. A. Vasilkovskii (Chapter 2); Dr. V. N. Revva (Chapter 4); Dr. G. P. Starikov (Sections 4.4 and 4.5); and Professor A. E. Filippov (Section 5.4).

I am also thankful to readers for their patience in covering this complex material and for their comments and advice. I hope this volume will be useful for scientists, technicians, and engineers, and also for students, especially those pursuing mining careers.

Author

Anatoly Dmitriyevich Alexeev is the leading earth scientist in Ukraine. Dr. Alexeev has combined a career in physics, chemistry, and engineering with teaching, research, and writing more than 300 publications, 14 monographs, and 50 patents. He developed the theory of gas absorption and emission in porous media and an applied approach to extreme states of rocks and rock massifs.

Dr. Alexeev's areas of interest in basic and applied research include the physics of rocks and mining processes, radiophysics, extreme states of solids, and physico-chemical mechanics. He studied the effects of surficants on stress states of rocks and implemented surfagents to prevent outbursts in coal mines. He also applied nuclear magnetic resonance (NMR) to studies of methane–coal systems and developed new methods and technologies for predicting mine outbursts.

Dr. Alexeev is a corresponding member of the National Academy of Sciences of Ukraine, a member of the Academy of Mining Sciences of Ukraine, and the Russian Academy of Natural Sciences. He graduated from Donetsk Polytechnical Institute in 1963 and joined the institute as an assistant professor. He was named head of the laboratory in 1967. In 1992, Dr. Alexeev became the division director of the Donetsk Physical and Technical Institute of the National Academy of Sciences of Ukraine. He was named the director of the National Academy's Institute for Physics of Mining Processes in 2001. In recognition of his accomplishments, Dr. Alexeev was awarded the State Prize of Ukraine (1996), the A.A. Skotchinsky prize (1990), and the I. Puluy prize (2001). He was designated an Honored Miner of Ukraine (1996) and received several President's medals in all three classes for "Valiant Labour," "Miner's Glory," and "Miner's Valour."

1

Coal Structure

Fossil coal is a "living witness" of prehistoric times (from the Carboniferous to Jurassic), when the Earth was covered with impenetrable jungles, and equisetums, lycopodiums, and ferns thrived. Due to seismic cataclysms, the jungles were first buried under water and later under various rock layers. That started the transformation of wood material into coal—a process known as metamorphism. Meanwhile, the world's oceans witnessed a similar process—the transformation of plankton into petroleum.

Petroleum is a natural accumulation of hydrocarbons. Its deposits occur in sedimentary rocks of various ages. Crude oil is an insoluble, dark brown, oily liquid with a density of 0.75 to 0.95 g/cm^3. It contains 83 to 87% carbon, 11 to 14% hydrogen, and minor amounts of nitrogen, oxygen, and sulfur. Cyclic hydrocarbons are dominant in some oil grades and chain polymers are dominant in others.

The chemistry of petroleum formation has not yet been fully elucidated, but most likely the source material was provided by life remains in shallow sea basins. Rapid development of primitive algae resulted in a rapid evolution of animal life. In view of the enormous reproductive rapidity of protozoa under favorable conditions, it is not surprising that great amounts of their remains accumulated at the bottoms of bodies of water. In the absence of air, the remains slowly decomposed in near-bottom waters and were gradually covered with clay and sand, steadily transforming into petroleum over millions of years.

Two competitive energy carriers emerged: coal and petroleum. They differ in form (one is solid, the other liquid). This difference is certainly important. Petroleum is easily transported via pipelines; coal must be repeatedly trans-shipped from origin to user. Furthermore, petroleum recovery is far less complicated. It requires only wells instead of mines. However, the crucial difference between these energy carriers that determines their recovery and use is caloric efficiency (approximately 8 kcal/g for coal and 11 kcal/g for petroleum). These values correspond to the transformation levels of the electron shells (0.4 and 0.6 eV/atom, respectively).

The second important feature is the amount of the world's reserves of these traditional energy resources. The estimates indicate more than 1.5×10^9 tons for coal; $0.17 \times 10^9 \, m^3$ for petroleum, and $1.72 \times 10^{11} \, m^3$ for gas. With respect to the amount and total energy content, petroleum ranks second to coal by an order of magnitude. As energy demands increase by approximately 3% per year, energy consumption in 2025 will amount to 2.28×10^7 tons of equivalent fuel. If industry as the main energy consumer continues to grow at this rate, the world resources of coal, petroleum, and gas will be exhausted in approximately 100 years. Note, however, that this holds for explored reserves (and also recoverable ones with regard to coal). What geologists may discover in the future remains unknown although estimates are still in favor of coal. Gas energy in the Ukraine amounts to 41% (21% world average). Additional sources are coal and petroleum (19% each), and uranium (17%); other sources constitute 4%. Manufacture of a commercial product with a cost of $1 requires 0.9 kg of equivalent fuel or 2.5 times the world average.

The specified values of 8 to 11 kcal/g for coal and petroleum, respectively, are minor in comparison to the caloric efficiency of a uranium fission reaction that produces as much as 10^4 times the energy. Thus, many countries prefer nuclear fuel to coal and petroleum as sources of chemical raw materials. Russia, for example, satisfies 16% of its needs through nuclear energy because its nationwide cost per equivalent unit is 13 to 15% less than the average. Nuclear fuel is virtually the most feasible energy source to replace limited and nonrenewable reserves of petroleum and coal. This problem still remains open. After sufficient thermonuclear energy sources are developed, coal and petroleum as energy carriers will be regarded as insignificant, and pass to the domain of chemists. Then perhaps the reversion to Mendeleev's idea of underground coal gasification will become possible.

Comparing coal with petroleum, we have passed over one general question: why do coal and petroleum liberate heat during combustion? More precisely, why exactly do they burn? The answer is simple. The C–C and C–H bonds that characterize hydrocarbons possess less energy (81 kcal/g-atom) than the C=O bond (85 kcal/g-atom) after combustion. This benefit of 4 kcal/g-atom does ensure caloric efficiency. Furthermore, combustion involves surmounting a certain energy barrier—somewhat low for petroleum and higher for coal. Under natural conditions, spontaneous ignition may occur due to the sulfides in beds that decrease this barrier.

What predominant problems does any physicist encounter when beginning studies of coal? The first step is to establish the general physical properties and create a certain model of its structure. Then one may proceed to analyze its behavior during interactions with various surrounding materials (gases, liquids, rocks, etc.) in different stress fields.

1.1 Genesis of Coal Substance

During carbonization, the content of the basic coal-forming elements in the organic matter of coals changes considerably.* Coal properties are predeter-

---

* A very complete systematic survey of coal typology, physicochemical properties, and structure can be found in a monograph by D.W. Van Krevelen [22]. We also employed studies of coal constitution to solve physical problems encountered in coal recovery.

mined by the manner of accumulation, surrounding conditions of sedimentation, and plant material required to form coal from peat. The first stage of carbonization is biochemical. Peat is the less metamorphosed product of the transformed plant bodies. It is a heterogeneous, porous (up to 70%) mixture consisting of decayed parts of dry land and marsh land vegetation that underwent different stages of transformation. It contains 55 to 65% carbon, 5 to 7% hydrogen, 30 to 40% oxygen, and 1 or 2% nitrogen and it approximates the composition of timber (50% carbon, 6% hydrogen, 44% oxygen, and 0.1 to 0.5% nitrogen on average). The composition of the most recent deposits of peat closely resembles the composition of timber. Under close examination, peat may reveal bands characteristic of brown and even black coals.

Because of limited oxygen presence, diagenesis occurs on peat surfaces. Decomposition of lignin provokes synthesis of humic acids that undergo carbonization and generate vitrinite. During the biochemical stage of carbon formation, decayed vegetation initially liberates colloids (heterogenic unstructured systems). Coal coagulation structures are not strong originally but their strength increases as water leaves the solidifying colloid and the coal substance consolidates.

Formation of the condensed coal structure is complete at the point of transition from the biochemical to the geochemical stage. Because of the loss of functional groups and simultaneous compaction of the aromatic ring in the crystallization lattice (both events typical for coal transformation during carbonization), black coal, in contrast to peat or brown coal, is harder, always black, less moist, and more compact. The formation of a condensed coal structure arises from increasing pressure and temperature and dehydration and compaction of colloid substances in the absence of oxygen.

A variety of geological and genetic factors led to formation of coals with different physical, chemical, and technological properties. Within a single coal basin, it is possible to categorize coals according to their maturity. On average, black coals of the Donets Basin constitute 75 to 80% vitrinite groups, 4 to 12% fusainite groups with large amounts of condensed aromatic fragments, and 6 to 12% liptinite groups, the latter having the longest aliphatic chains and providing aliphatic structural parts for the coal material.

Carbonization results from the influence of certain physical factors affecting the entire coal-bearing stratum and embraces all the beds. At the stage of diagenesis, an average of vitrinite reflection (R_0) amounts to 0.4%. During catagenesis, its value dramatically increases, reaching 2.6 to 5.6% at metagenesis (of anthracites). It follows that petrographic composition and metamorphism degree are two independent variables that determine coal properties. As the extent of carbonization rises, the ratio of chemical elements alters and the carbon contents in the aromatic ring networks and quantity of aromatic rings increase. The lattices of their clusters become more rigid and the coal eventually acquires the properties of a solid. The microhardness of vitrinite in brown coal-anthracite (B–A) series, for instance, becomes five times as great. This change is most apparent at lean coal-anthracite (T–A) stage [1].

Coal is a product of decayed organic materials from earlier geological ages. Carbonization progress depends on external conditions and the duration of their influence and results in multiple metamorphism products such as coals (brown, black, anthracite), sapropelites, pyroschists, and petroleum. Anthracites and graphites are considered the most perfect structures among coals and represent completion of the metamorphic line. Representative anthracites reveal properties unusual for bituminous coals (enhanced electrical conduction, for instance). It is believed that under certain conditions anthracite can pass through the intermediate stages to transform into graphite. In this case graphite concludes the succession from peat to lignite to bituminous coal to anthracite.

In its free state, carbon exists in three modifications (1) cubic lattice (diamond) (2) hexagonal graphite, and (3) hexagonal lonsdaleite extracted from meteorites or synthesized. A black diamond with graphite lattice was found first on the Moon, and later on Earth. Amorphous carbon does not occur in natural conditions. Graphite is composed of linked carbon networks (layered lattices) that account for the softness and flakiness observed while drawing with a graphite pencil. Each atom of carbon is bonded to four others. Distances to three of them are similar (0.142 nm), whereas the fourth atom is more distant (0.335 nm) and contributes a weaker bond. This feature manifests itself in the easy splitting of graphite into separate layers.

Attracting force between the layers is small, enabling the layers to glide over one another. Graphite peculiarities may result from a short-term (epoch—no more than 1,000 years) but powerful impact of an intrusive thermal field (up to 1,000°C) on already formed coal materials. In response to such an effect, graphite and vitrinite assume fine-grained structures, the spacing between carbon lattice planes decreases, carbon content increases, and ordering of the whole structure is enhanced. Graphite preserves the cellular structures of coal-forming plants [2].

In nature, solids have well defined macromolecular structures that may be regarded as molecules whose atoms are covalently bonded.

Comparatively few substances in a solid state can generate atom lattices. One example of such a lattice is the diamond—one of the free forms of carbon. Each atom of carbon forms four equal bonds with four adjacent carbon atoms to create an enormous, very durable, and consolidated macromolecular structure at the expense of common electron couples by strong covalent bonds that enable high hardness and high boiling and melting temperatures. Diamonds naturally originated from a combination of hard-to-reproduce conditions that liberated free carbon. While natural diamonds formed along with the Earth's crust by crystallization of carbon dissolved in melted magma, artificial diamonds are made under specific conditions within kimberlite pipe, and subjected to high pressures and temperatures for a long period; carbonization terminates at the stage of graphite. Contemporary physics laboratories succeeded in synthesizing fullerenes and nanotubes not yet found in nature.

Despite the experimental evidence on artificial carbonization, the significance of geological time must be regarded as a long-term impact on any factor (temperature, for instance) that may produce different results from organic material transformation under different timing conditions.

1.2 Elemental Composition of Coals

It may seem unusual to open a study of the physics of mining processes with a chapter related to the chemical composition of coals. This paradox, however, is absolutely natural because all the macroscopic properties of coal are dependent on microscopic ones that are in term conditioned by atomic properties; thus the boundary between physics and chemistry disappears.

The basic coal components are carbon, hydrogen, and oxygen. Carbonization is associated with regular and specific alterations of the amounts of these elements. The many variations among carbon compounds account for its peculiar properties, the most significant of which is ability to form strong atomic bonds. Due to the strength of bonding, molecules containing chains of carbon atoms are adequately stable under ordinary conditions, whereas molecules with similar chains of other elements are not nearly so strong.

The energy of the C–C bond (83 kcal/mole) is sufficient to provide stability for chains of virtually any length. x-ray examinations have shown that the carbon atoms in such compounds with linear structures are arranged not along a line, but along a zigzag. This results from carbon's four valences directed in a certain way toward one another; their relative position corresponds to lines drawn from the center to vertices of a tetrahedron. Carbon in organic compounds may exist in an aromatic state. This is defined as a special type of atom bonding into plane cycles (closed chains) in which all the atoms in a cycle participate in formation of a uniform electron system of bonding. Benzene is the simplest aromatic compound that forms a structural unit that determines the properties of all aromatic compounds.

In studies of coals via various physical and chemical techniques, it was found that an average of condensed aromatic cycles per structural unit ranges (subject to metamorphic grade) from 4 to 10 cycles with 80% of total carbon in cyclic components [3]. The high molecular structures of coals that include aromatic, hydroaromatic, heterocyclic and aliphatic components, and functional groups, are generally recognized. However, no uniform view indicates exactly how these fragments are connected, how the structural units are built, and how they communicate with the spatial system (see Section 1.3).

Organic matter of coal may be generally described as a mixture of condensed molecules whose C=C double bonds are represented only by aromatic bonds, and some carbon, hydrogen, oxygen, nitrogen, sulfur, and other heteroatoms form side aliphatic groups at the condensed aromatic nuclei. The oxygen-containing components in the aliphatic segments of the organic coal substance allow connections of the organic and non-organic contents. Various types of interaction are possible, for example, adsorption (electrostatic adsorption, hydrogen bonding, chemisorption), complexing (coordinate bonding), and formation of metal–organic compounds via the inclusion of heteroatoms or homopolar bonding. Essentially, the relative carbon content in the metamorphic line increases according to the pattern:

$$C–OH \rightarrow CH_3 \rightarrow CH_2 \rightarrow CH_{arom}.$$

For coals with V^{daf} = 28%, C–OH = 1.8%; CH_3 = 7.6%; CH_2 =12.0%; CH_{arom} = 20.8%; $C_{arom} - C_{arom}$ = 57.8%.

With increasing grades of metamorphism, concentrations of oxygen-containing groups decline, structural fragments become more homogeneous and ordered, and transformation of various components of the coal substance (those having undeveloped bonds and considerable oxygen and aliphatic components) proceeds into a three-dimensional homogeneous structure with a large poly-condensed system containing minimal amounts of hydrogen and heteroatoms.

As the degree of carbonization increases, reactions remove oxygen and hydrogen from coal as CO, CO_2, CH_4, and H_2O. At the middle stage of metamorphism, loss of hydrogen in such coals increases with rising oxygen presence, helping enhance the flexibility of macromolecules in the coal substance on the one hand and forming strong bonds between the fragments of the structure at the expense of the oxygen-containing functional groups on the other hand. During the formation of anthracites, with pressure as the predominant metamorphic factor, nitrogen, sulfur, and other heteroatoms undergo redistribution from the condensed systems of the molecular structure to the periphery. Classification of coals is traditionally based on grades of metamorphism (see Table 1.1).

Oxidation leads to self-heating and spontaneous combustion of coals. Oxidation initially involves aliphatic hydrogen and carbon; aromatic ones

TABLE 1.1

Rank Classification of Coals

Description	V	G	F	C	LB	L	A
Yield of volatiles *V* (%)	37	35–37	28–35	18–28	14–22	8–18	8
Carbon content (%)	78	81–85	85–88	88–89	90	91–94	94–98
Hydrogen content (%)	5.3	5.2	5.0	4.7	4.3	3.9	3.2

Source: Kizilshtein L.Y. 2006. *Geochemistry and Thermochemistry of Coals.* Rostov University Press, Rostov on Don.
V = volatile. *G* = gaseous. *F* = fat. *C* = coking. *LB* = lean baking. *L* = lean. *A* = anthracite.

are oxidized thereafter. Coals susceptible to spontaneous combustion contain hydrogen and oxygen amounting approximately to 4%; non-susceptible coals contain 4.8 to 5.2% and 7.3 to 7.8%, respectively. The former have average yields of volatiles of 10 to 15%, whereas the latter yields are 7.2 to 7.6%.

Combustion reactions allow conversion of one substance into another but also involve non-chemical (physical) processes such as diffusion, convection, heat migration, etc. One of the most dangerous phenomena of coal-mining—spontaneous combustion of fossil coals—is predetermined by the chemical activity and internal structure of the specific coal substance [4].

Oxidation rates of natural coals increase with rising oxygen, hydrogen, ferrous iron, and pyrite and yields of volatiles, and decrease with rising carbon quantity and metamorphic grade. During the transformation of peat into coal, carboxyl and hydroxyl group content declines, leading to a change of hygroscopic ability.

Moisture content, depending on origin and carbonization grade, can reach 30 to 45% for peats and 5 to 25% for brown coal. Moisture content in Donbass coals is up to 9% for gaseous and fat (GF), 5% for gaseous (G), 3.5% for fat (F), 2.5% for coking (C), 1.5% for lean baking (LB), 1% for lean (L), 4% for anthracite (A) coals, and 4 to 5% for bituminous shales. The amount of hygroscopic moisture in a coal indicates the quantities of polar fragments in a molecule. Water amount in coal is of great practical importance because it may enhance or impede oxidation and self-heating, change friability, and cause freezing that reduces caloric value.

Epshtein et al. [5] found that the coals of a similar metamorphic grade have different elemental compositions and coking capacities. They based this conclusion on the difference between the amounts of total and pyrite sulfur that occur in all coals regardless of their origin and metamorphic grade. Total sulfur in peats ranges from 0.5 to 2.5%, in brown coal from 3 to 7%, in Donbass coals from 0.5 to 9.3%, and in anthracites from 0.6 to 6.3%. Sulfur occurs in coals as an organic sulfur (an organic component of fuel), sulfides and pyrite sulfurs (including metallic sulfides and biosulfides), sulfate sulfur (as metallic sulfates), and elementary sulfur (in free state).

Most metallic sulfides form with the assistance of sulfate-reducing bacteria that catalyze reduction. This is impossible under ordinary temperatures and in the absence of ferments. Sulfide mineralization utilizes the biochemical properties of the material structure of the vegetation. Pyrite crystals are dispersed in the coal substance. More particularly they are regular octahedrons arranged in layers. Different coal beds may reveal different grades of sulfide mineralization. Oxidation of sulfur and its compounds leads to serious technological and environmental consequences. In pit refuse heaps, it causes an increase of temperature that may lead to spontaneous combustion of coal [6]. Mineral components in coals have different origins and fall into four fundamental groups [7]. Clay minerals and quartz of the first group are typical inclusions for most coals. They occur as interlayers and lenses or are evenly distributed in the organic substance, forming high-ash coal.

The second group includes minerals liberated from solutions that form peat during all the stages of transformation of sediments into solid rocks. The typical minerals of this group are sulfurous metallic compounds (pyrite, marcasite) and carbonates of calcium, magnesium, and iron (calcite, dolomite, ankerite, siderite, etc.). Minerals of the second group may be finely dispersed in the organic substance of coals. Pyrite, for example, is barely extractable via enrichment.

The third group consists of minerals brought to already-formed coal beds in solutions from enclosing rocks. The group includes alabaster, melanterite (ferrous sulfate solution), epsomite (magnesium sulfate solution), halite (common salt), and secondary sulfates of iron, copper, zinc, and quartz. Minerals of this group are confined in coal fissures and form balls and contractions. When coals are ground to a certain degree, these minerals are exposed and their efficient extraction becomes feasible.

The fourth group includes minerals as fragments of enclosing rocks, brought up with the coal while recovering it. They are various clay minerals, such as kaolinite, hydrous micas, mix-layered silicates, montmorillonite, chlorides, quartzes, micas, and feldspars. A carbonate roof allows the introduction of calcite and dolomite into coal. Minerals of the fourth group are not related to coal; they are fragments of soils and roofs that enter the coal and represent aggregates of multiple minerals.

The amounts of mineral impurities are usually characterized by an indirect factor known as ash content, estimated by combusting a coal sample of a certain weight under specified conditions according to the current standards. Ash basically consists of oxides of silicon, aluminum, iron, calcium, magnesium, and other elements. Its quantity varies considerably for different coals. At present, 98% of Ukrainian coal is recovered from underground mines. It is important to know the structures and properties of fossil coals to achieve certain industrial tasks and ensure miner safety.

1.3 Structure of Organic Mass

Since the structures of coals and their porosities are studied by various approaches using traditional methods of physics, Sections 1.3 through 1.5 are closely connected and their subdivision in this chapter is merely symbolic.

Fossil coal is a product of biochemical synthesis due to the functioning of microorganisms in decomposition of plants, and represents a complex structural system formed under virtually thermodynamic equilibrium conditions with multiple varieties of bonds both in its matrix structure and between the matrices and fluids. The distinguishing features of structural and chemical transformations of coals under natural conditions and the thermodynamic equilibrium nature of the process are high selectivity and screening of transformation courses in proceeding reactions with minimal activation energies. Under a series of chemical reactions, a continuous cluster of coal structures originates and displays the most enhanced thermodynamic stability under physical conditions of metamorphism [8]. A coal bed is a natural equilibrium system containing components such as coal, gas, water, and mineral impurities. Gases and water should not be regarded as filling or passive agents; they are components that exert considerable and even critical effects on the related reactions.

Mineral substances essentially include compounds of silicon, aluminum, iron, calcium, magnesium, sodium, potassium, and other elements. Impurities considerably influence coal characteristics, but their effect on physical and mechanical properties remains disputable. Note that coal processing properties have been well studied by chemical methods.

Water content of coals ranges from 2 to 14% and depends on grade, type of metamorphism, and bedding conditions. Water has a considerable impact on all the characteristics of fossil coals, particularly on their physical and mechanical properties [9].

Gases in coal deposits include nitrogen, methane, heavy hydrocarbons, carbon dioxide, carbon monoxide, hydrogen, sulfur dioxide, and rare gases. The principal gases are methane, carbon dioxide, and nitrogen. In the coal beds of Donbass, methane is predominant for black coals. The organic masses of coals are super-heterogeneous. Coal reveals visible banded arrangements of alternating matte and lustrous layers with multiple inclusions of irregular shape. Porosity and heterogeneity levels vary from millimetric (lithotypes) to micrometric (macerals) and even smaller scales, resulting in arbitrary, subjective choices of macromolecule sizes of organic masses. The six fundamental groups of petrographic constituents are vitrinite, semivitrinite, fusainite, liptinite, alginate, and mineral impurities. Although the total number of constituents approaches 20, only three morphologically homogeneous groups constituting most coals are considered: (1) vitrinite, the basis of lustrous coal; (2) fusainite, consisting of soft matte layers, and (3) liptinite, which includes exinite (spores), cutinite (cuticles), resinite (resin cores), and suberinite (cutinized fibers). Black coals of Donbass contain 75 to 80% average constituents of vitrinite groups, 4 to 12% fusainite, and 6 to 12% liptinite.

Coal properties are determined by both source material and by conditions of transformation during carbonization. Petrographic composition and metamorphic grade are considered independent variables that determine coal properties. As a result, structures and properties of a particular coal may be described only as statistical averages.

Understanding coal's natural structure took a long time because of considerable changes in experimental techniques. The structural models obtained via chemical experimental techniques such as an organic solvent extraction and thermal destruction indicated that the elementary compositions of coals and the chemical structures of fossil coal reaction products differed in nominal structural units of coal substance and molar mass.

The great adsorption ability of coals, their ability to expand in polar solvents and produce colloids, and their physical and optical qualities spread the concept that fossil coal had a colloid nature. That concept led to recognizing the micellar structure based on supramolecular entities (micellae) arranged as aggregates of fine macromolecules bonded by Van der Waals force. The shape of micellae was accepted as globular (spherical or resembling a pear). Differences in coal properties were attributed to degrees of nucleus and shell polymerization, construction of micellae, and their interacting forces [10]. However, the idea of micellar structures of black coal entities failed to explain the nature and mechanism of chemical processes arising from metamorphic transformation and heating of coals.

The first attempts to explore scale ordering were made by applying a new technique known as small angle x-ray scattering to organic substances (see Section 1.4). Krishimurty [11] found scattering in the areas of small angles, leading to his assumption that coal was an intermediate between true crystalline and amorphous states. Mahadevan [12] explained small angle scattering as the result of the colloid natures of diffracting entities. When investigating lignites, Brusset et al. [13] discovered bands in the areas of small angles; the dimensions of the elementary particles ranged from 0.4 to 0.6 nm. They proved the thesis that small angle scattering does not depend on atom position within scattering particles. Blayden, Gibson, and Riley [129] associated the presence of intensity maximum with the dimensions of fundamental particles and spacing between them. Scattering was characterized by both a steady increase in the area of small angles and progressive decrease in the presence of wide maximum.

Hirsch [14] obtained a scattering curve in the area of very small angles. He observed two types of scattering: (1) diffuse scattering in the area of 2.0 nm that rapidly inclined with angle narrowing and (2) discrete scattering characterizing the recurrence of the coal structure in the range of 1.5 to 4.0 nm (bands of 1.6, 2.0, and 3.4 nm). Diffuse scattering was explained by Hirsh as the result of large structural elements or distribution of pores in the solid phase.

A two-phase physical and chemical model of coal is generally accepted at present [15]. It was presented and described by Rouzaud and Oberlin [16,17] and Larsen et al. [18]. According to this model, the coal structure consists of two components—a macromolecular network and a molecular component that is a complex mixture of dissolved molecules [19]. The macromolecular network is formed of intercrossing aromatic structures that consist of turbostratic or randomly distributed layered lattices. The basic components of the macromolecular network are carbonic structures known as basic structural units (BSUs). With increasing carbonization grade, they accumulate two to four aromatic layers [20,21] and form so-called crystallites or aggregates of aromatic rings bonded to multiple diverse functional groups, aliphatic and hydroaromatic fragments, cross-linked into irregular macromolecules by various bonds.

This approach, used in models by Kasatochkin [8], Van Krevelen [22], and others is accepted by most researchers. The vitrain macromolecule model of incompletely metamorphosed coal by Fuks-Van Krevelen (Figure 1.1) is widely distributed. It consists of a central aromatic part containing 6 to 11 condensed rings bonded to one another and hydroaromatic groups with heteroatoms and side substituents surrounding the rings. Structural cells are not completely similar as they are in regular polymers. In Figure 1.1, the aromatic rings are cross-hatched and points of possible macromolecular decomposition (along ether bonds) are shown with dashed lines. Weak areas may also occur at the points of bonding aromatic rings to peripheral hydroaromatic groups.

More recent models describe the non-aromatic structure of the organic mass in greater detail. Average bituminous coal molecules proposed by Wiser, Given, Solomon, Heregy, and Lazarov are markedly similar despite differences in their skeletal structures and distributions of spatial groups. They include elongated planes interrupted by aliphatic and hydroaromatic protuberances of 0.2 to 0.5 nm. These models describe a fossil coal "molecule" as a two-dimensional layer or plane [23–28].

Shinn's research [29] on coal liquefaction led to revisions of scientific conceptions of a number of aspects of coal structure. For the purpose of creating the model, the data on compositions and reacting capabilities of bituminous coals of the United States and distribution of liquid products under one-stage and two-stage processing were collected. The coal compositions and liquid products were compared to gather data on chemical reactions and define bonding types that are destroyed under liquefaction.

A considerable contrast between the products of two-stage processing and mere coal dissolution reveals that coal contains predominantly active small molecular units. Relatively large molecular products of a single-stage thermal processing may, in fact, appear to be the products of condensing fine fragments, and not of depolymerizing source coal. The model under discussion—an example of coal behavior under liquefaction—reveals a number of ambiguities in its structure. They are associated with the nature of aliphatic coal, the abundance of active bonds, the dimensions of subunits within a cross-linked matrix, and macromolecule three-dimensional aspects.

More recent hypotheses based on C^{13} NMR and EPR data show that the organic mass consists of a rigid molecular network containing a mobile molecular component [30,31]. In their model, Larsen and Kovach [32] distinguish the molecular and macromolecular phases. The macromolecular phase forms a three-dimensional skeleton consisting of cross-bonded macromolecular fragments, whereas the molecular phase is distributed within the pores of the macromolecular phase or on its periphery. The model does not establish the nature of the connection of these phases.

On the basis of the data obtained from research on the solvent extraction of coals, Marzec et al. [33,34] concluded that the molecular and macromolecular phases are connected by electron donor and acceptor interactions. The latter lead to the existence of molecular complexes since electron donor and acceptor centers occur in both stages. The molecular network consisting of aromatic

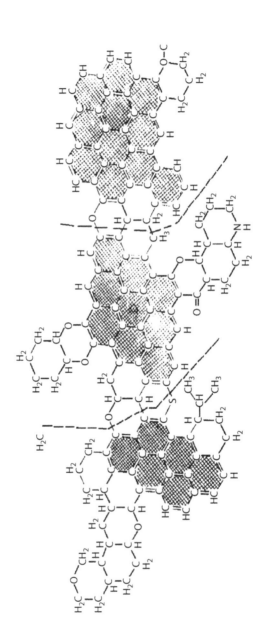

FIGURE 1.1
Fuks-Van Krevelen structural model. (Adapted from Van Krevelen D.L. 1961. *Coal: Typology, Chemistry, Physics, Constitution.* Amsterdam: Elsevier.)

fragments bonded by aliphatic and alicyclic units is understood as nonplanar. Cross links within the network form due to isolated methylene, ether, thioether groups, and large fragments [35].

The molecular phase consists of relatively small molecules (compared to "macromolecules" with cross links) that are not identical in chemical composition or physical state. More than 300 components have been identified in the extracts of highly volatile bituminous coals by high resolution mass spectroscopy. The basic components are aromatic hydrocarbons, phenols, pyridine, pyrrole, and benzene homologues.

Mass spectroscopy in ionization fields established that the volatiles liberated at room temperature consist mainly of substances with molecular masses ranging from 200 to 600 atomic units [36]. Based on the model, the extraction mechanism turns into a substitution reaction of the molecules of the molecular phase for the solvent molecules that have significant electron donor and acceptor forces. Some components of the molecular phase may not extract due to the dimensions of pore openings limited by the macromolecular network.

Duber and Wieckowski [31] interpreted the electron paramagnetic resonance (EPR) spectra of deaerated coals of different metamorphic grades in terms of the structural model of Larsen-Covach. They concluded that paramagnetic centers in the macromolecular phase give a narrow EPR line, whereas the molecular phase spins give a broad line.

Because chemical models have only two dimensions, they provide little information about the physical, mechanical, and other characteristics of fossil coals. Disregarding structure volume, the selection of macromolecule dimensions is subjective and can thus ignore such complex issues as molecular mass. It is very difficult to find solutions for gas static and gas dynamic mining and processing tasks at coal bed developments solely on the basis of chemical composition.

The most widely known model among the complete versions of the three-dimensional organic structure of coal was proposed by Kasatochkin [37]. In this model (Figure 1.2), the vitrain substances are regarded as space polymers in a glassy amorphous structure consisting of a congregation of plane hexagonal atom networks (lamellae) of cyclically polymerized polymers bridged by lateral radicals as molecular chains of linearly polymerized carbon atoms. Several lamellae or condensed carbon rings are spatially united into so-called crystallites—formations similar to graphite lattices with few parallel and disturbed layers. Such structural units have dimensions of up to tens of angstroms and are bonded to one another by alkyl chains. Free valences in the side chains may be linked to hydrogen, oxygen, or other atoms or atomic groups within the given substance.

High resolution microscopy produced a considerable effect on the development and improvement of coal structure conceptions. Transmission electron microscopy can generate precise pictures of the structural cells of the substances under investigation. Electron microscopy research in 1954

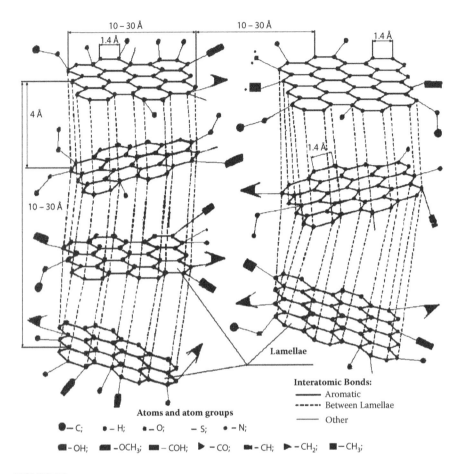

FIGURE 1.2
V. I. Kasatochkin's three-dimensional coal structural model. (Adapted from Larina I.K. 1975. *Structure and Properties of Natural Coals.* Moscow: Nedra.)

showed closed pores with dimensions of hundreds of angstroms along with aggregations into narrow chains and channels. At that time, use of electronic microscopy to investigate coal structures encountered considerable and even insurmountable procedural difficulties [38]. Research [39] using a YSM-35 electronic microscope showed that the vitrinite pores had rounded spherical shapes in GF, G, and F coals and spindle or bunch shapes in C, LB, and L coals. The pores formed evenly scattered aggregations within a homogeneous vitrinite layer, occurred layer-by-layer or at layer junctions, and congregated in groups at the points of forming microfissures. By magnifying the view of a scanning electron microscope, pore systems of even smaller dimensions were revealed. Alteration of pore dimensions and shapes under metamorphism attests to complex transformation processes that form vitrinite molecular structures via metamorphism and generation of gases.

Lipoid microcomponents such as sporinite, cutinite, and resinite feature individual architectures and sufficiently capacious intracavitary surfaces. Due to insufficient quantity, liptinite does not substantially affect gas generation. Heavy methane homologues form at the expense of latent lipid–humic bituminoids of vitrinite and liptinite at 2G through 4F stages. A special type of porous capacity is presented by interlayer spaces with distributed compression, slide, and dislocation zones and fissure cavities characteristic of junction zones of vitrinite inclusions in homogeneous helium-treated masses. The junction planes are essentially available for multiple fissure systems under developing microcleavage.

Ajruni [40] also observed a series of porous structures in fossil coals at various scales including pores of $1 \div 10^5$ nm whose bonding types and gas contents differed considerably. Oberlin [41] proposed a dark field transmission electronic microscopy (TEM) method that provided contrast images for investigating minimally ordered carbon substances such as fossil coals. The method produced direct pictures of basic structural units and their general spatial orientations (substance microtexture) not visible with other methods. The macromolecular network is represented by particles bearing a resemblance to solids and the molecular component appears similar to liquid units. Both types contain basic structural units of approximately similar dimension (<1 nm) disregarding the solid fuel metamorphic grade. The analysis of low metamorphosed coals by dark field TEM showed that the basic structural units are randomly oriented due to a significant number of functional groups. Highly metamorphosed coals and pyrolysis products feature basic structural units of similar dimensions, but their orientations are locally parallel within the molecular orientation domains (MODs), with each MOD forming a pore membrane. The substance microtexture resembles a crumpled sheet of paper (Figure 1.3).

Oberlin and Rouzaud described the model evolution under pyrolysis and metamorphism in detail [42–44]. Under carbonization to a semi-coking state, carbon substances undergo a more or less evident transition into a plastic state in which basic structural units rearrange and form domains. Oxygen in this case accounts for cross linkage formation and prevents development of large domains. Hydrogen conversely acts a plasticizer, generating large domains. Thus, chemical composition determines the nature of basic structural unit orientation.

Dark field technique was used to analyze 65 coals of various metamorphic grades with different chemical, geological, and geographic characteristics [41]. The data obtained agree with those from NMR and EPR spectroscopy and organic solvent extraction aimed at recovering the molecular component. In conclusion, it is reasonable to note that in view of the diversity of the (sometimes contradictory) conceptions discussed in this section, every

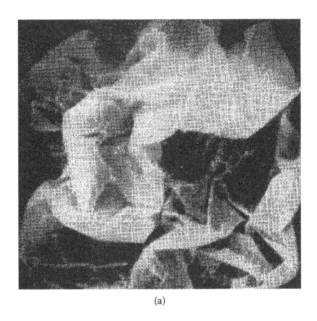

(a)

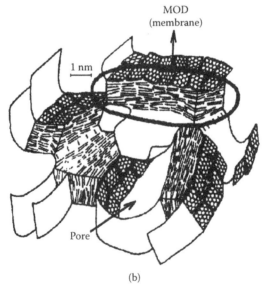

(b)

FIGURE 1.3
Models of molecular oriented domains. (a) Dark field transmission electron microcopy image. (b) Simulated image. (Adapted from Oberlin A. 1989. *Chem. Phys. Carbon* 22, 1–143.)

researcher is entitled to select his own point of view of the coal structure. We followed the spatial model by Kasatochkin and we believe that it allowed us to obtain a number of pioneering fundamental and applicable results that will be presented in the subsequent sections.

1.4 X-Ray Studies

This section covers in detail the history of x-ray studies of coals since no consensus yet exists regarding the interpretation of experimental results on the coal structure. As shown below, on the basis of experimental data obtained at the Institute for Physics of Metals of the National Academy of Sciences of Ukraine, x-ray techniques allow reasonable judgment about the dimensions of graphite-like crystallites.

1.4.1 Literature Review

Coal and its derivatives are sources of energy and carbon and the extensive use of these materials warrants thorough studies of the coal microstructure. Studying the atomic structure of coal by x-ray analysis commenced in the early twentieth century. Laue took the first x-ray photograph (showing what is now known as the Laue pattern) in 1912. It is fair to say that studies of coal structure began at the dawn of x-ray photography. Even the early works proved that diffraction photographs always feature three "blurred" maximums whose angle positions correspond to the lines on graphite x-ray patterns. The first maximum matches the graphite line (002), the second one (100), and the third one (110) according to Figure 1.4.

Also, under coal annealing, especially at temperatures exceeding 1000°C, the widths of the maximums gradually reduced and new reflections corresponded to graphite lines as well. x-ray patterns of some amorphous carbons that underwent annealing at 1000 to 3000°C showed all the graphite lines. These experiments enabled the authors to believe that a "carbonic substance" consisted of fine-grained graphite crystals whose shapes and dimensions were established by widths of diffraction maximums. The height increase and width decrease of the latter were associated with increasing dimensions of the crystals.

Over the past 50 years, x-ray techniques have undergone dramatic progress. Simple devices were replaced with powerful machines. For example, x-ray goniometers with high precision detectors and x-ray tube (or sample) positioning led to the creation of various x-ray diffractometers for general and special purposes. The use of x-ray tubes and x-ray power generators has increased significantly. Modern x-ray emission sensors allow increased resolution and measurement accuracy of the emission intensities of objects.

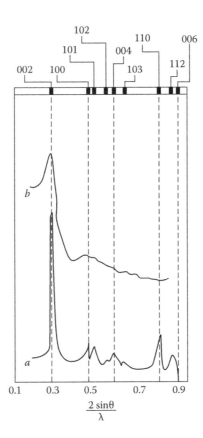

FIGURE 1.4
X-ray patterns of graphite (a) and vitrains of black coal (b).

Personal computers achieve automation of diffraction experiments, sharply improving accuracy and accelerating processing of data.

Unfortunately, this progress in instrument engineering and methodology has significantly promoted research only on crystalline objects. The achievements related to amorphous materials are not very impressive, mainly because the diffraction pattern from a disordered material (liquid or amorphous) is very simple—a continuous curve with only three or four peaks (diffusion scattering), among which only one—the first—is distinct. Diffraction patterns differ insignificantly even for objects with considerably different physical characteristics that can be seen in phase diagrams.

These features of experiment functions resulted in the lack of a reliable structure theory for both liquid and amorphous objects. Even now, studies of their structures have reached only model approximation. The defined characteristics of intensity function led to a great number of structure models of both liquid and amorphous objects, most of which lack a strict mathematical formalism and do not allow analysis on the quantitative level. Furthermore, diffraction methods are most often used to study ordered crystalline materials whose atomic structures may be unambiguously determined from crystalline phase symmetry. However, the loose atomic order of disordered solids and liquids yields wide diffraction peaks that carry less information than peaks derived from crystalline objects. This makes the analysis of diffraction patterns of disordered materials more difficult. That is why the researchers use material structure modeling (similar to common crystallographic modeling). Estimated likelihood studies were performed by comparing theoretical diffraction patterns with experimental ones. However, in disordered materials and liquids, model results may be in accord with experimental spectra.

As presented in this work, interpretation of coal structure differs from the methods described in the literature. An overview of the existing methods of building and estimating physically realistic structure models that are consistent with experimental diffraction patterns is given below. The methods were developed from those devised for analysis of disordered carbon forms.

According to most researchers, amorphous materials are characterized physically as frozen liquids. At the beginning of the twentieth century, a theory of x-ray radiation scattering by single-component melts was developed and later elaborated and extended to multi-component melts. The same theory is still used successfully to examine amorphous material structures.

The intensities of coherently scattered radiation of same-kind atoms aggregate and, reduced to electronic units, can be written in the following form [45–49]:

$$I(s) = \left\langle \sum_i \sum_j f^2 \exp\left(is(r_i - r_j) \right) \right\rangle, \tag{1.1}$$

where f^2 is the atomic factor; s is a diffraction vector with a scalar value $|s| = 4\pi\sin\theta/\lambda$ (2θ is the scattering angle); and $(r_i - r_j)$ is the vector interval between i- and j-atoms. To calculate the intensity, it is necessary to know the total interatomic spacing in an object, which can be described by a static integral. With liquids or amorphous materials, it is impossible to calculate a static integral. That is why, since atomic position statistics usually show spherical symmetry, spherically symmetric functions of atomic density $\rho(r)$ or probability $g(r) = \rho(r)/\rho_0$ that depend only on magnitude r_{ij} are introduced. The sum (1.1) is replaced by continuous distribution and angle averaging is carried out. Finally, intensity distribution, depending on the scattering angle takes the form:

$$I_k^{e.u}(s) = Nf^2 + Nf^2 \int_0^\infty 4\pi r^2 \left[\rho(r) - \rho_0 \right] \frac{\sin(sr)}{sr} dr, \tag{1.2}$$

where $\rho(r)$ is the function of atomic density, that determines the number of atoms in a volume unit at the distance r from the origin atom; and ρ_0 is the average atomic density of an object.

The $i(s) = I(s)/Nf^2$ is called the structural factor or interference function. Nf^2 is the sum scattering from isolated atoms (free "gas scattering"), which is easy to take into account during the calculation. Using Fourier transform, we calculate the total correlation function that characterizes atomic radial distribution:

$$4\pi r^2\rho(r)=4\pi r^2\rho_0 +\frac{2r}{\pi}\int\limits_0^\infty\left[i(s)-1\right]s\sin(sr)ds. \qquad (1.3)$$

With the help of the function of atomic radial distribution, we determine the average coordination number N (equal to the first maximum area) and the correlation radius. According to the structural factor, with the help of the Fourier transform, the relative function of atomic radial distribution $g(r)$ is calculated:

$$g(r)=1+\frac{1}{2\pi^2\rho_0 r}\int\limits_0^\infty\left[i(s)-1\right]s\sin(sr)ds. \qquad (1.4)$$

It should be noted that the information about the geometry of scattering centers distribution in a liquid structure contained in $g(r)$ functions is not comprehensive, as this function does not allow determination of the probability of distribution of three, four , or more particles. With alloys, the finite equation of the theory of x-ray radiation scattering becomes more complex because of the need to consider the difference in the scattering powers of atoms of different kinds.

Warren in his 1934 work [46] carried out the first quantitative x-ray diffraction analysis of carbonaceous materials. He suggests a model of the structure of imperfectly layered substances of soot types as lattices consisting of randomly oriented atomic layers. Warren derives the intensity equations for the line sectors in diffraction bands conditioned by such lattices. To accomplish this task, he introduces the average on a scattering volume function of scattering by one electron and effective quantity of electrons in each atom. These assumptions allow use of the Fourier transform to calculate the functions of electron radial distribution and later on the basis of model assumptions to derive the atomic radial distribution function.

Warren [46] calculated the curves to determine atomic concentration at any distances. As a result, the areas immediately under the peaks show the number of atoms in the given distance interval since the calculations are numerical and in absolute values. Thus, in the interval around 0.15 nm, the peak contains 3.2 atoms; at 0.27 nm, the peak contains 10.2 atoms, and crowding appears at 0.405 and 0.515 nm. With the help of these calculations, Warren showed that atomic configuration in the soot was not at all diamond-like. The diamond structure reveals 4 nearest neighbors at 0.15 nm, 12 neighbors at 0.252 nm, and another 12 at 0.295 nm. Warren surmises that distribution on the received curve corresponds to the atomic configuration of one graphite layer, even taking into account errors in defining atom quantities and peak arrangements (conditioned by overlapping of the latter). According to Warren, if the neighboring layers had particular orientations and positions relative to the first one, atomic configuration would completely correspond to graphite structure; otherwise coal soot would represent some sort of mesomorphic form of carbon (ordering only in two dimensions). However, he does not consider either of these two conditions a priority.

The comparison of the experimental curves of radial distribution of soot and graphite atoms allowed Warren to make a conclusion that the coal soot consists of either extremely small graphite crystals or roughly parallel (quasi-parallel) graphite monolayers, or, finally, heterogeneous mixture of both. These two distributions are not different to the degree where one of them can be positively excluded. Warren concludes from this that if it is possible to explain satisfactorily experimental x-ray data by quasi-parallel graphite monolayers, this serves as a warning against hasty conclusions while deriving diffraction spectra with the help of two or three indistinct peaks only.

At the same time, Warren determined a characteristic unusual for homogeneous solids and liquids that appeared on soot roentgenograms. Scattering intensity increased to a value above the highest peak instead of diminishing to small values with a decrease of scattering angle. This phenomenon known as small-angle scattering has also been noted by many other researchers who have suggested more complicated explanations.

Warren associated this effect with the fact that the material consists of very finely dispersed particles that are loosely packed in a sample due to their heterogeneous form. Thus, every atom is surrounded by a high density area that covers the space an order higher than the particle size, then a medium density area that is less condensed due to loose particle packing. Therefore the scattering intensity at small angles can have significant values, depending on the excess density amount. Assuming part of the material is in the form of multi-layered small graphite crystals, the difference between graphite and coal soot densities can be easily connected with the observed scattering intensity at small angles.

Warren expresses confidence in the fact that the decrease of small-angled scattering that occurs during angle increase depends on the dimension of the high-density area and thus represents a rough estimate of particle sizes. He surmises that small-angled scattering indicates that at least a minor part of the material is made of small crystals. Warren's works [46–49] assured him that soot definitely contained graphite monolayers. The particle sizes change from sample to sample based on the method of gathering, but coal soot includes continuous series of structures ranging from mesomorphic to crystalline states.

One way to determine aromaticity (sizes of concentrated aromatic layers and sizes of layers packed in a coal material) is also based on the equations introduced in Warren's works [47–49]. Using Sherrer's equation, he applies the general diffraction theory to statically homogeneous objects of arbitrary structure and determines the interdependence of the form and the intensity *hk0* stripe on plane size (later the formula for determination of L_c was derived [45]):

$$L_a = \frac{1.84\lambda}{B_{10}\cos\theta}\,; \Delta(\sin\theta) = \frac{0.16\lambda}{L_a}\,, L_c = \frac{0.9\lambda}{B_{002}\cos\theta}\,; \qquad (1.5)$$

where L_a and L_c are the width (diameter) and height of crystallites (packs containing several parallel layers), respectively, B is the half-width of the corresponding stripe, λ is the length of the scattering wave, B_{002} is the half-width of the graphite peak (002), θ_{002} is the peak position (002), and $\Delta(\sin\theta)$ is the difference in angular disposition of *hk0* stripe. This method of graphitized carbon analysis continues to be used.

Franklin [50] used Warren's constructions [45–47] to prove that all soot x-ray peaks correspond to graphite reflexes (001) and (*hk0*). She surmises that the main contribution to dependence of back-scattered radiation intensity $I(s)$ on the scattering vector s in soot is made by little vertical packs of completely ordered hexagonal graphite planes. In the same work [50], Franklin suggests a number of reasons for the observed broadening of radial distribution function $g(r)$. Having analyzed the disorder of carbon aromatic atoms in graphite layers, she concludes that this contribution to broadening $I(s)$ is small because the theoretical function $g(r)$ for an ideal graphite plane with introduced artificial decay corresponds to the experimental dependence only qualitatively. Franklin [50] notes the possibility of "disorganized carbon" in the purely gaseous state that may explain the monotonous background of the observed spectra. However, although she derives the model of very small reflecting aromatic layers (crystallites) in carbonaceous substance structures, ideal crystalline planes plus gas carbon later served as bases for building carbon structures.

To further develop the idea of plane crystalline interpretation of disordered carbon structures, Franklin later [51] devised methods of analysis for several carbonaceous materials, demonstrating diffraction patterns intermediate between cases of completely disordered structures with reflexes only (001) and (*hk0*) and crystalline graphites with complete sets of reflexes (*hkl*). Particularly, she analyzed the structures of carbonaceous materials with highly ordered crystallites consisting of a set of *c*-planes 0.344 nm from each other but absolutely not otherwise correlated (so-called turbostratic carbons). For such graphitized carbon-based materials (derived from coke, natural graphite, or polymers by thermal treatment at 1700 to 3000°C), the number of parallel *c*-planes in the crystallite was 30 to 150 and corresponds to ordered structures.

Franklin's [50,51] fraction of disorganized carbon was later used by Alexander and Sommer [52] as a parameter to determine soot structure. They calculated this fraction with the help of a theoretical expression for the expected peak (002) value for an ideal crystallite with a specified number of aromatic layers. The best approximation of an experimental peak (002) by this expression was achieved by subtracting a constant from the measured intensity $I(s)$ to determine the fraction of amorphous carbon; at this point, the peak must stay as symmetrical as possible and fall to zero at both sides of the central maximum.

Intensive research on fossil coals and cokes started in the early 1950s. The above described development of the theory of imperfect crystal lattices and x-ray diffraction in amorphous bodies on the basis of carbon-based material research seemed to allow the quantitative interpretation of coal roentgeno-grams. In fact, scientists differed on the interpretation of the results. Some agreed with Warren, Franklin, and Alexander [45–52] that coal consisted of small graphite crystallites. Others thought that direct crystallographic proof of this structure was lacking.

Kasatochkin [53,54] expressed his opinion about the inconsistency of con-sidering natural coals and cokes to be superfine graphite-structured materi-als because fossil coals, unlike other amorphous carbons, consist of a number of components with different morphologies, physical properties, and chemi-cal composition, for example, vitrains, fusains, and others. These constitu-ents differ sharply in their fine morphological and molecular structures. Kasatochkin assumes that the asymmetric maximum (002) is conditioned by non-aromatic side hydrocarbon groups in the peripheral parts of coal materi-als [54]. This approach allowed Kasatochkin [8,53–55] to express qualitatively the idea that hydrocarbon structural fragments surrounding the aromatic layers interconnect the layers, thus forming a spatial polymer. This repre-sents the first explanation of the elementary structure of coal (Figure 1.5).

Structural units that serve as the bases of vitrain macromolecules are het-erogeneous and differ from one coal rank to another in carbon lattice sizes and side chains. Kasatochkin observed interference fringes, coincident with graphite lines ($hk0$), as a result of x-ray interference on a flat hexagonal carbon atomic lattice, similar to the graphite carbon monolayer. Fringes correspond-ing to graphite lines with indices (001), are considered interlattice (intermolecu-lar) interferences. That is why it was assumed that the reason for the widening of interference fringes (001) was the reciprocal disorder of atomic carbon lat-tices. Lessening of interference fringe ($hk0$) and (001) width corresponds to the increase of micromolecules (carbon lattices) and an increase of their mutual orientation level. Structural changes of coal carbonaceous materials with meta-morphism involve two parallel processes: (1) increase of atomic carbon lattices due to side group decomposition and (2) mutual orientation of carbon lattices. Coalification, according to Kasatochkin, should be the direct transition of sp^3 hybrid atoms into $(sp^2\pi)_{ar}$, omitting the stage $(sp^2\pi)_{chain}$. This opinion presup-poses that at all stages of coalification only the non-aromatic hydrocarbon part

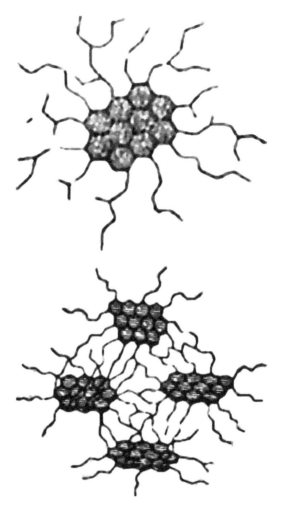

FIGURE 1.5
Vitrain of fossil coal according to V. I. Kasatochkin. (a) Structural unit. (b) Spatial polymer vitrain material. (Adapted from Kasatochkin V.I. 1952. *Dokl. akad. nauk. SSSR* 4, 759–762.)

links condensed aromatic nuclei in a spatial polymer. Heteroatoms that are essential parts of younger coals and play linking roles in polymer coal structure were not taken into account.

According to radiographic examinations [14] carried out by Hirsch and supported by other works [1,56], with the increase of aromatic layer ordering, the space-linking role of the non-aromatic part changes. Figure 1.6 shows Hirsch's structural model. The "open" structure (*a*) is a characteristic of black coal with a carbon content of approximately 85%. Aromatic layers rarely group crystallites, although there is a slight orientation relative to one plane. Linking of layers by hydrocarbon chains leads to linkage of the layers and such coals can

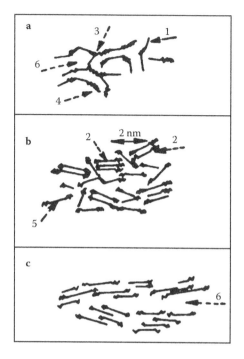

FIGURE 1.6
Structural model according to Hirsch. 1. Carbon layer. 2. Crystallite. 3. Spatial linkage.
4. Amorphous material. 5. Monolayer. 6. Pores. (a) Open structure. (b) Liquid structure.
(c) Anthracite structure. (Adapted from Hirsch P.B. 1958. *Proc. Res. Conf. on Use of Coal*, Sheffield,
pp. 29–35.)

be considered spatial polymers. Vitrains with 85 to 91% carbon contents are
considered liquid structured (*b*).

At a certain orientation in a liquid structure, the crystallites of two or three
lamellae are formed, although single layers may form. According to Hirsch,
coals are not characterized by large molecular nature (reticulation) and it
is possible to find ways to decompose their structures without breaking
valence bonds. Hirsch uses this approach to explain changes of coals in the
plastic or soluble state. Even for highly oriented and layered anthracites (*c*),
Hirsch assumes the absence of spatial graphite structures.

Diamond [57] further developed the ideas of Warren [45–48] and calculated a theoretical x-ray diffraction for carbon using Warren's method of
pair correlation functions with the Debye equation for aromatic monolayers. In addition, Diamond built diffraction spectra for various structures
with various numbers of atoms grouped in corresponding classes and
discovered that even significant variations of bond distances in atomic
planes barely influence the calculated spectra, at least from the onset of
noticeable broadening of correlation peaks due to small sizes of aromatic
layers.

After Diamond, Ergun, and Tiensuu [58] calculated pair distribution functions for planes that have various sections of diamond-like lattices. Design peaks of carbon with tetrahedral coordination are similar to those of aromatic carbon. Using these data, Ergun and Tiensuu tried to explain the difficulty of interpretation of disordered carbon diffraction intensity curves. They claim that although it is difficult by diffraction to determine the presence of diamond-like structures, these structures may be present in large quantities in amorphous carbon. Also, the work of Grigoriev [59] on coals to achieve the correspondence of experimental and theoretical diffraction curves suggests that models of coal structures include both the aromatic carbon layers and the diamond-like structures. Ergun [60] objects to Franklin's [51] assumption about the obligatory presence of some parts of carbon atoms in gaseous state. Ruland [61] and Rietveld [62] note that the share of "disorganized carbon" does not depend on gaseous atoms and may be explained by disordering within aromatic planes due to including penta element rings and side carbon and non-carbon groups in the configuration. A review [63] notes that Ruland and Ergun et al. found it necessary to take into account lattice imperfections such as aromatic plane bending, holes, and implemented and out-of-plane atoms.

Shi and colleagues [64] improved the crystallographic approach of Ruland [61] based on the least-squares method by developing a calculating algorithm similar to Rietveld's [62] with 11 parameters to analyze crystallites in experimental x-ray diffraction diagrams. Shi's model [64] assumes that the disordered carbon represents layers of completely ordered aromatic planes displaced relative to adjacent planes only in parallel or orthogonal directions. While the distance between the neighboring atoms is constant within each plane, the planes may be extended by changing the lattice parameters of the current plane to allow extension of theoretical peaks $I(s)$. To account for the "deformation zones" in the model crystallites, defined as packets of parallel planes with non-graphite steps along the c-axis, the authors presented the distance between aromatic planes as a normal crystallographic parameter d_{002} of graphite plus deformation. During the serial input of more planes parallel to the initial planes in the model crystallite, the deformation increased due to the stochastic choice of the inter-plane distance whose deviation from the ideal value d_{002} is subject to Gaussian distribution. Having defined from this model various crystallite parameters of disordered carbonaceous materials, Shi and colleagues [64] showed agreement with the experimental data. Although the suggested algorithm does not account for peculiarities such as interstitial atoms, aromatic plane bending, and defects, the algorithm is fast and simple.

In a recent study of the influence of thermal treatment on Australian semi-coke [65], Shi's algorithm was used to determine structural parameters. Crystallite parameters increased during thermal treatment, but based on algorithm calculations were at least five times more than those calculated by Sherrer's equation [46]. As the algorithm is in good agreement with existing experimental data, we surmise that Sherrer's equation may not work with strongly disordered carbonaceous materials that have wide diffraction peaks. However, the same authors [66] later achieved good agreement with the experiment and found that Shi's algorithm [64] yielded false results because in some cases the calculated share of disordered carbon was negative. Other researchers who use Shi's algorithm [64] reported the flaws in Sherrer's equation for analyzing strongly disordered carbonaceous materials. In conclusion, all the methods need further research. Nevertheless, these methods of diffraction pattern analysis are still used to analyze coals and its derivatives.

Yen and colleagues [67], using the supposition about ideal crystallites, expressed the idea that peak (002) asymmetry is conditioned not by Franklin's variations of ordered crystallite sizes [51], but by structural defects and this allows calculation of so-called aromaticity levels. The authors separate the diffraction peak (002) into two separate symmetric peaks and surmise that the smaller side peak to the left of the true normal peak (002) reflects the diffraction from non-aromatic carbon so that the ratio of peak areas allows determination of the number of aromatic carbon atoms in the material. They present experimental and theoretical proofs connected with research on the diffraction of a soot and saturated polycyclic aromatic compound mixture.

This method was used to study Australian coals [21]. Although the authors noted the simplification of structural models, they considered that this method could yield maximum data about the structures of carbonaceous materials if the data were taken from the areas of medium and wide angles. Further work [68] based on the concept of ideal crystallites and Franklin's methods [50,51] attempted to calculate the rates of graphitization of some Pennsylvania anthracites with measured peak (002) positions.

Until now, fine coal structures were determined directly from x-ray intensity profiles at medium and wide angles [21,68–74]. Such structural parameters as fraction of amorphous carbon (x_A), aromaticity (f_a), crystallite sizes, and characteristics of their distribution (L_a, L_c, d_{002}) are defined. The main issue is the need to draw the background line correctly. That is why the main stage of wide-angled diffraction pattern processing involves background subtraction and definition of an accurate diffraction profile taking into account the γ component and distortions from mineral impurities [73]. The diffraction patterns transformed by these methods [Equation (1.5) and References 47 and 49] provide information about coal structures. The definition of interplanar spacing d_{002} is carried out by the Bragg equation $[d = n\lambda/2\sin\theta]$ where n is the order of reflection. Figure 1.7 shows diffraction pattern of lean baking coal $(V^2 = 15.1\%)$. Figure 1.8 shows the same curve after background line subtraction. The portion of amorphous carbon (x_A) can be found by making the left side of stripe 002 as asymmetric as possible (Figure 1.9). Figure 1.10 shows x-ray diffraction patterns of all phases of metamorphism for different coal ranks.

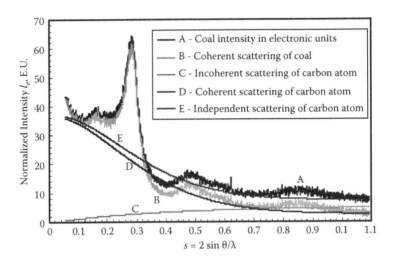

FIGURE 1.7
Normal curve of intensity (Ie) for probe 3JD (lean baking coal). (Adapted from Lu L. Sahajawalla C.K. and Harris D. 2001. *Carbon 39.*)

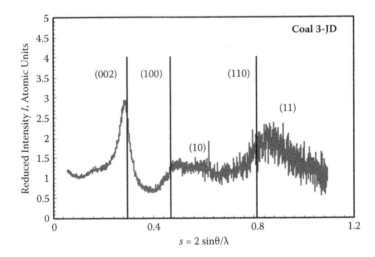

FIGURE 1.8
Typical curve of intensity (I) for probe 3JD (lean baking coal). (Adapted from Lu L. Sahajawalla C.K. and Harris D. 2001. *Carbon 39.*)

As shown in Figure 1.10, the diffraction maximum 002 has an asymmetric view but its position and half-width vary based on stage of metamorphism. These special features of diffraction patterns indicate that (1) maximum displacement is a result of change of interplane distance d_{002} of a graphite-like phase; i.e., the distance between graphite nettings is changed in a crystallite,

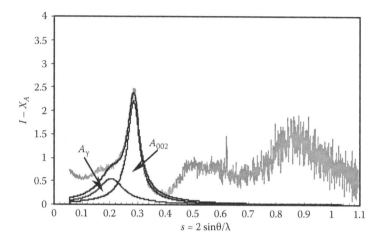

FIGURE 1.9
Subtraction of given intensity for probe 3JD (lean baking coal). A_γ represents γ stripe from coal aliphatics. A_{002} is coal aromaticity. (Adapted from Lu L. Sahajawalla C.K. and Harris D. 2001. *Carbon* 39.)

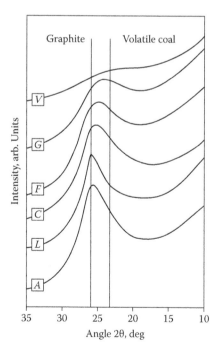

FIGURE 1.10
X-ray diffraction patterns of coal samples (diffraction maximum 002). Coal ranks are denoted by letters. V = Volatile. G = Gaseous. F = Fat. C = Coking. L = Lean. A = Anthracite. (Adapted from Alexeev A.D. et al. 2002. *Fiz. tekhn. vyso. davl.* 12, 63–69.)

and (2) it is possible to define sizes of particles in direction 002 and consequently determine the number of nettings in a crystallite according to the
half-width of the peak. Asymmetry of diffraction maximum 002 has been
explained by the presence of a γ phase that indicates the availability of side
hydrocarbon chains. Researchers [64–74] explain these results as follows:

- Coal contains graphite-like zones of crystalline carbon in the form
  of excessively tiny particles. Crystalline carbon produces three indistinct peaks and its structure is transitional between the graphite
  and amorphous states. This structure is stochastically layered or
  turbostratic.
- In addition to graphite-like particles, coal also contains a significant
  quantity of disordered substance giving rise to background intensity; the authors call it amorphous carbon.
- The asymmetric property of stripe 002 is caused by the second (γ)
  stripe on the left. The γ stripe is explained by the presence of aliphatic side chains joined to the edges of coal crystallites.
- High intensity in a small-angled part of a spectrum is conditioned
  on interparticle scattering on densely packed particles, i.e., by stripe
  [20] and also by scattering of microcracks, micropores, and other
  adhesion faults in coal samples [64].

According to these assumptions, two types of structures appear in coal
roentgenograms: crystalline and amorphous carbons. The authors think that
crystalline carbon that has a turbostratic structure is the main structure and
amorphous carbon is not spread on aromatic sections and contributes only
to background intensity. The simplified coal structure proposed in [21] is
shown in Figure 1.11. Graphite crystallites in the corner are very tiny and

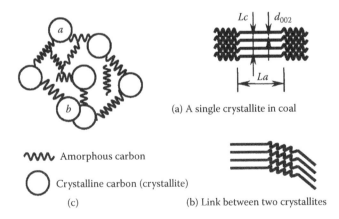

(a) A single crystallite in coal

ww Amorphous carbon

◯ Crystalline carbon (crystallite)

(c)

(b) Link between two crystallites

FIGURE 1.11
Simplified coal structure. (Adapted from Lu L. Sahajawalla C.K. and Harris. D. 2001. *Carbon* 39.)

surrounded by many aliphatic side chains (a). The crystallites are connected with each other by their side chains in so called macromolecules that in turn capture amorphous carbon [(b) and (c)].

1.4.2 Structural Peculiarities of Coal–Methane System

After reviewing the relevant literature, we suggest that coal substance consists of high dispersive crystallites of graphite, the shapes and sizes of which are determined by widths of diffraction maximums using Equation (1.5) and by the γ stripes of aliphatic compounds. A height increase and a width decrease of a diffraction maximum are linked with an increase of crystallite size.

During coal metamorphism, the number of rings and lamellae in crystallites increases and this leads to the appearance of a graphite-like structure in later stages of coalification. Interplanar spacing of the two described phases based on carbon concentration changes within $d_{002} = 0.35$ to 0.404 nm and $d_\gamma = 0.40$ to 0.82 nm [1,21,55]. However, overlapping of these ranges for d_{002} and d_γ that are peculiar to different structural formations indicates very free interpretation of these experimental data. In our opinion, the value of $d_{002} = 0.404$ nm [27] for volatile coal with 80.2% carbon content is too high to be part of graphite-like structure (graphite requires only 0.336 nm). The radius of a Van der Waals interaction does not exceed 0.35 nm for all known aromatic combinations of pure carbon. Second, an increase of distance between graphite-like layers would weaken their interactions and led to a decrease of mechanical properties of coals with low carbon concentrations. In ranks of coking coals, those with $d_{002} = 0.35$ nm have the weakest strength and contain 90% carbon.

In addition, the formulas for calculating L_c and L_a [Equation (1.5)] cannot be very accurate because of the approximate values they yield. To get more accurate values of L_c and L_a, we should correlate the theoretical curve with the experimental one by trying different values of L_c and L_a. It is worth noting that these formulas give only overstated values [45].

The current interpretation of diffraction presents many disadvantages and does not allow construction of an adequate coal structure model. Considering the studies discussed above, we tried to observe the given experimental data from another view during x-ray studies at the Galkin Institute for Physics and Engineering of the National Academy of Sciences of Ukraine in Donetsk. An external view of a roentgenogram of a volatile coal is similar to a view of a linear polymer like cellulose [45]. Based on that, we assumed that a basic mass of volatile coal is not comprised of graphite-like aromatic crystallites and represents another form of molecular carbon. This suggestion stimulated us to carry out the third phase of our research [75,76].

To analyze the form of the diffraction maximum, we used the superposition of Lorentzian methods on the supposed phases in different proportions. The initial roentgenogram processing showed that to get the best approximation of experimental profile it was necessary to introduce one

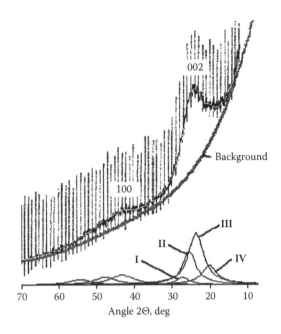

FIGURE 1.12
Decomposition of basic peak 002 into Lorentzian phases I through IV.

more (graphite) phase. Figure 1.12 shows the location of the peak of volatile coal. We subtracted background of non-coherent amorphous component scattering from the diffraction curve. Results indicate that coal substance composition is a superposition of four forms (phases) of pure carbon. Phase I is graphite—supposedly found in small quantities in samples with carbon content above 83%. Phase II is a graphite-like component with distance $d_{002} = 0.355$ nm, typical for defective layers packing of aromatic carbon. With some assurance, we can represent Phase II as both graphite-like and fullerene-like because the fullerenes C_{60} $d_{002} = 0.355$ nm and the sphere has a diameter of 0.71 nm. This does not mean that the corner is pure crystalline fullerene. A number of fullerenes (coaxial nanotubes, nanospheres, polymers of different types) are natural components of coal.

Phase III may have a chain-like composition of carbon molecules and act as an analogue to carbyne. The hexagonal cells of carbyne have lattice parameters of $A = 0.508$ nm and $C = 0.780$ nm. Thus, within the framework of accuracy of the experiment, $d = 0.385$ nm in volatile coal conforms to the reflection of carbyne that equals 002. Phase IV is characterized by interplanar spacing $d = 0.46$ nm. The γ stripe corresponds to that value and indicates hydrocarbon ordering of the coal mass. However, the given assumption has been disproved by the appearance in roentgenograms of pitches and cokes [71]. A carbon composition that is steady at any temperature, similar to the γ

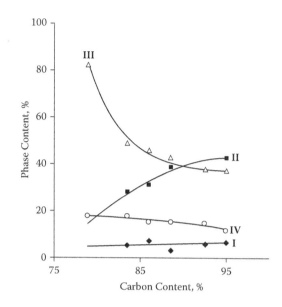

FIGURE 1.13
Phase coal composition depending on level of carbon content: Phases I through IV.

phase, and called a graphyne has now been synthesized. Aromatic rings are tied in lattices by carbonic chains of carbyne.

An elementary cell of a graphyne is triclinic. Values of $d_{100} = 0.508$ nm, $d_{010} = 0.458$ nm, and $d_{001} = 0.509$ nm were noted; $d = 0.46$ nm and $d_y = 0.47$ nm represent average values of three maximums located near each other; this is why Phase IV may be a graphyne. Figure 1.13 shows roentgenogram results. The content of each phase was estimated by comparing the integral intensity of the corresponding peak to general integral intensity. As we can see from the increase of C^{daf}, the number of Phases III and IV decreases accordingly while the number of Phases I and II increases, in accordance with coal graphitization via metamorphism. Changing the phase ratio and carbon content entails changes of diffraction maximum configuration. Phase III dominates in volatile coals and moves the aggregate maximum to the side of smaller angles. The result looks like displacement of reflection, explained previously as change of interplane distance d_{002} during metamorphism.

From several diffraction spectra of coal, we singled out the part that shows aromatic carbon and estimated level of aromaticity. The linear character of dependence f_{ar} (Figure 1.14) on carbon quantity conforms to the data yielded by various methods for examining coals during intervals of coalification [27]. The angle of the right line slope coincides with the slope of dependence shown in this work [1]. We found that the absolute values f_{ar} were understated in comparison to those cited in the literature. This can be explained by the specific character of x-ray experiments. Discrete maximums of medium

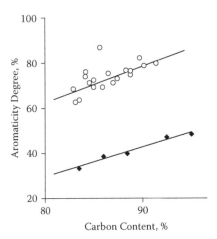

FIGURE 1.14
Levels of coal aromaticity based on carbon content. Triangles = our results. Circles = literature data. (Adapted from Korolev Y.M. 1989. *Him. tverd. topl.* 16, 11–19.)

angle ranges of diffraction pattern show scattering of large crystalline zones in the pure carbon phase. The contribution of the fine crystalline amorphous phase has not yet been documented.

This interpretation of phase coal composition covers well examined substances and it does not require other interactive models of carbon clusters that are difficult to calculate [57–66,69]. The discovery of fullerenes led to advances in carbon chemistry and it became clear that unique features of fullerenes helped them form big molecules and clusters without having to lock chains broken by hydrogen or heteroatoms. Considering the results of our research within the framework of a polymeric model of coal composition, aromatic kernels in coal can be linked in a polymer by both hydrocarbon chains and pure carbon chains. A graphite component consisting of aromatic rings and carbyne chains provides a transition between structural units. The lack of constant quantities of Phase I shown in our patterns can be explained by the availability of graphite particles in the form of a mechanical admixture.

The model proposed is useful for studying the influences of external conditions on coal properties, since the values of d_{002} for all phase components are fixed accurately. This allows us to record the results of external influences on of each structural component separately.

X-ray structural research carried out on coal samples simultaneously with nuclear magnetic resonance (NMR) studies made it possible to explain changes in roentgenograms of coal with high methane content during methane desorption [77]. In this work [75], we assumed that four-phase structural black coal can be found in any coal substance. Phases I and IV remain constant when affected by external conditions having no dependence on the stage of metamorphism and they can be considered admixtures.

We singled out the two main phases because of their high concentrations in coal. Phase II is similar to graphite and contains aromatic carbon with a typical interplane distance $d = 0.35$ nm that is used to pack defective layers. Phase III is a hydrocarbon phase consistent with the dominant $d = 0.38$ nm structure of new coals. As carbon content increases during coal formation, Phase II increases and Phase III decreases (Figure 1.13). Phase II is a hydrocarbon of aliphatic and alicyclic character. Aliphatic compounds corresponding to CH_3-$(CH_2)_n$-CH_3 consist of carbon atoms combined in unlocked chains. In alicyclic hydrocarbons, the carbon atoms form locked chains with a single combination. The relative positions of flat, condensed, aromatic carbon kernels and aliphatic and alicyclic atoms of hydrocarbons are shown in the scheme of Van Krevelen (Figure 1.1) [78]. In interpreting the properties of coal, we have focused on aliphatic chains because their mobility can help explain non-monotonous changes of coal properties in cases of linear dependence of structural features. We received data from several authors on atomic ratios of basic coal mass components (C, H, O, and N) and tried to value the atomic ratio of H to C for Phase III. We assumed that each atom of oxygen and nitrogen added an atom of hydrogen and the remaining hydrogen belonged to Phase III. The values calculated ranged from 1.49 to 1.54, which corresponds to an H:C ratio of 3:2. This proportion is typical for an alicyclic chain that consists of rings with a single combination of carbon and alternating groups of H and H_2. These fragments have a short-range order that is sufficient for two-dimension diffraction. That is why they reach the top in a roentgenogram similar to aromatic kernels. Therefore, if we consider Phase III to consist of alicyclic hydrocarbons, we can then describe the process of metamorphism.

Alicyclic rings lose hydrogen, form double connections, and finally form aromatic rings. Quantitative spectroscopic analysis shows a transition of carbon of alicyclic groups into aromatic carbon with a growth of coalification rate [1,78]. This conforms to the data obtained from ^{1}H NMR wide line spectroscopy that revealed the availability of significant quantities of alicyclic methylene groups. For coals at the medium stage of metamorphism, the relative quantity of hydrogen from the alicyclic groups constitutes 30 to 40% [1]. As to aliphatic groups, according to infrared spectroscopy, their numbers decrease based on the stage of metamorphism [79]. Chains differ in size and may be sterically disoriented when various terminal groups attach. x-ray diffraction of linear disordered objects does not yield discrete maximums.

x-ray studies of methane desorption were carried out using coking coal samples extracted from the Zasyadko mine. This experiment was performed with a DRON-2 diffractometer and results were assembled and processed by a computer. A lump of coal was saturated for several days, removed from the saturation material, and x-ray images were made sequentially at certain time intervals. NMR spectra and measurements of mass changes were received simultaneously from the same coal samples. The results of x-ray data processing are shown in Figure 1.15.

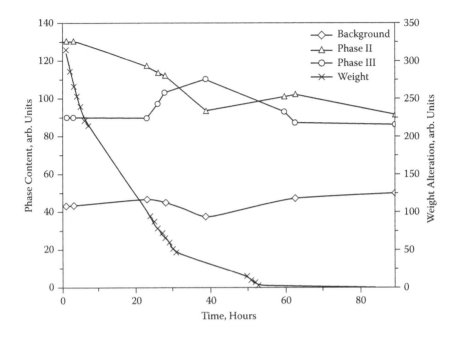

FIGURE 1.15
Phase content: Graphite-like phase (II) and carbon phase (III). Background and mass of coal sample depending on duration of methane desorption.

In comparison to the initial x-ray spectra of decontaminated diffraction patterns, coal samples with high methane concentrations featured an increase of Phase II and a decrease of intensity of background scattering ($I_{background}$) while the content of Phase III was invariable. Widths of peaks and sizes of zones of coherent scattering of both crystalline phases remained invariable during methane absorption. Consequently, an increase of peak intensity 002 in Phase II did not occur at the expense of enlargement of the available aromatic kernels; it occurred at the expense of new ones.

The decrease of background intensity indicates that Phase II increases at the expense of the amorphous component (aliphatic group). A great concentration of methane molecules can bring carbon atoms together at the distance of formation of double connections. Some aromatic kernels will arise and some free hydrogen will be released. Since pressure is removed, the distance between atoms of carbon increases. The process of hydrogenation (joining of hydrogen atoms) then starts. Some metastable aromatic fragments under pressure lose double connections and become alicyclic and Phase III begins to grow. A further decrease of pressure entails disappearance of single connections between chains. The quantity of Phase III decreases to its initial value. Some experimental curves shown in Figure 1.15 confirm this assumption. At a certain time and pressure, the quantity of Phase III grows, reaches its maximum, and then decreases again. A change of the mechanism of mass loss is shown in the graph of phase correlation. After 90 uninterrupted hours, the x-ray view coincided with the picture of the initial state before methane absorption; this verified the termination of the desorbing process.

Despite the complexity of the coal–methane experiment, the proposed interpretation of the results was relatively simple and very useful.

1.4.3 New Approach to Analyzing Coal via Scattering X-Ray Investigation

Warren's suggestion [46] to use small-angled scattering to study diffraction patterns of fossil coals and investigations of metallic material structures in liquid and amorphous states enabled A. P. Shpak and A. G. Illinsky of the Institute for Physics of Metals in Kiev to correctly determine crystal-lite dimensions in coal. Early x-ray studies of fossil coals clearly established strong scattering in the zone of small angles (small-angle scattering). Despite the primitive chambers and low power of early x-ray generators, the early conclusions [45–51] were basically correct.

We think that most scientists simplified their conclusions and in some cases produced contradictory or not quite correct conclusions about coal structures. Thus, some scientists began to consider significant small-angle scattering as a background to be "used" correctly and then eliminated. We consider this practice incorrect. The intensity function must be used correctly.

In 1966, A.S. Lashko [80] offered a more direct method for calculating atomic distributions of multi-component systems. In his calculations, the function of radial distribution of atoms is a linear combination of special functions of atomic density $\rho_{ij}(r)$ (later called *partial*). Function $\rho_{ij}(r)$ is proportional to the relativity of location of atoms of i type at distance r from atoms of j type. The basic equation for the intensity of scattering x-ray radiation by multi-component objects is as follows:

$$I(s) = \left\langle \sum_i \sum_j f_i f_j \exp\left(is(r_i - r_j)\right) \right\rangle, \tag{1.6}$$

where summing relates to all the atoms of scattering volume and f_i and f_j are atomic factors of the components. In case of a two-component object, the right section of Equation (1.5) is given in the form of three totals:

$$I(s) = \left\langle f_1^2 \sum_n \sum_k \exp\left(is(r_n - r_k)\right) \right\rangle + \left\langle f_2^2 \sum_n \sum_k \exp\left(is(r_n - r_k)\right) \right\rangle$$

$$+ \left\langle f_1 f_2 \sum_n \sum_k \exp\left(is(r_n - r_k)\right) \right\rangle. \tag{1.7}$$

The averaging of the sums is similar to the method for one-component objects. After the corresponding transformations, Equation (1.6) will be given as

$$I_k^{e,u}(s)=\left\langle F^2\right\rangle+\left\langle F^2\right\rangle\int\limits_0^\infty 4\pi r^2\left[\sum_i\sum_j n_iK_iK_j\rho_{ij}(r)-\left(\sum_{i=1}n_iK_i\right)^2\rho_0\right]\frac{\sin(sr)}{sr}dr, \quad (1.8)$$

where $\left\langle F^2\right\rangle=\sum_{i=1}n_if_i^2$, $K_i^2=\left\langle f_i^2/F^2\right\rangle$ are coefficients independent of the scattering angle and f_i^2 and n_i are the atomic factor and atomic concentration of *i*-component, respectively. Using the Fourier transform, we find a complete function of radial distribution of atoms from Equation (1.8):

$$4\pi r^2\sum_i\sum_j n_iK_iK_j\rho_{ij}(r)=\left(\sum_i n_iK_i\right)^2 4\pi r^2\rho_0+\frac{2r}{\pi}\int\limits_0^\infty\left[i(s)-1\right]s\sin(sr)ds. \quad (1.9)$$

From Equations (1.8) and (1.9), we conclude that in binary objects, in addition to three-dimensional atomic distributions (topographic short-range ordering), the relative positions in the spaces of atoms of different types (compositional short-range ordering) is essential as well. The information about the spaces is given by the partial functions of atomic density $\rho_{ij}(r)$. These methods for obtaining diffraction patterns apply to metallic objects in liquid and amorphous stages. x-ray studies of the structures of materials in liquid and amorphous stages follow three steps:

1. Experimental measurement of angular dependence of intensity of x-ray radiation scattered by specimen (diffraction function)
2. Mathematical treatment of experimental data, calculation of structural factor and functions of radial distribution of atoms
3. Interpretation of results

As indicated above, to obtain quantitative information on an amorphous material structure, a high degree of accuracy is required to obtain the diffraction pattern, calculate the structural factor and function of radial distribution, and determine the basic structural characteristics. Two types of uncertainties of the obtained functions must be noted: experimental uncertainties and uncertainties of mathematical treatment.

The experimental uncertainties are based on a number of factors: (1) the efficiency of monochromatization of x-ray radiation, (2) the technique of sample alignment, (3) the ability to provide and maintain experimental conditions for a long period (steady operation of x-ray and electronic equipment, temperature, centering of sample, etc.) of experimentation, (4) the method of x-ray radiation recording, (5) the geometry of the diffraction pattern, and

(6) the shape of the exposed surface of the sample. The potential experimental errors are described in detail in the literature [80–87].

It is not possible to estimate accurately the errors of some parameters. To determine the errors in experimental diffraction patterns, for example, initial treatment, calculation of basic structural parameters, and other factors over a long period (usually a year), we repeatedly obtain diffraction patterns of some components. Usually at this time the x-ray tubes are renewed and the diffractometer is realigned. We realize that only occasional errors can be found this way; it is impossible to estimate systematic errors.

Criteria for estimating obtained results are simpler: (1) changes of structural factors (how clearly they oscillate) and (2) the sizes and forms of false oscillations (in the area of short distances) with the full functions of radial distribution. Figure 1.16 illustrates two structural factors determined according to the diffraction patterns for massive (curve 1) and thin (curve 2) anthracite samples. In curve 1, the structural factor clearly oscillates; curve 2 shows nothing of the kind. It is better to compare the intensity functions reduced to electronic units or the conclusions may be incorrect. Figure 1.17 shows two experimental diffraction patterns obtained in different months. While the intensities expressed in impulses (a) significantly differ from each other, those expressed in electronic units (b) coincide completely.

Equations (1.1) through (1.9) lead to two important conclusions: (1) experimental diffraction patterns should only be obtained for liquid or amorphous objects in monochromatic radiation; and (2) Equations (1.2) through (1.9) are integral; integration should be performed when calculating all the functions of atomic distribution according to the theory of the vector interval of diffraction from 0 to ∞.

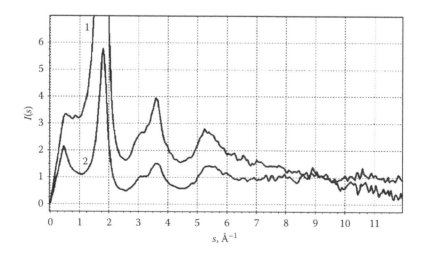

FIGURE 1.16
Structural factors of anthracite obtained by diffractometry of massive (1) and thin (2) samples.

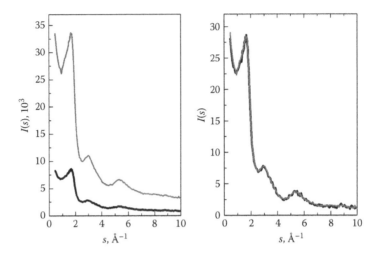

FIGURE 1.17
Functions of intensity of a volatile coal, expressed in impulses (left), or electronic units (right).
Diffraction patterns were obtained in different months.

It is impossible to conduct measurement over a very large range; the available range is limited by the size of the cover of the x-ray tube and the detection unit and also by the radiation wavelength. This limitation is obvious and the dependence of the value of diffraction value on the wavelength of x-ray radiation is shown as $s = 4\pi\sin\theta/\lambda$. Indeed, s will be maximum when $\sin\theta = 1$, i.e., $s = 4\pi/\lambda$. If copper radiation is used, the maximum value s will be less than 8 Å^{-1}. Substitution of an infinite limit of integration by a final value results in a miscalculation known as the "cut-off effect"—the shorter the diffraction vector, the larger the miscalculation and the narrower the diffraction range (function of intensity). If the task is to calculate correctly the functions of atomic distribution, the use of copper radiation is undesirable; a more acceptable radiation type is K_α molybdenum.

We suggest using one more function while interpreting the data of diffractive experiment. It can be given as follows:

$$f(r) = \int_0^r 4\pi r^2 \left[\rho(r) - \rho_0\right] dr. \tag{1.10}$$

The argument of this function is the maximum level of integration. In [82], the physical meaning of the given function is defined and the whole area around the atom which is at the origin may be divided into spherical layers whose average atomic densities will be the same as that of the whole object. This fact allows us to focus on all the coordination spheres (instead of one) within the correlation radius according to common physically grounded positions.

Bregg's equation can be related to the simplified theory of x-ray radiation scattering of crystalline objects. It is based on geometry and enables us to define quickly only the positions of diffraction maxima. It does not work for liquid and amorphous materials. The d symbol indicates interplanar spacing that does not always coincide with spacing between some atoms.

The authors of some works add Muller indices to the secondary peaks on the diffraction patterns of amorphous coals (Figure 1.10). This is in disagreement with the Laue theory according to which it is not possible to obtain a reflection characteristic of a two-dimensional object from a three-dimensional one. The common opinion is that long-range order cannot be found in amorphous materials; this is not in the agreement with the theory of diffuse peaks and no valid arguments prove the contrary.

Finally, most scientists consider the boosted scattering in the small angle zone as background and instead of checking the cause of its appearance, they eliminate it. Although this kind of scattering should be separated from the scattering on the independent cluster, we think that this practice is not correct.

1.4.3.1 Materials and Methods

The objects of the research to determine the atomic structures of coals of different types include volatile, gaseous, fat, coking, lean baking, lean, and anthracite types (Table 1.2). The lean coals were classed as outburst-prone and non-outburst-prone. The coal samples were taken from different lavas and horizons of 18 mines of the Donetsk Basin (Kommunist, Trudovskaya, 13-Bis, Yasinovskaya, Chaykino, and others). Diffraction patterns of volatile, coking, and lean baking coals were obtained for both original samples and those annealed at 100°C for 1 hour.

As noted earlier, diffraction patterns should be obtained in the widest interval of the diffraction vector and may be obtained only when using x-ray radiation with the shortest wavelength; the most optimum is K_α molybdenum radiation. As we expected, this radiation is slightly absorbed by coals and roentgen analysis of the free space using θ to 2θ geometry of filming is

TABLE 1.2

Results of Chemical Analysis of Coal Samples

Coal Rank	V^g (%)	C (%)	H (%)	W (%)	Ash Content (%)
Volatile	42.9	81.9	5.6	11.5	2.87
Gaseous	35.6	85.0	5.5	7.4	1.35
Fat	34.1	86.1	5.4	1.0	1.9
Coking	23.7	89.1	5.15	1.0	3.9
Lean baking	21.4	90.0	4.94	1.1	2.79
Lean	11.2	91.8	4.55	1.7	3.5
Anthracite	7.0	95.4	2.2	4.0	1.0

Note: V^g indicates quantity of volatile matter per gram of coal matter.

impossible because scattering volume changes considerably when the scattering angle changes and measurement becomes impossible.

In this connection, we suggest obtaining diffraction patterns for thin samples (in the form of wires) that are placed along the main axis of the diffractometer (the axes of the sample and goniometer are matched) in a "washing" x-ray (similar geometry was used in Debye chamber). Indeed, at all scattering angles, the scattering volume remains the same. We first tried that method on coals of several ranks and in all the cases got positive results. Then we used the same method to take diffraction patterns of all the objects listed in the table. The correct structural factor (curve 2) was obtained by taking a diffraction pattern of a thin cylindrical sample according to the described method but the false structural factor (curve 1) came from a massive flat sample (see Figure 1.16).

1.4.3.2 Results and Discussion

The diffraction patterns of the coals of all the ranks made into samples in the form described in the preceding section were made with two radiations (K_α molybdenum [K_αMo] and K_α cobalt [K_αCo]. The diffraction patterns were taken at a 2θ angular range: 3 to 90 degrees for K_α Mo radiation and 2 to 40 degrees for K_α Co radiation. Based on the diffraction patterns of the Mo radiation, we calculated the structural factors and all the functions of atomic distribution and determined the basic structural characteristics that were later used for simulating the coal structure. The Co radiation studies were performed to determine the physical nature of small-angle scattering. The experiment on coals of several ranks repeated as number of times within 1 month. No changes were found (see Figure 1.17b).

Figure 1.18 shows coal diffraction patterns. The patterns indicate that the functions for coals of different ranks differ significantly. After we calculated

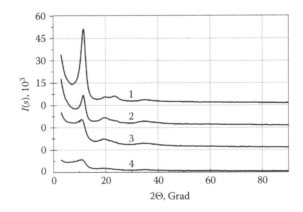

FIGURE 1.18
Diffraction patterns of anthracite (curve 1), lean (2), fat (3), and volatile (4) coals.

and drew graphs of the total functions of atomic distribution (Figure 1.19), we noted that the differences in those functions were not as considerable as those for the intensity function. (Later we will show how that abnormality can be explained.)

Because the basic element of all the coal ranks is carbon, we also studied spectra of pure graphite. One of the fat coals crystallized at room temperature. The first diffraction pattern of that coal was taken on July 21, 2004, and the second on October 25, 2004. When we obtained the first diffraction pattern, we noted a diffuse function of intensity characteristic of the amorphous phase (dashed line in Figure 1.20). The second pattern revealed a function characteristic of crystallized objects (solid line). The rank of the coal that

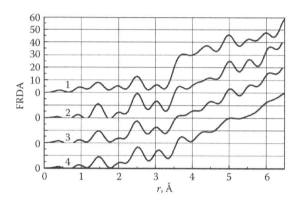

FIGURE 1.19
Radial atomic distributions of anthracite (curve 1), lean (2), fat (3), and volatile (4) coals.

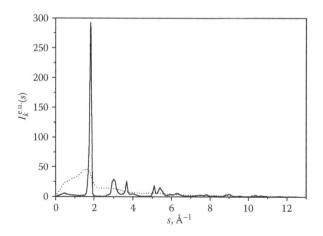

FIGURE 1.20
Diffraction patterns of fat coal taken on different dates. Dotted line: July 21, 2004. Solid line: October 25, 2004.

showed that crystallization was unclear and we have not been able to repeat the result on other fat coals. However, if we compare the roentgenograms of the crystallized coal and graphite, both diffraction patterns reveal the same set of diffraction peaks and their intensities are very similar (the difference represents experimental error), but their positions shifted. Diffraction peaks of coal are not always aligned to smaller values of diffraction vectors; this is connected to the increased parameters of the elementary cell.

After we studied the structures and crystallization peculiarities of amorphous alloys containing metals and metalloids, we obtained convincing evidence that the composition and atomic packing of clusters within amorphous strands are the same as those that appear in the early phases of crystallization [88]. This result enabled us to assume that the main structural constituent of an amorphous coal is a cluster with atomic packing of the graphite lattice type. We now discuss experimental details that clearly prove the validity of the assumption.

Note in Figure 1.21 the diffraction patterns of the crystalline fat coal (top) and spectral pure graphite (bottom). The same diffractive reflexive actions are revealed by both the coal and the pure graphite.

Furthermore, analyses of the elementary cells of graphite and the functions of atomic distributions of amorphous coals and crystalline graphite also prove the validity of our assumption. Indeed, the graphite lattice is hexagonal. The base of the lattice is a diamond pattern with parameters of $a = b = 0.246$ nm and $c = 0.671$ nm. The shortest atomic distance (between the atom is inside the base and three atoms of the diamond) was 0.142 nm (Figure 1.22, atoms c and d). The radius of the second coordination sphere equals the lattice parameters a and b (0.246 nm), that is, it is about 1.5 times the shortest distance. Apparently, because of the complete function of radial atomic distribution, the first coordination peak appeared completely isolated (see Figure 1.23)—an indication of a cluster structure of a coal. The figure shows complete functions of atomic distribution of pure graphite (crystalline) and amorphous lean coal. In both graphs, the first two coordinate spheres nearly match. This proves that the main structural constituents of amorphous coals are clusters with atomic packing graphite lattices.

This conclusion is also confirmed in Figure 1.24 that shows the complete function of radial distribution of lean coal. The vertical legs show coordinate spheres of crystalline spectral pure graphite (hexagonal lattice, $a = b = 0.2464$ nm and $c = 0.6711$ nm). Although coordinate spheres of graphite do not match the peaks of distribution function perfectly, they describe all the peaks in detail. The observed differences in the parameters of the graphite lattice and the coal may be explained by the fact that the clusters contain atoms of elements other than carbon; they likely are atoms of hydrogen. The atomic concentration of hydrogen is second to carbon in coals of all ranks. Atomic concentration of other constituents of coal content is too small to produce such considerable changes of parameters. It is possible that all coals except types containing clusters with atomic packing of solid solution also contain small amounts of different atomic packing.

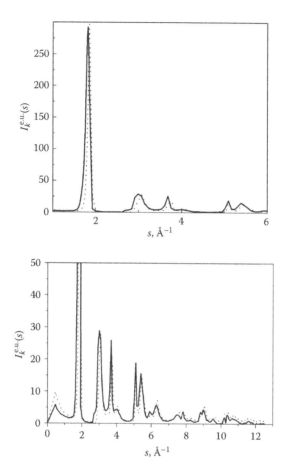

FIGURE 1.21
Diffraction pattern of crystalline fat coal (solid line) and spectral pure graphite (dashed line), (top) Stretched x axis. (bottom) Stretched y axis.

As noted earlier, in the diffraction patterns of all coals, at a scattering angle range of 3 to 7 degrees, even with Mo radiation, one can observe considerable scattering whose intensity increases when the scattering angle decreases. A common opinion is that this is a background phenomenon but we do not agree.

We carried out detailed measurements near the original beam with K_α Mo and K_α Co to define the nature of such scattering. Diffraction patterns were taken on an ordinary x-ray diffractometer (DRON-3) to exclude air scattering and influence of the "tails" of the original beam. We measured the intensity of scattering without a sample (Figure 1.25, solid curve) over the whole range before measuring the sample (dashed curve) and calculated the value of the first from the value of the second. Before we made an exposure of air scattering

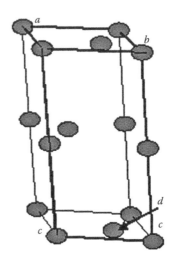

FIGURE 1.22
Elementary cell of graphite lattice.

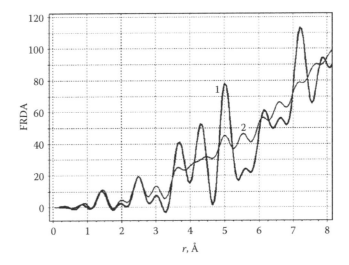

FIGURE 1.23
Radial atomic distributions of pure graphite (1) and lean coal (2).

of the original beam, we installed a special lead trap that completely absorbed the original beam but did not influence the intensity of scattering.

Figure 1.25 shows that after deduction of air scattering, the intensity in the zone of small angles is very high. Moreover, the diffraction patterns of most coal ranks taken in K_α Co radiation show clear peaks near the original beams

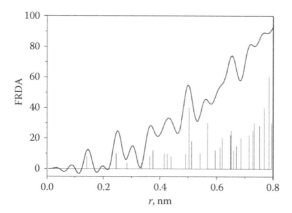

FIGURE 1.24
Radial distribution of atoms of lean coal. Vertical lines show coordinate spheres of graphite.

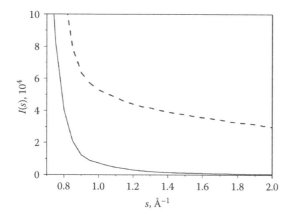

FIGURE 1.25
Fragment of diffraction pattern near original beam.

(Figure 1.26). The *d* symbol in Figure 1.26 represents cluster size, evaluated according to an equation introduced by A. F. Skryshevskiy [89]:

$$r_1 = \frac{7.73}{S_1} \tag{1.11}$$

Skryshevskiy used this equation to evaluate the most probable interatomic distances (r_1 or the principal maximum score of the function of radial atom distribution) of metal objects. In some cases, this simple equation enables high accuracy evaluation of the most probable distance according to the principal maximum of diffraction pattern.

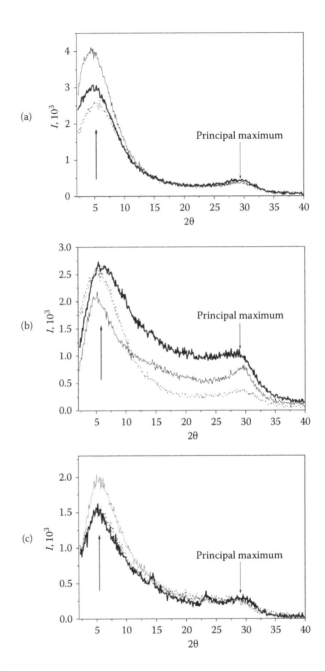

FIGURE 1.26
Diffraction pattern fragments for K_α cobalt radiation, $\lambda = 0.179$ nm. (a) Anthracite, $d = 0.257$ nm (thick line), 0.230 nm (dotted line), 0.274 nm (thin line). (b) Fat coal, $d = 0.197$ nm (thick line); lean baking coal, $d = 0.245$ nm (thin line); anthracite, $d = 0.230$ nm (dotted line). (c) Coking coal, $d = 0.229$ nm (thick line); lean baking coal, $d = 0.223$ nm (thin line); lean coal, $d = 0.238$ nm (dotted line). (d) Volatile coal (thin and dotted lines represent samples from different mines).

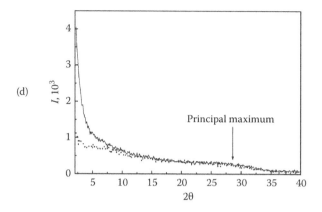

FIGURE 1.26 (CONTINUED)
Diffraction pattern fragments for K_α cobalt radiation, $\lambda = 0.179$ nm. (a) Anthracite, $d = 0.257$ nm (thick line), 0.230 nm (dotted line), 0.274 nm (thin line). (b) Fat coal, $d = 0.197$ nm (thick line); lean baking coal, $d = 0.245$ nm (thin line); anthracite, $d = 0.230$ nm (dotted line). (c) Coking coal, $d = 0.229$ nm (thick line); lean baking coal, $d = 0.223$ nm (thin line); lean coal, $d = 0.238$ nm (dotted line). (d) Volatile coal (thin and dotted lines represent samples from different mines).

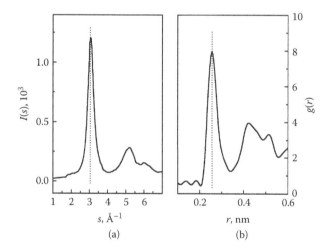

FIGURE 1.27
Intensity function (a) and relative function of amorphous alloy distribution (b) for $Fe_{73}Si_{15.8}B_{7.2}Cu_1Nb_3$. (a) $s_1 = 0.307$ nm^{-1}; $0.0773/0.307 = 0.252$ nm. (b) $r_1 = 0.255$ nm.

Figure 1.27 shows a successful application of Equation (1.11). (a) represents a fragment of a structural factor. The principal maximum score S_1 is deduced and the size (average) of the atom is rated according to Skryshevskiy's method. (b) represents fragment of relative function of atom distribution; the principal maximum score is r_1. According to Skryshevskiy's method the rated value r_1 is only 0.003 nm less than the experimental value.

Whether Equation (1.11) can be used in our case and what the measurement error of the d parameter will be cannot be answered clearly now. It is obvious that the error can be estimated only after correct retracting of the scattering present in the diffraction patterns of all coal grades in a vector area of little value. That scattering of roentgen radiation cannot be background, as noted in many articles. In our opinion, it is due to cluster property of coal. The clusters can be treated as large molecules, then the reason for full insulation of the principal maximum of radial atom distribution functions becomes clear.

As to radial atom distribution functions (Figures 1.19, 1.23, and 1.24), large oscillations were observed in the areas of small distances (before the principal isolated maximum). They were caused by the "cut-off effect" mentioned earlier. False oscillations rarely occur in amorphous metal materials. (Figure 1.28).

The cut-off effect (change of infinite integration limit for finite one) and other errors of experimental intensity functions influence the values and forms of these oscillations. We already mentioned the factors that cause errors of intensity functions. The results of coal analysis prove that multiple errors affect experimental intensity functions and calculations of atom distributions based on them. In our opinion the source of error is small-angle scattering (SAS).

SAS occurs on the significant area of the diffraction vector. This explains the peculiarities of diffraction patterns and full functions of radial atom distributions. Actually, the whole diffraction situation consists of scattering both in independent clusters (cluster factor) with the atomic packing according to the type of graphite lattice and in aggregates of clusters (SAS). The vector diffraction area where cluster factor occurs is equal to radius of the cluster but SAS circulates in an insignificant area where vector diffraction radiation activity diffused by the sample is equal to the sum of cluster factor and SAS. This causes the distortion of both the cluster factor and SAS in the

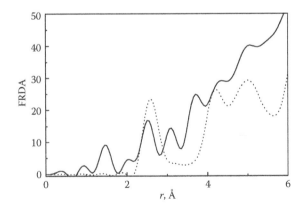

FIGURE 1.28
Fragments of full functions of radial atom distribution. Solid line = fat coal. Dotted line = $Fe_{80}Si_6B_{14}$.

given area of vector diffraction. In the vector diffraction area where SAS is missing, the cluster factor is not distorted. This can be observed when considering the structural factors of fat coal and anthracite (Figure 1.29). Both structural factors consist of two parts. One part of the structural factor (near the origin of coordinates) oscillates next to the horizontal line and the other next to the oblique line. The parts are divided by dotted lines in the figure. Figure 1.30 illustrates the same data.

Intensity functions should oscillate near the atomic factor. This is observed in metallic materials but not in coals. In one case the small-angle scattering is missing in the diffraction patterns; that is why the intensity function is not distorted. In the other case, SAS badly distorted the intensity function and most of the area does not comply with our assumption; this distortion also leads to errors in radial atom distribution.

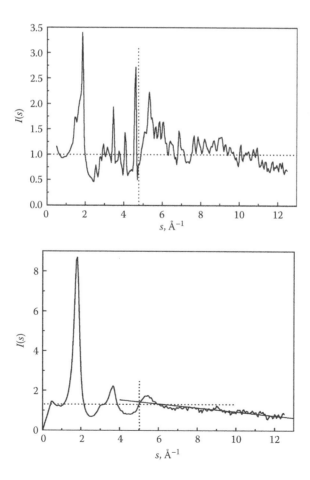

FIGURE 1.29
Structural factors of fat coal (top) and anthracite (bottom).

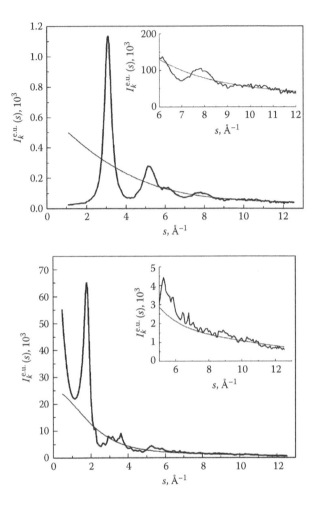

FIGURE 1.30
Intensity functions of (top) amorphous alloy $Fe_{80}Si_6B_{14}$ and (bottom) anthracite of 2-2 Bis mine (seam h_8, depth 493 m).

These observations made it necessary to find a way to isolate SAS from the general intensity function to determine more reliable intensity functions and radial atomic distributions. As noted, clusters can be treated as molecules with large numbers of atoms. Thus, to solve a problem of SAS diminution, one can use the James theory concerning scattering of coherent radiation on assemblies of polybasic molecules. Unfortunately, our attempts to solve the problem have not been successful yet, possibly because we find the function of scattering by independent clusters in one type of radiation and the total small-angle scattering in the other. We can convert only one function into electronic units and this affects the accuracy of SAS experiments.

1.4.4 Conclusions

The research discussed in this section [90] enables us to come to the following conclusions:

- The main structural component of coals of all ranks is the cluster with atomic packing of graphite lattices.
- Intensive small-angle scattering of roentgen radiation occurs in the diffraction patterns, deduced from examining coals of all ranks, and it distributes on the considerable area of the diffraction vector.
- Distinct maximums can be observed on SAS of the most coal ranks, proving the certainty of the first conclusion and confirming the dominating cluster sizes of coals.

1.5 Porosity

Sorption properties of coal and gas contents of coal seams are determined primarily by free spaces in coal materials. Volumes of cleavages and pores—classification by size under changeable conditions of carbonization—define the total amount of gas in coal. Methane and carbon dioxide are the most prevalent gas components in fluids that fill pores and cleavages. The most hazardous mine gas is methane. A sudden emission of methane can cause outbursts and explosions—and fatalities.

Under natural conditions, layers of methane accumulate in geological fault lines. These zones create the sudden and hazardous coal and gas outbursts during mining operations. Important steps for predicting outbursts include fault line prediction and drilling pilot holes to ensure coal seam drainage. Determination of volume is necessary to predict methane retention capacities of coal layers. Classifying pore structures based on their size in coals at different stages of development can determine methane content under various physical conditions.

Three types of methane conditions have been classified according to their bonds with coals: (1) free methane in cleavages and pores; (2) methane adsorbed at the sides of pore spaces; quantity is defined by the surface area of the sides of pore spaces and the active molecule sorption points of the adsorbate; and (3) methane incorporated into a coal substance similar to solid solutions or occupying very small pores comparable with methane molecules in size. In the last case, the methane molecules occur in the force fields of solid coal molecules. The methane molecules in the crystallite interlayer spaces have very specific properties and bonding energy since the distance between the aromatic layers (see Section 1.4.1) is equal to the diameter of the methane molecule: $d_{CH4} = 0.414$ nm. The sorption capacity of coal is defined by its opening-and-pore structure and by the complex physical and chemical interactions of the coal–gas–liquid system.

1.5.1 General Characteristics

All pores are divided into groups based on typical size:

Molecular pores	< 1 nm
Micropores	< 10 nm (d_m = 2 to 3 nm)
Mesopores	10 ÷ 100 nm (d_m = 25 nm)
Macropores	100 ÷ 10^4 nm

The larger voids in which one of the space coordinates is considerably superior to the two others (cylindrical and tunnel forms) relate to cleavage. The number of openings significantly influence coal mechanical properties, particularly its strength because the cleavages expand under the stress of the material and thus lead to coal destruction. Coal cleavage (failure) expands significantly during coal face movement in mines and during explosions.

Porosity classification is based on research with sorption methods. Voids in a volume of pore solid may be connected to the surface by a cylindrical or winding channel system. Closed pores may lack ways to reach the surface and only solid diffusion allows fluid transition to the pores [91,92]. In fact, this diffusion takes place through molecular-size channels in solid coal. Closed pores can be large and contain large volumes of gas (Figure 1.31). The detailed model explains the large volume of methane in coal and mass transitions of fluids in porous media.

The pore classification can be related to gas transition mechanisms in pores. Folmer diffusion takes place in the pores of the smallest diameter

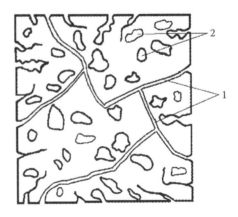

FIGURE 1.31
Models of pore structures of coal species. 1. Open porosity. 2. Closed porosity.

when gas molecules migrate along the pore sides, because the transfer in three-dimensional space is limited. The Folmer diffusion proceeds under the pore width (diameter), which is much less than free path of the molecule (d_p $\ll\lambda$) in which molecules of gas diffuse along the pore sides in adsorption layers of little density. In the first adsorption layer, molecules are bonded strongly. Some parts of molecules are involved in diffusion in which energy exceeds the necessary activation energy:

$$D_f = \text{const} \cdot f(T) e^{\frac{Q-E_a}{RT}}, \tag{1.12}$$

where Q is adsorption energy and E_a is activation energy. $E_a < Q$, so the exponent is positive in the equation and the coefficient of Folmer diffusion is decreased by the increase of temperature. Another limiting case is free diffusion carried out when the diameter of a coal pore is larger than the length of the free gas molecular path ($d > \lambda$). The coefficient of diffusion is described by Einstein's formula:

$$D_{free} = \frac{1}{3}\lambda \upsilon_{m.aver} = \frac{\text{const}}{Pd_m \sqrt{M}} \frac{T^{2.5}}{T+c}, \tag{1.13}$$

where λ is the average value of the free molecular path of methane ($\lambda = 10^2$ nm); $\upsilon_{m.aver}$ is the average speed of molecules; d_m is diameter of the molecule (for methane $d_m = 0,416$ nm); M is the molar mass for methane (0.016); P is gas pressure; T is temperature; and c is Sutherland's constant. Knudsen diffusion occurs at pore diameter, which is less than the free molecular path ($d < \lambda$) [93]:

$$D_K = \frac{1}{3}d_p \upsilon_{m.aver} = \frac{1}{3}d_p \sqrt{\frac{8RT}{2\pi M}}, \tag{1.14}$$

where d_n is pore diameter and R is the universal gas constant. Pressure does not influence the coefficient of Knudsen diffusion, but it depends on temperature T. D_K is usually one order higher than at free diffusion, D_{free}. In Knudsen's pores (10 to 100 nm), gas flows in the area of long-term laminar filtration. The number of collisions of gas molecular with sides of pores exceeds the number of collisions of molecules. In this case, the Knudsen number (ratio of mean molecular range to diameter of channel) is more than one.

Diffusion in solids occurs in pores of small diameter, where adsorption areas superimpose on one another. The coefficient of diffusion in solid body depends on activation energy:

$$D_S = D_0 e^{\frac{E_a}{RT}}, \tag{1.15}$$

where D_0 is a pre-exponential factor whose value depends on properties of both the adsorbed substance (gas) and the adsorbing substance (coal). The macropores and cracks of natural genesis sized 10^3 to 10^4 nm are characterized by the Knudsen number less than 0.01. Mixed laminar and turbulent filtration works in that situation. The gas flow becomes viscous and is described by Poiseuille's equation.

The concept of molecular-sized micropores in a diffusional field implies that the sorption field occurs in the total area of micropores at sorption interactions of any origin. The sorption in micropores is characterized by three-dimensional filling of all their spaces. Therefore the main parameter characterizing the sorbent becomes micropore volume. The proportion of micropores in the total is not large, as proven by later studies [39].

For a theoretical description of the sorption processes, the theory of three-dimensional filling of micropores (TTFM) was applied along with Langmuir theory and written clarifications and modifications. The authors [94] consider adsorption in coal to be monomolecular. Attempts were made to divide the process of sorption into stages. The initial stage is described by the Langmuir equation for multilayer coating; subsequent stages are described by the Flory theory for bulking and polymer solutions [95].

Dissolved methane is hardly suitable for ordinary commercial use, but it has some practical use based on its gradual long-term blowing and theoretical utility due to swelling caused by its presence in carbon matrices. The ability to differentiate absorption and adsorption was devised early in the twentieth century [96]; differentiation is now performed by processes of transfer [97] based on mathematical modeling.

The absorption of molecules occurs in free spaces of macromolecular lattices due to the three-dimensional filling mechanism and may be interpreted on the basis of the micropore adsorption theory according to Dubinin's equation. Isotherm via adsorption in micropores represents the fourth type of methane–coal bond; absorption in macromolecular structures of carbon matrices constitute the fifth type of bond of methane in coal. The sorption of methane in coal based only on TTFM can be identified as a form of adsorption in micropores only if curves of adsorption and desorption are reversible. If desorption is measured simultaneously, the irreversibility of the isotherm of sorption is shown clearly, and certain presuppositions are used, the joined isotherm can be determined and divided into adsorbed and occluded segments [98].

The most successful explanation of gas in coal is the theory of closed pores developed in 1992 [39]. Pore capacity is characterized by size, form, and number of pores. Pores of coals and other solids [99,100] are classified as open (vesicles and channels connected to surfaces) and closed (not connected to surfaces). The division is apparent after a comparison of gas diffusion coefficients in the tightest filtration channels to the coefficient of solid diffusion; the difference equals four to five orders of magnitude. Open pores are connected to surfaces by a system of cleavages that allow gases and liquids to quickly enter into and exit from coals. Infiltration of fluids in closed areas not connected to the surface can be realized only by solids diffusion, which explains essential process time. This concept explains pore values up to 0.3 cm³/g determined by gas sorption under pressure; most other methods reveal pore values up to 0.1 cm³/g. The distribution of porosity during metamorphism is represented in Figure 1.32. The closed pores act as accumulators and allow long-term gas storage. Based on results, closed pores constitute more than 60% of pore volume. The tendency toward closed porosity increases the likelihood of dangerous outbursts.

The methods of coal porosity determination are based on the open porosity method and may be applied to other sorbents [27], for example:

1. A mercury injection method is based on mercury meniscus curvature radius R determination at its indention in pores after a pressure boost ΔP. With a known mercury tension ratio, the radius of transport link of the open pore is defined by an equation obtained from the Laplace formula $\Delta P = 2\sigma/R$. This method calculates a correction for compressibility of coal. The maximum coal porosity based on this method under 2 GPa pressure is 15%. The shortcoming is that atoms of mercury cannot enter channels smaller than 10 μm in diameter under low pressure.

2. Adsorption of nitrogen and argon at low temperatures allows definition of the surface areas of pore spaces, but micropores about 0.5 nm in size are unavailable for these molecules.

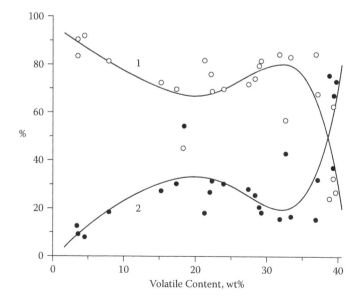

FIGURE 1.32
Dependence of closed (1) and open (2) porosities on volatile content.

3. The helium approach is the most appropriate for calculating real density and porosity because the helium atom is far smaller than atoms of other sorbates. Helium is considered *not* to be sorbed at pore sides and this allows accurate determination of pore volume.

It has been stated that open porosity determination approaches cannot yield values of complete porosity because they do not consider closed porosity. It is obvious also that at low temperature adsorption, N_2 molecules at 77 K cannot infiltrate space areas (< 0.4 nm) between aromatic crystallite scales (lamellae), and so they also represent the properties of molecular compounds. This also explains low values ($\leq$ 10 m^2/g) for the defined specific areas in comparison to the high values obtained in micropore analysis (100 to 300 m^2/g) [101]. Micropore analysis is usually performed with carbon dioxide at 273 K. Although the molecule sizes are equal, carbon dioxide at 273 K can reach ultra-micropore areas (< 0.4 nm) of coal. N_2 at 77 K is unable to do that because of diffusion limits [102,103].

The methods of porosity evaluation mentioned above introduce values less than 0.15 cm^3/g although the summary volumes of closed pores may exceed 0.3 m^3/g (more than the value of open porosity of deposit coal). Sorbent and bottle methods allow better approximations of total porosity. The first results of sorbent properties of deposit coal were published in 1941 [104]. Coal absorbed sufficient quantities of methane and carbon dioxide. Porosity values that exceeded the values found by different methods were obtained, but error rates for determining total porosity reached 100 to 200%. The results of more effective determination of the total porosity for coals of the Donets Basin are introduced in a later monograph [105].

Langmuir's theory proposed for describing adsorption on the interphase boundary suggests that the surface of an adsorbent presents a number of equivalent adsorbent centers. During absorption at one adsorbent center, only a single molecule may be absorbed (one-layered character of absorption). No interaction occurs between the adsorbed molecules. Such conditions describe an ideal process of absorption and their application to real systems is limited [99].

Brunauer, Emmet, and Teller [130] suggested an equation of absorption (BET) after refusing to use Langmuir's postulate of the one-layered character of absorption. Their multi-molecular BET method, which is usually in agreement with other methods of surface distribution showed fundamental differences between deposit coal and other sorbents for the first time. The specific surface obtained at room temperature based on wetting heat and absorption with different substances constituted hundreds of square meters for coal. At temperatures of 77 and 90 K, the area was several square meters or less [106]. Such divergence appeared in a comparison of the BET method and small-angle neutron scattering [15].

Analysis based on the above theories indicates that for the evaluation of any porous body as a sorbent it is necessary to study the parameters of its porous structure in detail. Furthermore, differential porosity is a very important aspect of pore distribution; the size of a sorbent determines the appropriate theory for describing the sorbent process.

1.5.2 Neutron Scattering

It is more reliable to evaluate total porosity using small-angle x-ray scattering (SAXS) and neutron scattering (SANS) [107,108]. SAXS can trace the development of coal porosity based on degree of metamorphism. Hirsch originally used the SAXS method during his work on structural models of coal to determine the parameters of the porous structures of coal (see Section 1.4).

The methods of small-angle scattering are non-destructive; samples are not subjected to chemical interactions and the methods are effective for open and closed pores. SAXS may be used to study micropores of diameters from 1 nm to 2 μm. This range coincides with the range of pore sizes in coal.

Polydisperse systems of porous carbon materials like coal scatter in small angles. This scattering is created by a difference of electronic density between the voids and the hard carbon matrix; we ignore mineral contents and regard coal as a two-phased substance consisting only of voids and densely packed carbon material. Pore distribution is statically isotropic. Thermal neutrons used in SANS are scattered non-uniformly. The regime of small-angle scattering infers that the scattering angle $\Theta < 5$ degrees. Thus, the scattering vector is $K = (4\pi / \lambda)\sin(\Theta / 2) \approx 2\pi\Theta / \lambda$ where λ denotes wavelength. To calculate pore distribution based on sizes obtained from SAS, the Fourier transform is used. The pores sizes studied via SAS are determined by $d = 2\pi/K$.

At present, SAS methods are powerful tools for studying mining deposit structures containing complex systems of pores [109–111].

According to the theory of small-angle x-ray scattering, if we take the random distribution of scattering pores, the intensity in the range of small angles scattered by spherical particles is described by Guinier's method of approximation [108]:

$$I = I_0 \exp\left(-\frac{(RK)^2}{3}\right), \tag{1.16}$$

where I is the intensity of scattering; I_0 is the intensity in the zero angle of scattering; and R is the radius of inertia of scattering particles. The value of R is determined by the slopes of the straight lines in the coordinates: ln $(I) = f(K^2)$.

The analysis of the internal part of the scattering curve yields data about the inertia radius (R) of a pore. The external part of the curve of small-angle scattering indicates pore shape. It is inversely proportional to the fourth degree of a scattering vector for three-dimensional pores that are restricted by smooth surfaces according to Porod [112], but inversely proportional to the second degree of a scattering vector for random-oriented flat pores [113,114]. For random-oriented fine, narrow pores, the scattering curve is inversely proportional to a scattering vector [108].

For coal showing very small angles for extrapolation of a scattering curve [16], Guinier's law is applied to extrapolate the experimental data to small values; Porod's law may be applied for big angles. A coal may reveal a wide area where intensity is proportional to a non-integer of a negative degree of a scattering vector K [42] that contradicts the traditional SAXS theories [108]. This behavior may be characteristic of independently scattering micropores distributed according to size based on the power law [114] or for macropores with fractal surface boundaries.

The pore distribution in coal samples varies based on stage of coalification. Thus, the function $P(r)$ for the coal at the lowest degree of metamorphism reflects non-uniformity of structure, mainly mesoporosity with typical pore size of 6.0 to 8.0 nm. Pore distribution according to $P(r)$ values for Mericourt mine coal (Figure 1.33) shows significantly transient porosity or mesoporosity not exceeding 20 nm (mean size about 8 to 10 nm). For coal of the highest degree of metamorphism from the Escarpelles mine, $P(r)$ is characterized mainly by mesoporosity in the range of 8 to 10 nm and 4 nm and also by micropores smaller than 2 nm.

Studies of the coals of two American basins were conducted [111]. Applied methods included SAXS, SANS, and infrared spectroscopy for samples of a wide range of coalification (Figure 1.34). The coefficient of vitrinite reflection was $0.55 \div 5.15\%$; carbon content was $66.7 \div 92.75\%$. One typical feature of pore size distribution (differential porosity) was power dependence:

$$f(r) = Ar^{-B}. \tag{1.17}$$

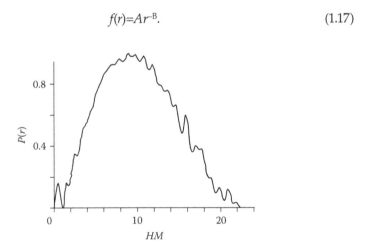

FIGURE 1.33
Pore radius density distribution derived from small-angle scattering for coal showing mean degree of coalification. (Adapted from Rouzaud J.N. and Oberlin A. 1990. *Characterization of Coals and Cokes by Transmission Electron Microscopy.* Amsterdam: Elsevier, pp. 311–355.)

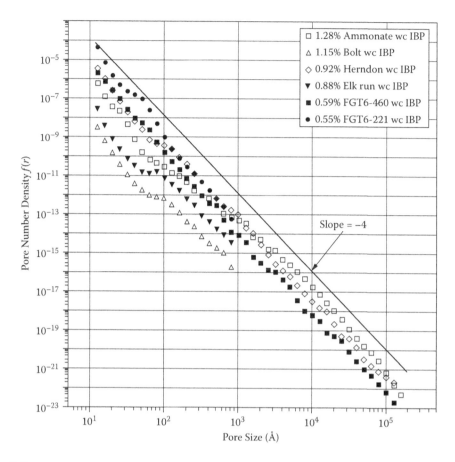

FIGURE 1.34

Distribution of American coal based on pore sizes determined by SAXS and SANS. (Adapted from Radlinsky A.P. et al. 2004. *Int. J. Coal Geol.* 59, 245–271.)

Power value B varies in the range of $3.53 \div 4.79$, suggesting fractality of the specific surfaces of pores of sizes $D_s = B - 1 = 2.53$ to 3.0 (see Section 1.5.3). The number of micropores primarily determines the specific surface. The distribution of total porosity (closed pores and pores connected to the surface) was minimum in the coal metamorphic sequence coinciding with carbon content of 87 to 90%.

The contradictions among measurements of coal porosity via SAXS, SANS, and sorption experiments at low temperatures are ascribed to different values of the measured surfaces of the samples. If in the process of adsorbing nitrogen using the BET method, the $S/V \approx 4{,}7 \times 10^3$ cm^2/cm^3 equation was obtained, SANS suggests the specific surface values are five to ten times greater [111], especially for micropores (Figure 1.35). The results of a comparison of both methods using coal from the Ruhr Basin are shown in Table 1.3 [115].

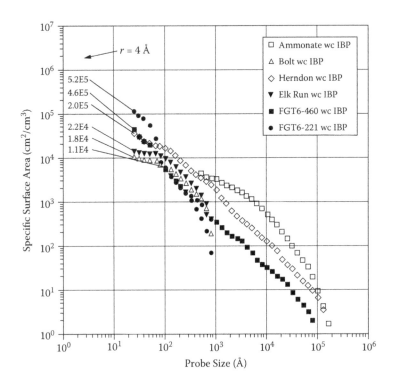

FIGURE 1.35
Specific surface depending on pores size for coal samples oriented parallel with bedding.
(Adapted from Radlinsky A.P. et al. 2004. *Int. J. Coal Geol.* 59, 245–271.)

TABLE 1.3

Parameters and Specific Surfaces Obtained via
Small-Angle Neutron Scattering and Adsorbing
Experiments

Sample	D_S	S/m(cm²/g), SANS	S/m (cm²/g), Adsorption
2	2.40	24.62	2.24
3	2.33	55.82	1.48
4	2.27	93.18	0.90
6	2.38	18.46	1.14
8	2.22	129.11	0.53
9	2.13	363.48	0.33
10	2.29	43.88	0.59

Source: Prinz D. et al. *Fuel* 38, 547–556.

The reason for deviations of the figures from neutron and x-ray scattering in sorption experiments is the presence of closed pores in complex coal structures. They are included in the macromolecular lattices. The adsorbing experiments allow determination only of open pores.

In fact, the results of adsorbing N_2 and SANS experiments on the Ruhr basin coal showed essentially positive correlations. In the initial stages of coalification, the data from N_2 and SANS experiments start to diverge as coalification decreases since neutrons reach all the pores in the ranges of the resolving power while the N_2 molecules do not reach the pores at 77 K. We remind you [116] that the pore surface (square centimeters/gram) is determined by the formulae for cylindrical forms that have a mean radius $\bar{r}_p$:

$$S = \frac{2{,}0 \cdot V_p}{\bar{r}_p} 10^4.$$ (1.18)

For spherical forms:

$$S = \frac{3.0 \cdot V_p}{\bar{r}_p} 10^3,$$ (1.19)

where V_p represents specific pore volumes expressed in cubic centimeters/gram, and $\bar{r}_p$ indicates mean radii of pores in nanometers. This means that power distribution of pores in a wide range of sizes over most of a specific surface will be determined by micropores. In fact, the specific surfaces of macropores are sufficiently small in comparison with those of micropores [99,116].

Only low metamorphic coal with a small content of vitrinite (<1.1% *VRr*) shows wide distribution according to the sizes in mesopore and macropore ranges. In highly metamorphic coal (>1.90 to 2.23% *VRr*), sufficient mesoporosity is not observed.

The structural parameters obtained from adsorbing and SANS experiments show exponential decay with increasing of coalification degree (*VRr*). It is obvious that the structural conversion in early coalification occurs mainly because of loss of oxygen of functional groups. Using a specific surface as a parameter, we found out that mesoporosity and macroporosity are one to three times less than microporosity. Distribution of pores according to radii is fractal, with fractal dimension function $2.13 < D_s > 2.78$. Study results indicate a two-phase model coal substance structure is considered to have a structural organization similar to a macromolecular matrix (Section 1.3).

1.5.3 Fractality

The fractal peculiarities of coal structures demand special studies because the theory of fractal geometry is widely used in natural sciences [117], particularly in geology and geophysics [118–120]. Methods of studies of coal porosity yield different kinds of pore distributions according to sizes. Due to the limitations of sorbent methods at low temperatures and mercury porometry. the density of pore distribution based on size is obviously unimodal and bimodal in cases of two pore systems. Figure 1.36 shows such distribution. The limitations of these methods mean that the numbers of molecular pores and micropores are underestimated. The calculation involves the fractal dimension function D (fractional exponent for probable density) and the integral function of pores distribution according to size:

$$P(r) \sim r^{-D} \quad \text{and} \quad f(r) \sim r^{-(D+1)}. \tag{1.20}$$

These functions are linked by

$$f(r) = \frac{d}{dr} P(r) \tag{1.21}$$

Taking into account the index B in probability density

$$f(r) = A \cdot r^{-B} \tag{1.22}$$

From (1.20) and (1.21) we obtain $B = D + 1$. After integration over the range of pores we obtain for the mean pore size [121]:

$$\bar{V}_P = \frac{B-1}{4-B} \pi^{\frac{4-B}{3}} \left(\frac{4}{3}\right)^{\frac{2-B}{3}} \cdot r_{min}^{B-1}\left(r_{max}^{4-B} - r_{min}^{4-B}\right) \tag{1.23}$$

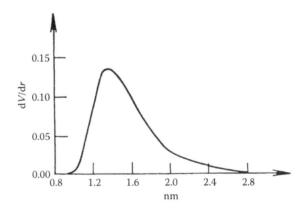

FIGURE 1.36
Distribution of micropore volume based on its diameter in coal containing 86.3% carbon. (Adapted from Medek, J. 1977. *Fuel* 56, 131.)

Two variants are possible:

$$\text{If } B < 4, \text{ then } \bar{V}_P = \frac{B-1}{4-B} \pi^{\frac{4-B}{3}} \left(\frac{4}{3}\right)^{\frac{2-B}{3}} \cdot r_{min}^{B-1}\left(r_{max}^{4-B}\right) \tag{1.24a}$$

$$\text{If } B > 4, \text{ then } \bar{V}_P = \frac{B-1}{4-B} \pi^{\frac{4-B}{3}} \left(\frac{4}{3}\right)^{\frac{2-B}{3}} \cdot r_{min}^{B-1}\left(r_{min}^{4-B}\right) \tag{1.24b}$$

These two equations give us the opportunity to determine variations of average pore volumes at the known index B in pore distribution according to their sizes. Besides, if total porosity θ is determined, we can find total number of pores in a unit of volume $\bar{n}_p$:

$$\theta = \frac{V_p}{V} = \bar{n}_p \cdot \bar{V}_p , \tag{1.25}$$

where Vp and V indicate pores and body volume, respectively; $\bar{n}_p$ is the mean index of pores in a unit of volume, and $\bar{V}_p$ is mean pore size. Small-angle methods of scattering examine the entire pore system including ultra-microscopic pores in the distribution. This distribution has a scaling organization. Based on the radii of pores in the distribution, we can determine distribution by volume [122]. We start with the relation known from the theory of relativity: $f(r)dr = f(V)dV$. For spherical pores we have

$$f(V) = A'\left(V^{\frac{B+2}{3}}\right)^{-1}. \tag{1.26}$$

That is why we can note that

$$f(V) = A' \cdot V^{-\left(1+\frac{D}{3}\right)}. \tag{1.27}$$

As $B > 1$ and $r_{min} \ll r_{max}$, it is sufficient to note that $A = (B-1) \cdot r_{min}^{B-1}$. Then we obtain for distribution according to the pore volume assuming its spherical forms:

$$f(V) = A'\left(V^{\frac{B+2}{3}}\right)^{-1} \tag{1.28}$$

where the normalizing constant:

$$A' = \frac{B-1}{3}\left(\frac{3}{4\pi}\right)^{\frac{2}{3}} \cdot r_{min}^{B-1} \tag{1.29}$$

For the scaling dependence of pore distribution according to the sizes, we can obtain the formula for the total surface S of pore area [117]:

$$S = C \int_{r_{min}}^{r_{max}} r^2 f(r) dr \tag{1.30}$$

where parameter C indicates geometrical forms of pores. After the integration we obtain:

$$S = \frac{C}{B-3} \left(\frac{1}{r_{min}^{B-3}} - \frac{1}{r_{max}^{B-3}} \right) \tag{1.31}$$

As shown by experiments, the conditions $B > 3$ and $r_{min} \ll r_{max}$ are always true. That is why it is sufficient in Equation (1.31) to restrict the components in the brackets. Note that the value $B = 4$ is critical for distribution of pores according to sizes [123]. Under the condition $B < 4$, fractality for pore surfaces is realized because $2 < D < 3$. In accordance with Equation (1.24a), the volume of pore area equals minimum r_{min} and maximum r_{max} pore sizes. At $B > 4$, the pore volume will be determined by pores of minimum sizes. During uploading of a coal layer near the face area or breaking a layer, the number of large pores and cracks increases; methane enters more easily in layers containing smaller pores. As a result, rapid explosion or slow methane efflux may occur, depending on the index of power distribution of pore sizes.

In accordance with [124], when the pressure reaches 20 MPa, the volume of macropores changes 27 times; the volume for micropores changes 2 or 3 times and remains almost the same for pores of 1.25 to 7.5 nm size. These data were experimentally substantiated by evaluating mean pore size using Equation (1.23). The indices B obtained by the SANS method to determine uniaxial loading of T (lean) rank coal up to 1.6 GPa [125] changed in the range of 3.95 to 4.56. Under conditions of minimum and maximum pore sizes [Equation (1.23)], it follows that after loading the pore volume was 0.12 of the initial volume.

Evaluation of pore volume via scaling distribution with the index $B = 3.5$ substantiated the conclusions mentioned above about the dominance of macropore volume in the total coal pore area. Table 1.4 shows the results of the evaluation [121]. Total porosity $V_p/V = 0.116$.

From Table 1.4, it follows that most pores are smaller than 10 nm and thus specific surface characteristics are determined by micropores. In fact the specific surface of macropores is sufficiently small in comparison with that of micropores [116,126]. However, the volume of pore area is mostly determined by macropores of diameter $d > 10^3$ nm.

Because pore geometry varies based on degree of metamorphism, it is essential to consider this factor when evaluating porosity during loading

TABLE 1.4

Pore Characteristics for Scaling Distribution with Index B = 3.5

Number	Size Range (nm)	Relative Content	Mean Pore Size (nm³)	Total Pore Volume (Range = 1 m³ × 10⁴)
1	1 to 5	0.982	3.1	9.3
3	5 to 10	0.0147	157	6.9
4	10 to 100	3.15×10^{-3}	5.4×10^3	51.2
5	10^2 to 10^3	1.0×10^{-6}	5.4×10^8	162
6	10^3 to 10^4	3.1×10^{-8}	1.0×10^{10}	927

changes. One factor that may cause changes of pore distribution slope may be pore volumes of different geometries. Micropores are spherical. Open pores are more extended in one dimension. Some cracks have slot-like shapes. If spherical and cylindrical pores are present, they will change their volume when stress level changes in a rock massif. Taking into account Hooke's law and isotropic compression, we obtain relative changes of volume of spherical and cylinder pores:

$$\frac{\Delta V}{V_0} = -\frac{3 \cdot \Delta \sigma}{K} \tag{1.32a}$$

$$\frac{\Delta V}{V_0} \cong -\frac{2 \Delta \sigma}{E} \tag{1.32b}$$

where E is Young's modulus, $K = E / 3(1 - 2v)$ is the module of all-round compression, and v is Poisson's coefficient.

Modeling of real distribution of pores is possible if contributions of different kinds of distribution are approximately known: (1) the scaling invariant distribution for the whole range of pores and (2) Weibull or lognormal distribution of some macropores and cracks. These kinds of distribution are shown in Figure 1.37. When the range of scales is quite large (about five times), the differential porosity of geomaterials is described by scaling distribution [119]. However for some experimental data, it is preferable to describe porosity by logarithmical normal distribution and Weibull distribution.

Study data [27,127] can be considered as examples of two-component distributions of pores according to volumes. Two kinds of distribution of pores observed are described by power dependence with different indices (Figure 1.38).

In conclusion, each researcher, despite thoroughly analyzed (and sometimes inconsistent) views, may choose his or her representation of coal structure. We have been guided by the spatial model of Kasatochkin and obtained a number of fundamental and applied results that we will discuss in more detail in this and other chapters.

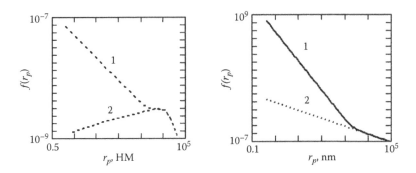

FIGURE 1.37
Kinds of pore radius r_p distribution of spherical (1) and cylinder pores (2) Scaling distribution (power law) is implied for spherical pores and Weibull distribution for cylindrical pores.

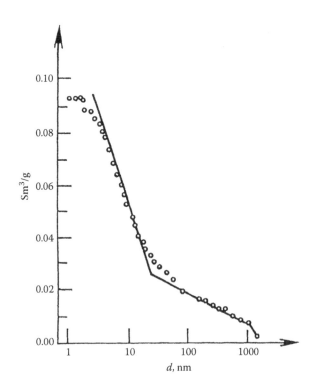

FIGURE 1.38
Distribution of total pores in coal with 77.8% carbon content. (Adapted from Gan H. et al. 1972. *Fuel* 51, 272–277.)

1.5.4 Closed Porosity of Donbass Coals

Figure 1.39 represents the distribution of molecules of gas in a coal substance near closed pores. One can think that the division of pores into closed and opened is rather theoretical due to gas possible migration into closed pores. However, such division becomes clear when comparing the diffusion coefficient values for the narrowest filtration channels (not less than 10^{-4} cm²/s) and for solid-state diffusion (~10^{-9} 10^{-8} cm²/s.)

As the results of electronic and microscopic studies show, pores of vitrinite, a gel-like substance, have roundish, spherical forms (maturing stage volatile; gaseous and fat coals) or roundish, spindle, or cluster shapes (coking, lean baking, and lean coals) 0.9 to 1.0 nm. Pores form regularly diffused congestions (layer by layer) within homogeneous layers of vitrinite and in areas where microcracks form group congestions.

Lipoid microcomponents (sporinite, cutinite, resinite) are characterized by specific structures and capacious intracavitary surfaces. The porous capacity of interlayer spaces show widespread zones of crumpling, sliding, and moving along with cavities typical of contact zones for vitrinite inclusions in a homogeneous gelified mass. As a rule, the numerous systems of split cracks of developing microcleavage run normally along the interlayer contact plane.

The opened pores are linked to the external surface of a coal sample by a system of cracks and other channels that allow various liquids and gases to enter and leave the coal mass [128]. The closed porosity of fossil coals is defined by a system of cavities of various sizes and configurations that are not connected with the external surface of the coal by transport channels. The entry into (or evacuation from) these cavities of gas molecules can be

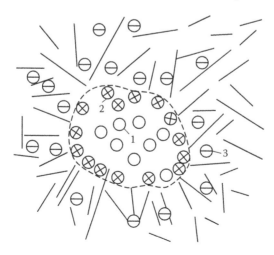

FIGURE 1.39
Distribution of molecules of gas in a coal pore. 1. Free molecule. 2. Adsorbed molecule. 3. Molecule in solid solution.

carried out exclusively by solid-state diffusion that determines the duration
of this process.

The technique was developed at the Institute for Physics of Mining
Processes of the National Academy of Sciences of Ukraine to define closed
porosity by changing the pressure of fluids absorbed by coal samples at saturation. This approach was used earlier for the definition of full porosity. To
measure closed porosity, it was necessary to

- Develop quality monitoring of drying and degassing efficiency
- Study kinetics and thermodynamics of absorption
- Perform sorbate experiments

The proposed methodology was based on the results of theoretical research
[91]. For simplicity, we assumed that fossil coal is a porous solid body (PSB)
in the shape of a sphere with volume V [$(4/3) \pi R^3$] filled with spherical pores
of radius r_p at regular intervals with density N_p and separated from each
other by distance l, considerably larger than r_p ($r_p \ll l \ll R; l \sim N_p^{-1/3}$).

The PSB was placed in a sorption ampoule (SA) with volume V_A, filled
with a gas at a constant temperature T. Gas distribution was characterized
by the density of molecules in a solid solution with $c(r, t)$, pores $\rho(r, t)$ and
free volume of chamber $n(t)$. The diffusion movement of gas in the volume
of a solid phase can occur by migration of both gas molecules and migration
of both gas molecules and molecule-vacancy or molecule-interstitial complexes or in other ways, which can be described by the magnitude of the corresponding diffusion coefficient D. The equation of diffusion of molecules
on PSB volume in the presence of pores as sinks (sources) is

$$\frac{\partial c(r,t)}{\partial t} = Dr^{-2} \frac{\partial}{\partial r}\left(r^2 \frac{\partial c(r,t)}{\partial r}\right) - 4\pi r_p^2 N_p I(r,t), \qquad (1.33)$$

where $c(r, t)$ is the density of molecules in a solid solution; t is time; D is the
diffusion factor; r is distance; r_p is the pore radius; N_p is the density of distribution of pores in volume; and $I(r, t)$ is the density of a gas stream into a pore.
The gas density in pores is changed with the speed:

$$\frac{\partial \rho(r,t)}{\partial t} = \frac{3}{r_p} I(r,t) \qquad (1.34)$$

The boundary condition of Equation (1.33) at $r = R$ is the condition of a continuity of a stream of particles through an external surface of a solid body:

$$D_0 \frac{\partial c(R,t)}{\partial R} = -I_R(t) \qquad (1.35)$$

connected directly with the speed of gas density change in free volume (in SA) $V_{fs}=V_A-V$:

$$\frac{dn(t)}{dt} = \frac{4\pi R^2}{V_{fs}} I_R(t)$$

(1.36)

The initial conditions of Equations (1.33), (1.34), and (1.36) define the initial distribution of gas molecules in the system:

$$n(0) = n_0; c(r,0) = c_0; \rho(r,0) = \rho_0.$$

(1.37)

The density of streams of molecules of gas into a pore $I(r, t)$ and through the external PSB surface I_R (t) is

$$I(r,t) = Dr_p^{-1}\left[c(r,t) - v\rho(r,t)\right],$$

(1.38)

$$I_R(T) = \sigma R^{-1}\left[c(R,t) - vn(t)\right],$$

(1.39)

where v is the constant characterizing local thermal balance of molecules of gas on the surface of a solid body defined by the equality of chemical potentials of gas molecules in a solid solution and free gas molecules and has a meaning of the specific volume of dissolution ($V = M \Omega_o \delta$); δ is the gas solubility [92]; Ω_o is the volume per molecule of gas in a solid solution; M is number of possible positions of molecules of gas in a unit of volume of a solid body (of the order of the number of atoms of a solid body in a volume unit: $M \sim 10^{28}$ to $10^{29}/m^{-3}$); σ is the diffusion factor through the PSB border, generally different from D.

The solution of Equations (1.33), (1.34), and (1.36) taking into account the conditions of Equations (1.35) and (1.37) by means of a Laplace transform with time t allows us to derive equations necessary for the further analysis of molecule density SA $n(t)$ and numbers of gas molecules $\theta(t)$ contained in a unit of PSB volume defined by

$$\theta(t) = \frac{4\pi}{V} \int_0^R drr^2\left[c(r,t) + \mu\rho(r,t)\right]$$

(1.40)

or, at very long times,

$$n(t) \sim n_\infty + \frac{V}{V_{fs}} 3Kv\exp\left(-\frac{t}{\tau}\right),$$

(1.41)

$$\theta(t) \sim (\mu + v)n_\infty - 3Kv\exp\left(-\frac{t}{\tau}\right),$$

(1.42)

where n_∞ is the equilibrium density of gas in the chamber:

$$n_\infty = \left[1 + (\mu + \nu)\frac{V}{V_{fs}}\right]^{-1} \left\{n_0 + c_0 \frac{V}{V_{fs}}\left[1 + (1-\gamma)\frac{\mu}{\nu}\right]\right\}; \qquad (1.43)$$

τ is relaxation time in the system;

$$\tau = \tau_D \xi^{-1} = R^2 (D\xi)^{-1}; \qquad (1.44)$$

$$K = f^{-1}\left[n_0 - c_0 \nu^{-1}\left(1 - \gamma\frac{\varepsilon_2}{\varepsilon_1 + \varepsilon_2 - \gamma}\right)\right];$$

$$f = \beta + \frac{(\alpha\beta - \xi)^2}{2\alpha}\left(1 + ctg^2 y - \frac{ctgy}{y}\right)\frac{\xi^2 - 2\varepsilon_1\xi + \varepsilon_1(\varepsilon_1 + \varepsilon_2)}{(\varepsilon_1 - \xi)^2};$$

ξ is the least positive root of the transcendental equation:

$$yctgy = 1 + \frac{\alpha\xi}{\alpha\beta - \xi}; y \equiv \left[\xi\left(1 + \frac{\varepsilon_2}{\varepsilon_1 - \xi}\right)\right]^{\frac{1}{2}}; \alpha = \frac{\sigma}{D}; \beta = \frac{4\pi R^3 \nu}{V_{fs}};$$

$$\tau_D = \frac{R^2}{D}; \mu = \frac{4}{3}\pi r_p^3 N_p; \varepsilon_1 = \frac{3\nu R^2}{r_p^2}; \varepsilon_2 = \frac{3\mu R^2}{r_p^2}; \gamma = 1 - \nu p_0/c_0. \qquad (1.45)$$

Analysis of the Equation (1.44) for relaxation time τ taking into consideration Equation (1.45) shows that

1. If the porosity of a solid body μ is small ($\mu \ll \nu$), according to Equation (1.45) $\xi = y^2$ and the time of relaxation τ is equal $\tau = R^2/(Dy^2)$. Thus, in case of $\alpha \ll 1$ $y = \sqrt{3\alpha}$, the speed of relaxation in a system is limited by diffusion through the PSB surface. If parameter α is not small ($y \sim 1$), relaxation is defined basically by volume diffusion.

2. For high porosity ($\mu \gg \nu$)$\xi = \nu y^2/\mu$ and relaxation time $\tau = \mu R^2/(\nu Dy^2)$ in comparison with the previous case, $\mu/\nu \gg 1$. It is easy to understand because the share of "cells" in the solid body filled with gas molecules increases in the same way. Thus $y = \sqrt{3\alpha}$ if $\alpha \ll 1$ and $y \sim 1$ if parameter α is not small. These conclusions about the dominating role of diffusion through a surface ($\alpha \ll 1$) or volume ($\alpha \geq 1$) of PSB remain valid.

The results show the defining influence of time on kinetics of transition in the balance of a gas–PSB system that is essential for conducting sorbate experiments on rocks and fossil coals. The accounts of substantial growth (in μ/v times) in highly porous ($\mu \gg v$) compared with minimally porous ($\mu \ll v$) materials may explain the discrepancy concerning isotherms based on experimental data from gas occlusion of the same porous substance (hysteresis). The discrepancy arises from the absence of a global equilibrium in the considered gas–PSB system during the final moment of time t.

Equation (1.41) shows that at a particular moment of time t $(t > \tau)$, it is possible to achieve infinitesimal value for the second component in comparison with the first one by replacing $n(t)$ for n_∞ in Equation (1.41). The experimental conditions to establish balance at a stage of gas occlusion originally degasified the PSB, so when $C_0 = 0$, we get from Equations (1.41) and (1.43):

$$n = \frac{n_0}{1 + (\mu + v)V/V_{fs}},$$

Taking into account the connection of concentration of gas molecules with pressure at temperature $T(p = nkT, p_0 = n_0kT)$, the expression for relative change of pressure is

$$\frac{p_0 - p}{p} = \frac{V}{V_{fs}}(\mu + v). \tag{1.46}$$

The estimations based on the laws of statistical physics show that $v < 10^{-4}$. For coals of $\mu \sim 10^{-1}$ order and rocks of $\mu \sim 10^{-2}$ order, rewriting Equation (1.46) is permissible and yields a simple formula:

$$\mu = \frac{(p_0 - p)V_{fs}}{Vp} \tag{1.47}$$

The size μ defined by this formula is dimensionless; it characterizes the ratio of closed pore volume to PSB volume. For practical purposes, it is more convenient to attribute the volume of pores to PSB weight, that is, to replace volume V by mass m in the denominator of Equation (1.47). The result is a basic formula useful for further research to define values of pressure p_0 and p, PSB sample mass m, and free volume V_{fp} to calculate the volume of closed pores per PSB mass unit μ (cm³/g):

$$\mu = \frac{(p_0 - p)V_{fs}}{pm} \tag{1.48}$$

Establishment time of sorption balance is different: for minimally porous PSB particles:

$$\tau_L = R^2/(Dv) \tag{1.49}$$

and for highly porous PSB particles:

$$\tau_H = \mu R^2 /(Dv) \tag{1.50}$$

where R is particle size; v is specific volume of dissolution, and D is diffusion factor. The condition of a solid substance (humidity in particular) defines the scales. An increase of humidity reduces v by blocking structural voids with molecules of water; this reduces the mobility of gas molecules in wet PSB and consequently reduces the diffusion factor.

A number of experiments have been carried out on coal samples of various ranks to study the effect of moisture content on the volumes of closed pores (μ). The results in Table 1.5 testify to the reduction in wet coal in comparison with dried coal. The table also shows the growth of saturation time (column 8). The moisture and gas contained in the organic substance of naturally wet coal partially block volumes of closed pores. Correct definition of saturation time requires preliminary drying and degasification of samples to allow the transformation of a wide line of a spectrum of a nuclear magnetic resonance (NMR) image to a Gaussian shape. The NMR spectrum of dry degasified coal (width 5 to 6 Oe) is Gaussian. The NMR spectrum of coal containing even small amounts of water and methane shows a superposition of wide and narrow lines (0.1 to 0.5 Oe). The change of character of the NMR to a Gaussian shape testifies to almost complete drying and degasification of a sample. The results cited above concerning the experimental determination of closed pores in fossil coal allowed us to formulate and implement a program for further research. Representative samples of fossil coal and rocks weighing more than 400 g were taken. The samples were crushed to sizes below 3 mm to reduce the time needed for controllable diffusion of methane molecules in coal particles [Equation (1.44)]. The time required to establish sorptive balance τ depends on linear particle size R. The reduction of the sizes of experimental particles also reduces drying and degasification times. The crushed samples were dried and degasified by heating up to 120°C and simultaneous pumping of moisture for at least 10 hours.

This research would not be possible without efficient sample drying and degasification. The wide variations in porous structures of the many ranks of coal require particular modes of drying and degasification and real-time information about humidity and gas content of samples. Only coals with identical porous structures obtained after drying and degasification can serve as standards.

The use of NMR (see Chapter 3) to monitor molecules of water and methane in pore spaces of coal and rock is invaluable because it provides quantitative information about water and methane at all stages of drying and degasification rather than after their removal (for example, by reduction of sample mass or by calculating the number of the molecules that left the samples which represents an error source). The NMR spectrum of wet gas-saturated coal is a superposition of two resonant lines: wide $\Delta H_1 = 5$ to 6 Oe from protons

TABLE 1.5

Influence of Moisture Content on Measured Values of Closed Pores Volume of Fossil Coal

Sample Number	Coal Rank	Moisture Content (W_A, %)	Sample Weight (m, r)	Free Space Volume (V_{fs}, cm³)	Sorption Ampoule Pressure (PA)		Saturation Time (τ, days)	Closed Pore Volume (μ, cm³/g)
					Initial $p_0 \cdot 10^5$	Final $p \cdot 10^5$		
1	Volatile	2.4 <0.1	320.5	224.07	11.3	10.9	10	0.027
			289	253.66	58.7	55.2	11	0.059
15	Fat	0.35 <0.1	349.9	177.68	65.7	48.75	6	0.197
			344.8	187.16	56.9	39.25	17	0.271
16	Fat	0.3 <0.1	326.4	194.63	56.5	41.8	13	0.24
			321.9	210.92	56.8	40.1	17	0.279
19	Lean	0.3	416.6	188.05	53.9	39.1	10	0.237
		<0.1	416.38	206.89	55.2	38.6	17	0.238

Note: Value for wet test is given in numerator; value for dry test is given in denominator.

of coal substance and the methane in a firm solution, and narrow $\Delta H_2 = 0.1$ to 0.5 Oe from water and methane protons and sorbate on the surfaces of pores and the cracks; the lines are of Lorentzian shape. The NMR spectrum of dry degasified coal consists only of a wide line of Gaussian shape. The moment of transition from a complex NMR spectrum consisting of two lines to a spectrum containing only one wide Gaussian line corresponds to the defined point of the dried and degasified coal.

The NMR technique for wet gas-saturated coal included a reference standard for dry degasified coal. Unlike other methods, NMR allows stage-by-stage control of the efficiency of degasification and drying of samples. The method is highly accurate; it can register 0.01% weights of sorbate, water, and methane molecules with a signal-to-noise ratio of 50, that is, the weight error does not exceed 0.0002%. The NMR spectrometer used in our studies was created and developed at the Institute for Physics of Mining Processes of the National Academy of Sciences of Ukraine.

The saturation of the dried degasified samples of fossil coal was conducted on a sorption installation with intermediate capacity (Figure 1.40). The first step was to load the sorption ampoule (SA) with the dried degasified sample of coal or rock with $m \approx 400$ g. The reservoir for the preliminary gas intake by the opening of valve K1 was filled from a transport cylinder with methane to pressure $(110 \div 130) \cdot 10^5$ Pa, registered by manometer M1. After opening valves K3 and K5 (with valve K4 closed and valve K2 opened), gas was fed into the SA. In most cases methane was forced to pressure $(60 \div 70) \cdot 10^5$ Pa. After that, valve K3 was blocked and in 15 minutes the indication of manometer M2 was registered. The volume of open pores in the sample of coal or the rock placed in the SA was filled for 15 minutes and yielded initial pressure p_0 [see Equation (1.48)].

The absorption of methane by the coal sample or rock and drop of pressure in the SA was registered by manometer M2. As sorption balance is established over infinitely long periods and is not technically realizable, we had to choose the moment of registration of final pressure in the SA. Studies of coal samples containing both closed and open pores established the presence of at least two times of relaxation: τ_1 (several minutes) and τ_2 (several days). The estimations show that τ_1 is connected with the kinetics of filling open pores and some closed pores near the coal or rock surfaces; τ_2 characterizes the process of filling all closed pores. A change of methane pressure in a sorption ampoule at t can be approximated by

$$p(t) = p_\infty + A \exp\left(-\frac{t}{\tau}\right) + B \exp\left(-\frac{t}{\tau_2}\right) \qquad (1.51)$$

where p_∞ is the equilibrium pressure and A and B are constants with pressure depending on coal sample characteristics.

The time scale τ_1 influenced the interval of time shown above by 15 minutes. To ensure the filling of methane closed pore volume and measure the

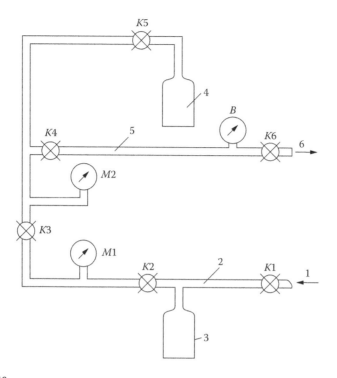

FIGURE 1.40

Apparatus to determine closed pore volume showing preliminary intake for compressed gas.
1. Channel for methane feeding from transport cylinder. 2. High pressure line. 3. Reservoir
for preliminary gas intake (to 20 MPA). 4. Sorption ampoule. 5. Low pressure line. 6. Channel
of methane feeding in metering unit and atmosphere. K1 through K6 = valves. M1 and M2 =
sample manometers (to 16 MPa). B = sample vacuum gauge.

volume with sufficient accuracy, the criterion for choosing the moment of registration of final pressure p of methane in the SA was experimentally determined: when the pressure decay rate reached 0.5 to 0.15% per day of the initial
pressure p_0 (Figure 1.41). Further saturation of sample by methane produced
only an insignificant increase in the maintenance of methane in the closed
pore volume; this may cause an error in closed porosity in the fourth decimal
place.

The release of methane from the SA and evacuation of the low-pressure
line through channel 6 was carried out within 30 minutes with valves K4
and K 6 open (Figure 1.40). Pressure line helium was fed into the SA through
channel 6 and the volume of SA free space V_{fs} was measured by the standard technique.

The calculation of closed pore volume of a mass unit of fossil coal or rock
based on experimental values of pressure p_0, p mass of sample m, and volume of free SA space V_{fs} followed Equation (1.48): $\mu = ((p_0 - p)V_{fs}/pm)$. The
volume of open pores was calculated by the known formula

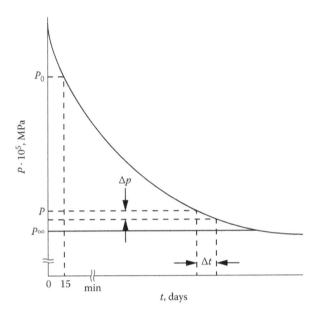

FIGURE 1.41
Curve of change of pressure in a sorption ampoule over time during study of closed pore volume, $\Delta p/\Delta t \sim 10^{-3}\, p_0$/day.

$V_p= 1/d_A - 1/d_R$ where d_A и d_R are apparent and real densities defined in accordance with State Standard Specification (GOST) 2160-82. The total number of pores was calculated by

$$\Omega = V_p + \mu \qquad (1.52)$$

where V_p и μ are volumes of open and closed pores, respectively. Experiments were conducted on 20 samples of fossil coal. The values and errors caused by ignoring the volumes of closed pores are shown in Table 1.6. This 90% error leads to the conclusion that the main contribution to general porosity is made by the volume of closed pores. The described way of determining closed porosity serves as a standard (GDS 10.1.24647077.002:2007) for the coal mining industry.

1.6 Conclusions

The sequential genesis of coal (peat → brown coal → coal → anthracite → graphite → diamond) has been described. The modern concepts of the structures of coals and their mineral components in relation to metamorphism

TABLE 1.6

Volumes of Fossil Coal Pores

Sample Number	Coal	Saturation Time	Coal Density (g/cm³)		Pore Volume (cm³/g)			Permissible Error from Ignoring Closed Pores (%)
			Real d_R	Apparent d_A	open V_P	close μ	total Ω	
1	Volatile	11	1.33	1.07	0.183	0.059	0.242	24.4
2	Gas	12	1.31	1.24	0.43	0.214	0.257	83.3
3	Gas	10	1.3	1.24	0.37	0.199	0.236	84.3
4	Gas	14	1.38	1.24	0.082	0.108	0.190	56.8
5	Gas	12	1.3	1.25	0.031	0.168	0.199	844
6	Gas	14	1.35	1.27	0.047	0.099	0.146	67.8
7	Gas	14	1.28	1.27	0.061	0.103	0.164	62.8
8	Gas	9	1.31	1.17	0.091	0.044	0.135	32.6
9	Gas	9	1.31	1.13	0.122	0.045	0.167	26.9
10	Fat	3	1.34	1.2	0.087	0.245	0.332	73.8
11	Fat	3	1.4	1.3	0.055	0.248	0.303	81.8
12	Fat	3	1.51	1.36	0.073	0.16	0.233	68.7
13	Fat	5	1.58	1.2	0.2	0.165	0.365	45.2
14	Fat	5	1.35	1.21	0.086	0.247	0.333	74.2
15	Fat	17	1.31	1.2	0.007	0.271	0.341	79.5
16	Fat	17	1.31	1.21	0.063	0.279	0.342	81.6
17	Coking	14	1.32	1.25	0.042	0.096	0.138	69.6
18	Lean	40	1.33	1.22	0.068	0.174	0.242	71.9
19	Lean	17	1.16	1.05	0.009	0.238	0.328	72.6
20	Lean	50	1.34	1.26	0.047	0.108	0.155	69.7

have been presented. The modern ideas of scientists about the structure of organic mass of coal have been analyzed. In view of the various viewpoints (including inconsistent views), each scientist, depending on the purpose of his or her research, can choose among the various models of coal structures. We have chosen the spatial model of Kasatochkin. x-ray studies from many countries have been described. Our original research conducted at the Institute of Metal Physics of the National Academy of Sciences of Ukraine demonstrates that

- The basic structural component of all ranks of coal is the cluster of atoms packed in a graphite lattice.
- In diffraction patterns revealed by all ranks of coal, a powerful, small-angle dispersion (SAD) of x-ray radiation spreads over a considerable range of a diffraction vector.
- SAD studies of most coals reveal clear peaks that testify to the primary size clusters.

The methods of estimation using a porous solid body as an absorbent have been described. Starting with the differential porosity of a sorbent, it is possible to choose a theory for a particular sorbent that can fully describe processes in the sorbate. Small-angle x-ray and neutron dispersion show that ratios of mesoporosity and macroporosity of coals vary largely. The distribution of pores due to radius size is fractal: $2.13 < D_s < 2.7$.

For the first time, a technique for finding the closed porosity of coal based on changes from saturation of pressure of absorbed methane has been developed. For all ranks of coal, the most common porosity is closed and the volume (for example, for anthracite) is higher than that of open porosity types.

References

1. Saranchuk V.I., Ayruni A.T., and Kovalev K.E. 1988. *Supramolecular Organization, Structure and Properties of Coal.* Kiev: Naukova Dumka.
2. Kizilshtein L.Y. 2006. *Geochemistry and Thermochemistry of Coals.* Rostov University Press, Rostov on Don.
3. Larina I.K. 1975. *Structure and Properties of Natural Coals.* Moscow: Nedra.
4. Saranchuk V.I. 1982. *Oxidation and Spontaneous Ignition of Coal.* Kiev: Naukova Dumka.
5. Epshtein O.I., Suprunenko O.M., and Barabanova P. 2005. Material composition and reactivity of black coal vitrinites with various degrees of reduction. *Him. tverd.topl.* 1, 22–25.

6. Lebedev N.V. 2004. Interplay between physicochemical and thermal processes under self-oxidation, self-heating and spontaneous ignition of coals and stored mining waste. In *Visti Donetskogo Girnychogo Institutu*. Donetsk: Nat. Tech. Univ., pp. 42–46.

7. Antipov I.V. and Lysenko O.N. 2005. Estimation of soaking ability of rock impurities in coals in terms of physical characteristics. In *Fiziko-tehnicheskie problemy gornogo proizvodstva*. Donetsk: Institute for Physics of Mining Processes, pp. 64–67.

8. Kasatochkin V.I. 1969. *Transitive Forms of Carbon*. Moscow: Nauka, pp. 7–16.

9. Alexeev A.D., Krivitskaya R.M., Pestryakov B.V. et al. 1977. NMR study of water adsorption by fossil coal. *Him. tverd, topl.* 2, 94–97.

10. Ruschev D.D. 1976. *Chemistry of Solid Fuel*. St. Petersburg: Himiya.

11. Krishimurti P. 1930. The structure of amorphous carbon. *Ind. J. Phys.* 5, 126–139.

12. Mahadevan C. 1935. Studies in coal problem with x-ray diffraction methods. *Q. J. Geol. Min. Met. Coc. Ind.*, 7, 1–24.

13. Brusset H. 1943. Study of carbon fuels by means of x-rays diffused at very small angles. *Cr. Hebd. Acad. Sci.* 216, 152.

14. Hirsch P.B. 1958. Structural model of coals. *Proc. Res. Conf. on Use of Coal*, Sheffield, pp. 29–35.

15. Prinz D., Pyckhout-Hintzen W., and Littkea R. 2004. Development of the meso- and macroporous structure of coals with rank as analyzed with small-angle neutron scattering and adsorption experiments. *Fuel* 38, 547–556.

16. Rouzaud J.N. and Oberlin A. 1990. *Advanced Methodologies in Coal Characterization*. Amsterdam: Elsevier.

17. Oberlin A., Villey M., and Combaz A. 1980. Influence of elemental composition on carbonization: Pyrolysis of kerosene shale and kuckersite. *Carbon* 18, 347.

18. Larsen J.W., Green T.K., and Kovac, J. 1985. The nature of the macromolecular network structure of bituminous coals. *J. Org. Chem.* 50, 4729.

19. Clarkson C.R. and Bustin R.M. 1996. Variation in micropore capacity and size distribution with composition in bituminous coal of the Western Canadian Sedimentary Basin: Implications for coalbed methane potential. *Fuel* 75, 1483.

20. Boulmier J.L., Oberlin A., Rouzaud J.N. et al. 1982. *Scanning Electron Microscopy*. Chicago, SEM, pp. 1523–1538.

21. Lu L., Sahajawalla C.K., and Harris D. 2001. Quantitative x-ray diffraction analysis and its application to various coals. *Carbon* 39.

22. Van Krevelen D.W. 1961. *Coal: Typology, Chemistry, Physics, Constitution*. Amsterdam: Elsevier.

23. Wiser W.H. 1978. *Organic Chemistry of Coal*. Washington: American Chemical Society, Series 71, pp. 147–150.

24. Given P.H. 1960. Distribution of hydrogen in coals and its relation to coal structure. *Fuel* 39, 147–150.

25. Solomon P.R. 1981. *Coal Structure and Thermal Decomposition: New Approaches in Coal Chemistry*. Washington: American Chemical Society, Series 169, *pp. 61–71*.

26. Heregy L.A. 1980. *Model Structure of Bituminous Coal*. Washington: American Chemical Society, pp. 38–41.

27. Lazarov L. and Angelova G. 1990. *Structure and Reactions of Coals*. Sofia: BAS.

28. Spiro C.L. 1981. Space-filling models for coals: molecular description of coal plasticity. *Fuel* 60, 1121–1126.

29. Shinn Y.N. 1984. From coal to single-stage and two-stage products: a reactive model of coal structure. *Fuel* 63, 1187–1196.
30. Derbishire F., Marzec A., Schulten H.R. et al. 1989. Molecular structure of coals: a debate. *Fuel* 68, 1091–1106.
31. Duber S. and Wieckowski A.B. 1982. EPR study of molecular phases in coal. *Fuel* 61, 433–436.
32. Larsen J.W. and Kovach J. 1978. Polymer structure of bituminous coals. *ACS Symposium on Organic Chemistry of Coal.* Washington: American Chemical Society, Series 71, pp. 36–49.
33. Marzec A., Juzwa M., Dotlej K. et al. 1979. Bituminous coals extraction in terms of electron-donor and acceptor interactions in a solvent–coal system. *Fuel Proc. Technol.* 2, 35–39.
34. Marzec A., Jurkiewich A., and Pislevski N. 1983. Application of ¹H pulse NMR to the determination of molecular and macromolecular phases in coals. *Fuel* 62, 996–998.
35. Marzec A. 1986. Macromolecular and molecular model of coal structure. *Fuel Proc. Technol.* 14. 39–46.
36. Bodzek D. and Marzec A. 1981. Molecular components of coal and coal structure. *Fuel* 60, 47–51.
37. Kasatochkin V.I. and Larina N.K. 1975. *Structure and Properties of Natural Coals.* Moscow: Nedra.
38. Taylor C.H. 1966. Electron microscopy of vitrinites. *Coal Sci. Chem.* 55, 274–283.
39. Alexeev A.D., Sinolitskiy V.V., and Vasilenko T.A. et al. 1992. Closed pores in fossil coals. *Phyz tekhni. probl. razrab. polez. nyhz.* 2, 99–106.
40. Ajruni A.T. 1987. *Prognostication and Prevention of Gas Dynamic Phenomena in Coal Mines.* Moscow: Nauka.
41. Oberlin A. 1989. High-resolution TEM studies of carbonization and graphitization. *Chem. Phys. Carbon* 22, 1–143.
42. Rouzaud J.N. and Oberlin A. 1990. *Characterization of Coals and Cokes by Transmission Electron Microscopy.* Amsterdam: Elsevier, pp. 311–355.
43. Rouzaud J.N., Vogt D., and Oberlin A. 1988. Coke properties and their microtexture I. Microtextural analysis: a guide for cokemaking. *Fuel Proc. Technol.* 20, 143–154.
44. Rouzaud J.N. 1990. Contribution of transmission electron microscopy to the study of coal carbonization processes. *Fuel Proc. Technol.* 24, 55–69.
45. Kitaygorodski A.I. 1952. *X-Ray Structural Analysis of Fine Crystalline and Amorphous Bodies.* St. Petersburg: Tehniko Teoreticheskoy.
46. Warren B.E. 1934. X-ray diffraction study of carbon black. *J. Chem. Phys.* 2, 551–556.
47. Warren B. E. 1941. X-ray diffraction in random layer lattices. *Phys. Rev.* 59, 693–698.
48. Biscoe J. and Warren B.E. 1942. X-ray study of carbon black. *J. Appl. Phys.* 13, 364–371.
49. Warren B.E. 1969. *X-Ray Diffraction.* Reading, MA: Addison-Wesley, pp. 116–150.
50. Franklin R.E. 1950. Interpretation of diffuse x-ray diagrams of carbon. *Acta Cryst.* 3, 107–121.
51. Franklin R.E. 1951. Structure of graphitic carbons. *Acta Cryst.* 4, 253–261.

52. Alexander L.E. and Sommer E.C. 1956. Systematic analysis of carbon black structures. *J. Phys. Chem.* 60, 1646–1649.
53. Kasatochkin V.I., Zolotorevskaya E.Yu., and Razumova L.L. 1951. Alteration of fine structure of fossil coals at different stages of metamorphism. *Dokl. akad. nauk. SSSR* 79, 315–318.
54. Kasatochkin V.I. 1952. On molecular structure and properties of black coals. *Dokl. akad. nauk. SSSR* 4, 759–762.
55. Kasatochkin V.I. 1969. The problem of molecular structure and structural chemistry of natural coals. *Him. tverd, topl.* 4. 33–48.
56. Hirsch P.B. 1954. X-ray scattering from coals. *Proc. Roy. Soc. London* 226, 143–169.
57. Diamond R. 1957. X-ray diffraction data for large aromatic molecules. *Acta Cryst.* 10, 359—364.
58. Ergun S. and Tiensuu V. 1959. Tetrahedral structures in amorphous carbons. *Acta Cryst.* 12, 1050–1051.
59. Grigoriev H. 1988. Interpretation of the pair function for laminar amorphous materials in the case of coals II. Structure of coals. *J. Appl. Cryst.* 21, 102–105.
60. Ergun S. 1967. X-ray studies of carbon. In: Walker P.L., Ed., *Chemistry and Physics of Carbon*. New York: Marcel Dekker, pp. 211–288.
61. Ruland W. 1959. Structural parameter determination of small aromatic systems in non-crystalline solids. *Acta Cryst.* 12, 679–683.
62. Rietveld H.M. 1967. Line profiles of neutron powder diffraction peaks for structure refinement. *Acta Cryst.* 22, 151–152.
63. Fischbach D.B. 1971. The kinetics and mechanism of graphitisation. In: Walker P.L., Ed., *Chemistry and Physics of Carbon*. New York: Marcel Dekker, pp. 1–105.
64. Shi H., Reimers J.N, and Dahn J.R. 1993. Structure refinement program for disordered carbons. *J. Appl. Cryst.* 26, 827–836.
65. Feng B., Bhatia S.K., and Barry J.C. 2002. Structural ordering of coal char during heat treatment and its impact on reactivity. *Carbon* 40, 481–496.
66. Feng B., Bhatia S.K., and Barry J.C. 2003. Variation of the crystalline structure of coal char during gasification. *Energy Fuels* 17, 744–754.
67. Yen T.F, Erdman J.G, and Pollack S.S. 1961. Investigation of the structure of petroleum asphaltenes by x-ray diffraction. *Anal. Chem.* 33, 1587–1594.
68. Atria J.V., Rusinko F. Jr., and Schobert H.H. 2002. Structural ordering of Pennsylvania anthracites on heat treatment to 2000 to 2900°C. *Energy Fuels* 16, 1343–1347.
69. Skripchenko G.B. 1984. Intermolecular orderliness in fossil coals. *Him. tverd. topl.* 6, 18–26.
70. Korolev Y.M. 1989. X-ray studies of humus organic matter. *Him. tverd. topl.* 16, 11–19.
71. Golovin G.S., Korolev Yu. M., and Lunin V.V. et al. 1999. X-ray studies of humus coal structure. *Him. tverd. topl.* 4, 7–27.
72. Wertz D.L. 1998. X-ray scattering analysis of the average polycyclic aromatic unit in Argonne premium coal 401. *Fuel* 77, 43–53.
73. Petersen, T., Yarovsky, I., Snook, I. et al. 2004. Microstructure of an industrial char by diffraction techniques and reverse Monte Carlo modelling. *Carbon* 4, 2457–2469.

74. Maity S. and Mukherjee P. 2006. X-ray structural parameters of some of the Indian coals. *Curr. Sci.* 91, 340–345.
75. Alexeev A.D., Shatalova G.E., Ulyanova E.V. et al. 2002. Allotropic forms of carbon in the natural coal. *Fiz. tekhn. vyso. davl.* 12, 63–69.
76. Alexeev A.D., Shatalova G.E., Ulyanova E.V. et al. 2003. NMR and x-ray studies of the coal–methane system. *Fiz. tekhn. probl. gorn. proiz.* 6.
77. Alexeev A.D., Shatalova G.E., Ulyanova E.V. et al. 2003. Structural features of the coal–methane system. *Fiz. tekhn. vyso. davl.* 13. 100–105.
78. Van Krevelen D.V. 1960. *Coal Science.* Moscow: Mir.
79. Rusjanova N.D. 2003. *Coal Fuel Chemistry.* Moscow: Nauka.
80. A.S. Lashko. 1955. *Problems of Physics of Metals and Metallurgy.* Kiev: Ukrainian Academy of Sciences Press, pp. 66–70.
81. Danilov I. 1956. *Structure and Crystallization of Liquid.* Kiev: Ukrainian Academy of Sciences Press, Kiev.
82. Temperly G., Roulinson, J., and Rashbruk, J., Eds. 1971. *Physics of Simple Liquids.* Moscow: Mir.
83. Dutchak Y.A. 1977. *Roentgenography of Liquid Metals.* Lvov: Vyshya Shkola.
84. Wagner C.N.J. and Ruppersberg H. 1981. *Atomic Energy Rev.* 1, 101–141.
85. Bek G. and Gunterodt G., Eds. 1988. *Metallic Glasses.* Moscow: Mir.
86. Romanova A.V. 1971. Collected articles. *Metallophyz. nauk. dumka* 36, 3–14.
87. Romanova A.V., Illinsky A.G. 1987. Structures of amorphous metal alloys. In *Amorphous metal alloys.* Kiev: Naukova Dumka, pp. 7–51.
88. Ilinskiy, A.G. Maslov, V.V., Nosenko, G.M. et al. 2006. Transformation of atomic structure in Fe-Si-B based amorphous alloys during annealing. *Fiz. i noveysh. tekhnol.* 28, 1369–1384.
89. Skryshevskiy A.F. 1966. *Roentgenography of Liquids.* Kiev: University Publishers.
90. Alexeev A.D., Zelinskaya G.M., Ilinskiy A.G. et al. 2008. Atomic structure of natural coals. *Fiz. i tekhn. vyso. davl.* 18, 35–52.
91. Alexeev A.D., Feldman E.P., and Vasilenko T.A. 2000. Alternation of methane pressure in closed pores of fossil coals. *Fuel* 79, 939–943.
92. Alexeev A.D., Vasilenko T.A. and Ulyanova E.F. 1999. Closed porosity in fossil coals. *Fuel* 78, 635–638.
93. Khodot V.V., Yanovskaya M.F., and Premysler Y.S. 1973. *Physicochemistry of Gas Dynamics Phenomena in Mines.* Moscow: Nauka.
94. Milewska-Duda J. 1987. Polymeric model of coal in light of sorptive investigations. *Fuel* 66, 1570–1573.
95. Gallegos D.P., Smith D.M., and Sterner D.L. 1987. Pore structure analysis of American coals. IUPAC Symposium (COPS1), April 26–29, 1987, pp. 509–518.
96. McBain J.V. 1909. The mechanism of the adsorption of hydrogen by carbon. *Phil. Mag.* 18, 916–935.
97. Milewska-Duda J. 1993. The coal–sorbate system in light of the theory of polymer solutions. *Fuel* 72, 419–425.
98. Weishauptová Z, Medek J., and Kovář L. 2004. Bond forms of methane in porous system of coal II. *Fuel* 83, 1759–1764.
99. Greg S. and Sing K. 1984. *Adsorption, Specific Surface, Porosity.* Moscow: Mir.
100. Cheremskoi P.G., Slezov V.V., and Betehtin V.I. 1990. *Pores in a Solid Body.* Moscow: Energoatomizdat.
101. Marsh H. 1987. Adsorption methods to study microporosity in coals and carbons—A critique. *Carbon* 1, 49–58.

102. Alcaniz J., Monge A., Linares S. et al. 2002. Water adsorption on active carbons: Study of water adsorption in micro- and mesopores. *J. Phys. Chem.* B 106, 3209–3216.
103. Kopp O.C., Bennett M.E. III, and Clark C.E. 2000. Volatiles lost during coalification. *Int. J. Coal Geol.* 44, 69–84.
104. Lidin G.D. and Airuni A.T. 1963. Gas flooding in coal mines of the central region of Donets basin. In *Gas Flooding in Coal Mines of the USSR*. Moscow, pp. 57–78.
105. Alexeev A.D., Zaidenvarg V.E., Sinolitski V.V. et al. 1992. *Radiophysics in the Coal Industry*. Moscow: Nedra.
106. Dubinin M.M. 1966. *Chem. Phys. Carbon* 2, 51–120.
107. Svergun D.I. and Feigin L.A. 1986. *X-Ray and Neutron Small-Angle Scattering*. Moscow: Nauka.
108. Guinier A., Fournet G., Walker C.B. et al. 1955. *Small-Angle Scattering of X-Rays*. New York: John Wiley & Sons.
109. Nikitin A.N., Vasin R.N., Balagurov A.M. et al. 2006. Investigation of thermal and deformation properties of quartzite at the temperature interval of polymorphic α–β transition by neutron diffraction and acoustic emission. *Part. Nucl. Lett.* 3, 76–91.
110. Radlinski A.P., Boreham C.J, Wignall G.D. et al. 1996. Microstructural evolution of source rocks during hydrocarbon generation: a small-angle scattering study. *Phys. Rev. B* 53, 14152–14160.
111. Radlinsky A.P., Mastalerz M., Hinde A.L. et al. 2004. Application of SAXS and SANS in evaluation of porosity, pore size distribution and surface area of coal. *Int. J. Coal Geol.* 59, 245–271.
112. Porod G. 1951. Die Rontgen Keenwinkelsteng Won Deshtgepasten Kolloiden Sistemen. *Kolloid Z.* 124, 83–114.
113. Schmidt P. and Hight R. 1959. Calculation of intensity of small-angle x-ray scattering at relatively large scattering angles. *J. Appl. Phys.* 30, 866–867.
114. Schmidt P. 1982. Interpretation of small-angle scattering curves proportional to a negative power of the scattering vector. *J. Appl. Cryst.* 15, 567–569.
115. Bale H.D. and Schmidt P. 1984. Small-angle x-ray-scattering investigation of submicroscopic porosity with fractal properties. *Phys. Rev. Lett.* 53, 596–599.
116. Linsel, B.G., Ed. 1973. *Structure and Properties of Adsorbents and Catalysts*. Moscow: Mir, pp. 25–26.
117. Gorobets Y.I., Kuchko A.M., and Vavilova I.B. 2008. *Fractal Geometry in Natural Science*. Kiev: Naukova Dumka.
118. Radlinski A.P., Radlinska E.Z., Agamalian M. et al. 2000. The fractal microstructure of ancient sedimentary rocks. *J. Appl. Cryst.* 33, 860–862.
119. Radlinski A.P., Radlinska E.Z., Agamalian M. et al. 1999. Fractal geometry of rocks. *Phys. Rev. Lett.* 82, 3078–3081.
120. Bernabé Y. and Revil A. 1995. Pore-scale heterogeneity, energy dissipation and the transport properties of rocks. *Geophys. Res. Lett.* 22, 1529–1532.
121. Alexeev A.D., Vasilenko T.A., and Kirillov A.K. 2008. Modeling of size distribution of pores under deformation of porous materials. *Phys. i tekhn. vyso. davl.* 18, 110–119.
122. Medek, J. 1977. Possibility of micropore analysis of coal and coke from the carbon dioxide isotherm. *Fuel* 56, 131–133.
123. Kirillov A.K. and Polyakov P.I. 2005. Methods of determination of fossil coal porosity. *Vist. donets. girnych. inst.* 2, 17–20.

124. Trubetskoy Y.N. and Ayruni A.T. 2000. *Fundamental and Applied Methods of Solution of the Problem of Methane in Coal Beds.* Moscow: Izd. Akademii gornykh nauk p. 519.
125. Vasilenko T.A., Polyakov P.I., and Slyusarev V.V. 2000. Investigation of physico-mechanical properties of coal under hydrostatic and quasi- hydrostatic pressure. *Fiz. i tekhn. vyso. davl.* 10, 72–85.
126. Petukhov I.M. and Linkov A.M. 1983. *Mechanics of Rock Bumps.* Moscow: Nedra.
127. Gan H., Nandi D., and Walker P.L. 1972. Nature of the porosity in American coal. *Fuel* 51, 272–277.
128. Alexeev A.D. and Sinolitskiy V.V. 1985. Kinetics of absorption and emission of gases by solid bodies. *Inzhen. fiz.* 4, 648–654.
129. Blayden H. E., Gibson J., and Riley H. 1944, An x-ray study of structure of coals, cokes and tars. *Proc. Conference on the Ultra-fine Structure of Coals and Cokes,* British Coal Utilisation Research Association, London, UK, pp. 176–231.
130. Brunauer S., Emmett P.H., and Teller E. 1938. Adsorption of gases in multimolecular layers. *J. Amer. Chem. Soc.* 60: 309–319.

2

Equilibrium Phase States and Mass Transfer in Coal–Methane Systems

---

2.1 Equilibrium and Dynamics of Mass Exchange between Sorbed and Free Methane

2.1.1 Gaseous State of Methane in System of Opened and Closed Pores

Methane in coal formed and accumulated over geologic time. The main methane mass is found in virgin rock in a system that consists of cracks, canals, and open pores (that determine filtration capacity) and closed pores (not connected by means of transport links with filtration capacity). The seam methane pressure in virgin rock is homogeneous because of the homogeneous structure constants according to Bernoulli's equation and its pressure correlates with rock pressure.

The gaseous methane pressure in a virgin seam is equal in the filtration capacity and closed pores. As seam depth grows, seam methane pressure increases with few exceptions. The stresses in the mountain mass and the seam methane pressure resist the rock pressure in a state of thermodynamic equilibrium. As an approximation, the energy (and other thermodynamic potentials) of the seam form from the elastic energy of the coal structure and the energy of gaseous methane. When examining the energy of a coal seam, it is necessary to use the thermodynamic Gibbs potential as the environment sets the overburden load on the coal seam and seam temperature.

It is usually possible to consider methane in coal as an ideal gas, in which case its thermodynamic potential (based on mountain massif capacity) can be written as [1]

$$\varphi_g = \gamma P \ln \frac{P}{P_T}. \tag{2.1}$$

P is the seam methane pressure; porosity γ takes into account that methane is only in pores of gas and coal bearing mass; P_T is dimension pressure; its temperature dependence is determined by the energy of a separate free molecule of methane. If in calculating the energy of the molecule we consider only progressive and rotation degrees of freedom and do not account for the atomic oscillation and electronic conditions, we will get for the P_T of methane the following correlation:

$$P_T = T\left(\frac{mT}{2\pi\hbar^2}\right)^{3/2}\left(\frac{T}{T_r}\right)^{3/2} \tag{2.2}$$

Here m is molecule mass of methane, $\hbar$ is Planck's constant divided by 2π, T is the gas absolute temperature in energy units, $T_r \equiv (18/\pi)^{1/3}\hbar^2/J$ is the so-called rotary temperature, and J is the moment of inertia of a molecule of methane. It is possible to consider gas ideal if the real pressure of methane reaches several tens of atmospheres and complies with the ideal gas law:

$$P = \rho T \tag{2.3}$$

where density ρ corresponds to the quantity of molecules of methane in the unit of volume of gas (not in the unit of volume of the coal mass).

If the integrity of the massif is broken, for example, by coal mining or geological displacement, the equilibrium is broken and gas under seam pressure overflows from seams to open areas where its pressure is lower than atmospheric pressure. The flow of gas in the filtration volume inside the seam corresponds to filtration through the system of narrow canals. It is acceptable to describe this viscous flow of gas (intrastratal dynamics) by Darcy's equation. The derivation of this equation is that the driving force of this process is the pressure gradient. The process is isothermal because of its slowness but not adiabatic. The gas is compressible which follows directly from the equation of state (2.3). Darcy's equation is

$$\frac{\partial Q}{\partial t} = \text{div}\left[\frac{k}{\eta}\rho\,\text{grad}\,P\right] \tag{2.4}$$

Here η is the dynamic viscosity of gas, k is the penetrability of coal, and Q is the quantity of gas in the unit of volume of the seam. After methane has escaped from a seam or from broken coal, its movement complies with the usual equations of aerodynamics of gas mixtures as it mixes with air. This phase of the process of mass transfer is called the external gas dynamic. The estimate of the coefficient of viscosity leads us to the well known gas-dynamic formula [2]:

$$\eta = m\rho\bar{\upsilon}l \tag{2.5}$$

where m is mass of gas molecule, $\bar{v}$ is the average thermal velocity of molecules, and l is the length of molecular free path. For free gas $l \sim 1/\rho\sigma$, σ represents the dispersion of methane molecules on each other; in this case the viscosity $\eta \sim m\bar{v}/\sigma$ does not depend on the density ρ and the filter equation is nonlinear. In this case the flow is called Poiseuille flow. If the gas moves in the system of narrow channels and the typical diameter of a channel d is less than the path length of molecule in free gas, the substitution of l on d should be made in Equation (2.5).

At room temperature, channel size is about 10^{-7}m, that is, a few hundred nanometers. In this case, $\eta = m\rho\bar{v}d$, Equation (2.4) is linear and concerns Knudsen flow.

We now offer a mathematical description of mass transfer in coal on the basis of Knudsen flow. We will consider the isothermal process and treat the seam temperature as homogeneous. Based on Equation (2.3) and the viscosity of gas, it is possible to simplify Equation (2.4):

$$\mathrm{div}\left[D_f \, \mathrm{grad}\rho\right] = \frac{\partial Q}{\partial t} \tag{2.6}$$

where the filtration factor is applied:

$$D_f = \frac{kT}{m\bar{v}d} \tag{2.7}$$

Order of magnitude $D_f \sim k\bar{v}/d$. The permeability coefficient k has a dimension of area and defines the degree of filling of coal with transport channels. It is a generalized parameter of the nanostructure and mesostructure of coal and depends weakly on temperature (according to the root law). The crude estimation gives $k \sim \gamma_0 d^2$, so that

$$D_f \sim \gamma_0 d\bar{v}, \tag{2.8}$$

where γ_0 is an open porosity, i.e., relation of the total volume of transport channels to the volume of the gas and coal mountain mass. The average speed of thermal motion is approximately the speed of sound in gas. With a crude approximation as in Equation (2.8), the filtration factor changes in the range $D_f \sim (10^{-8} - 10^{-4})$m/s. The first consideration for studying the mass transfer of methane in coal is the solution of the equation of the diffuse type (2.6) with a diffusion coefficient that depends on the coal structure and not on temperature.

2.1.2 Solid Solution of Methane in Coal: Absorption and Adsorption

We can imagine the coal structure on the macro and meso levels as a solid-state framework in contact with the filtration volume. Experimental data [3] and theoretical concepts led to the unique conclusion that methane goes from the filtration capacity into the solid-state framework. Inside the framework, the molecules of methane penetrate into the most energy advantageous places. The "fringes" of coal crystalline grains are aliphatic. The incorporation of molecules of methane into the coal framework usually takes place by their penetration into pores of different sizes. If we talk about micropores (about 10 nm), it is natural to talk about the solid solution of methane in coal. Methane that penetrates into pores 10 nm and larger is divided into free and adsorbed types. The condition of methane in pores of intermediate size is open to discussion and not examined here.

The main characteristics of a solid solution of methane in coal are the binding energy ψ of molecules of methane with coal and a number of "matching sites" for methane per unit of volume of solid material (inverse value is volume Ω that falls on one molecule of methane in the solid solution). In large closed pores that do not mix with the filtration volume, methane is generally a free gas. Critical characteristics of methane, which is adsorbed on the surfaces of both opened and closed pores, are the binding energy χ of a molecule of methane with the surface of coal and the area that fits one molecule of adsorbed methane.

Binding energies both ψ and χ are of course averaged values because of the complex hierarchical structure of coal. Present theoretic concepts and calculations suggest that the connection of methane with coal is a combination of Van der Waals forces that induce dipole and hydrogenous connections. Calorimetric measuring [4,5] confirms this qualitative idea by giving ψ a value of 8 to 35 kilojoules/mole subject to the rank of coal. Remember that binding energy is the difference between the energy of a molecule in the bound and free conditions. If coal contains methane, the binding energy is negative as well as for most other gases in the coal.

We will use the averaged quantities to construct theoretical models and different quantitative estimates because specific values of binding energies ψ and χ and volume Ω are very specific, that is, they depend on the rank of coal, the structure of the seam area, the degree of stress, and other natural and technological details. For this reason, we will consider a generalized view of the methane–coal system.

We know that methane appears in coal in three phase states: (1) free in the filtration volume and closed pores [6,7]; (2) absorbed in coal lumps (solid solution); and (3) on the surfaces of closed and opened pores (adsorbed). Naturally, the question of correlations between the quantities of methane in these three phase states occurs. To answer the question, we must consider the equilibrium between gaseous and sorbed methane. Statistical physics and thermodynamics research has proven that the temperature and the chemical potential must be homogeneous along the system; the chemical potential of free gas must coincide with the chemical potential of the solid solution. Remember that the chemical potential corresponds to the derivative of the thermodynamic potential according to the number of particles. The thermodynamic potential of gas is represented by Equation (2.1) without the constant γ. The thermodynamic potential of weak solid solution of concentration c has the following form:

$$\psi_c = c\psi + cT \ln c\Omega \tag{2.9}$$

While differentiating (2.1) and (2.9) according to density ρ and concentration c, respectively, and equating the obtained derivatives, we come to the correlation:

$$c = \nu\rho, \tag{2.10}$$

where solubility $\nu = (T / P_T\Omega)\exp(-\psi / T)$ is introduced. Equation (2.10) expresses the well known Henry's law. If we consider the correlation of Equation (2.2), as well as the negativity of binding energy ψ, we will get the following result for solubility:

$$\nu = \frac{1}{\Omega}\left(\frac{2\pi\hbar^2}{mT}\right)^{3/2}\left(\frac{T_r}{T}\right)^{3/2} e^{\frac{|\psi|}{T}}. \tag{2.11}$$

Note that solubility depends on temperature and increases when temperature drops. When calculating the equilibrium of gaseous and adsorbed methane, remember that the surface concentration of methane can be considerable enough to limit the number of "seats." This leads to the correlation called Langmuir's isotherm:

$$\rho_S = \frac{V_b\rho}{s\Lambda(1+V_b\rho)}. \tag{2.12}$$

where ρ_s is the number of "surface" (adsorbed) molecules of methane per unit of coal volume, s is the area of one adsorbed molecule, and Λ is an inverse specific surface of the pore and crack system, i.e., the inverse value of the relation of total area of the inner surface of coal to volume. We can call quantity:

$$V_b = \left(\frac{2\pi\hbar^2}{mT}\right)^{3/2}\left(\frac{T_r}{T}\right)^{3/2} e^{\frac{|x|}{T}} \tag{2.13}$$

a volume of incorporation of methane into the surface structure. The dimensionless parameter $V_b/s\Lambda$ is analogous to solubility ν (with solubility it represents incorporation into the lump volume; in this case is it incorporation into the lump surface). At room temperature, solubility ν calculated per Equation (2.11) varies over limits 10^{-3} to 10^{-1} (because of spread of values ψ and Ω), while $V_b/s\Lambda$ is two to three orders smaller. That is why we discuss only two phase states of methane: free and absorbed like solid solution gas.

Let γ_0 designate open porosity, i.e., the relation of filtration volume to the full mass volume and γ indicate closed porosity, i.e., the relation of the closed pore volume to coal volume. In equilibrium, the quantity of methane (per unit of mass volume) in the filtration volume is equal to $\gamma_0\rho$; in the closed pores is equal to $(1-\gamma_0)\gamma\rho$; and in the solid solution is equal to $\nu(1-\gamma_0)(1-\gamma)\rho$. The total quantity of free methane is $\rho[\gamma_0+\gamma(1-\gamma_0)]$. According to experimental data, the open porosity varies. The closed porosity can be at about 0.3 to 0.4 [6,7], i.e., 30 to 40%, and the solubility changes from 10^{-3} to 10^{-1}. The total quantity of methane in all the phase states (per unit of mass volume) is

$$Q = \rho\left[\gamma_0 + \nu(1-\gamma_0)(1-\gamma+\frac{\gamma}{\nu})\right], \tag{2.14}$$

and density $\rho(m^{-3})$ is defined by the seam pressure P according to Equation (2.3), i.e., $\rho = P/T$.

These correlations and estimates indicate that at room temperature the largest methane quantity, about 60 to 70%, is contained in closed pores and the balance is distributed between the open pores (filtration volume) and the solid solution in comparable quantities. When the temperature falls, a growing quantity of methane penetrates into the solid solution so that at a temperature of approximately 0°C, about 10% of the methane can penetrate into the solid solution.

Equation (2.14) shows methane quantity as the number of methane molecules per unit of volume of a coal–gas massif. In accepted practice, the formula to calculate cubic meters of gas (at atmospheric pressure) per ton of coal is

$$Q = \frac{P}{P_a n_c}\left[\gamma_0 + \nu(1-\gamma_0)(1-\gamma+\frac{\gamma}{\nu})\right], \tag{2.14'}$$

where P/P_a is the relation of the seam pressure to atmospheric pressure and n_c is coal density. For example, at seam pressure $P = 30$ atm, the open porosity $\gamma_0 = 0.05$; the closed porosity $\gamma = 0.3$; solubility $\upsilon = 0.01$; and the coal density. $n_c = 1.3$ t/m³ the methane content Q in the seam will be 7.5 m³/t or 0.5% of mass. The methane content in the seam increases according to depth based on increased seam pressure and some increase of binding energy and means of solubility.

2.1.3 Diffusion of Methane from Coal Lumps into Filtration Volume: Efficient Diffusion Coefficient

We now describe the process of methane mass transfer in coal. Consider a coal framework as a set of lumps of unbroken fragments of coal dipped into the filtration volume. The lump dimension R is defined by the structure of the coal massif on a mesoscale and is comparable to the inverse specific surface Λ of a filtration system [Equation (2.12)], i.e, $R \sim \Lambda$. Since coal has a fractal structure, the result of measuring of the area of the inside surface depends to an extent on the choice of the scale of measuring scheme [7].

R may be estimated ($R \sim 10^{-5} \div 10^{-4}$ cm) based on the literature method [8,9] for measuring the specific surface ($1 \div 20$ m^2 per 1 cm^3 of coal). If the seam equilibrium is broken, the conditions for methane mass transfer are created. The movement of methane inside the lump toward the filtration volume is a part of this process. We believe the methane mass transfer within the lump occurs via solid-state diffusion, i.e., a methane molecule jumps from one type of "seat" to neighboring ones. However, the picture of mass transfer in reality is far more complex one than a transfer in an ideal crystalline solid solution because of the complexity and hierarchical structure of coal at mesolevel and microlevel.

In discussing solid-state diffusion, we rely on the strong exponential dependence of the corresponding diffusion coefficient on the temperature established by experiment [9] that is typical for solid-state diffusion but not for infiltration through narrow channels. When examining diffusion within lumps, it is very important to consider the availability of the closed pores distributed throughout a lump. We know that methane mainly remains in closed pores. To escape from a lump, a methane molecule must move from a closed pore first into a solid solution and then diffuse through the solid solution to the lump boundary. In other words, gas in closed pores that increases its concentration in the solid solution. The reduction of gas density in the closed pores corresponds to the increase of the gas concentration in a solid solution. Here is the equation for methane diffusion inside the lump:

$$(1-\gamma)\frac{\partial}{\partial t}c(\vec{r},t)= D\Delta c(\vec{r},t)+q(\vec{r},t)$$

(2.15)

Here $c(\vec{r},t)$ is methane concentration at a place $\vec{r}$ of the solid solution at point of time t, D is a solid-state diffusion coefficient, Δ is a Laplace operator, and $q(\vec{r},t)$ is the source strength at a given lump place. The multiplier $(1-\gamma)$ takes into account the fact that the entire lump is not occupied by a solid solution; part γ of the volume consists of pores. Suppose that the medium size of pores is far smaller than the medium distance between them. It becomes possible to use the average density $\rho(\vec{r},t)$ to calculate source density:

$$q(\vec{r},t) = -\gamma \frac{\partial \rho(\vec{r},t)}{\partial t}.$$ (2.16)

Further we will consider that Henry's law [Equation (1.10)] works even in the absence of equilibrium. We thus rewrite Equation (1.15):

$$(1-\gamma)\frac{\partial c}{\partial t} = D\Delta c - \frac{\gamma}{v}\frac{\partial c}{\partial t}.$$ (2.17)

As a result, Equation (2.15) become the standard diffusion equation:

$$\frac{\partial c(\vec{r},t)}{\partial t} = D_{eff}\Delta c(\vec{r},t),$$ (2.18)

With substitution of the solid-state diffusion coefficient for the efficient diffusion coefficient:

$$D_{eff} = \frac{D}{1-\gamma+\dfrac{\gamma}{v}}.$$ (2.19)

At reduced solubility, $v \ll \gamma$, $D_{eff} \approx vD/\gamma$, i.e., as the solubility reduces, the efficiency of methane escape from a lump also reduces because methane is concentrated in the closed pores and its escape from pores into the solid solution is blocked. Note that this is the only way that the escape of methane from closed pores in the lumps into the filtration volume is possible. Despite the rise of the solid-state diffusion coefficient as the methane solubility reduces with the rise of temperature and gas penetrates the closed pores instead of escaping from them, the rise of temperature does not always intensify the escape of methane from coal.

2.2 Joint Flow of Filtration and Diffusion Processes in Coal Massifs

2.2.1 Double-Time Models of Mass Transfer: Fast and Slow Methane

Let us examine a macroscopic coal massif. It may be a seam or a piece of coal detached from a seam. It is clear that thermodynamic equilibrium is broken and the gas from the filtration volume, because of a difference between seam and outer pressures, will seep outside to the volume not filled with coal. The gas pressure inside the coal massif reduces and starts the methane diffusion process of lumps into the filtration volume. Gas filtration with simultaneous methane replenishment of the filtration volume takes place. In other words, lumps (microlumps) play the role of methane sources distributed over the whole coal volume.

In the coal, the filtration processes described by Darcy's equation (2.6) are interconnected with and interdependent on gas diffusion in lumps (2.18). We want to determine the type of connection. To maintain simplicity in formula writing without limiting generality, let us use Equation (2.18) for the spherical form of radius R:

$$\frac{\partial c(r,t)}{\partial t} = D_{eff} \left[\frac{\partial^2 c(r,t)}{\partial r^2} + \frac{2}{r} \frac{\partial c(r,t)}{\partial r} \right]. \tag{2.20}$$

In Equation (2.20), r is the distance from the given site to the sphere center. At $r = 0$, gas concentration must be final. Physical content is defined by the boundary condition on the surface of the border of the lump and filtration volume, that is, at $r = R$.

Suppose that time of incorporation of a methane molecule into the lump surface is far less than the time of its transfer from the lump interior to its surface. In this case, global equilibrium is not available on the lump border and a local equilibrium for methane molecule exchange between the lump and the filtration volume is being established. The equilibrium is expressed by Henry's law (2.10) where concentration and density are taken at a boundary of the border, that is

$$c(R,t) = v\rho(t), \tag{2.21}$$

where $\rho(t)$ is methane density in the filtration volume in the lump location and ρ changes slowly at distances of about R, in contrast to density $c(r,t)$. Now let us consider the average methane concentration $c(t)$ in the lump solid solution:

$$c(t) = \frac{3}{R^3} \int_0^R r^2 c(r,t) \ dr. \tag{2.22}$$

Change of average concentration over time can be found through the averaging procedure (2.22) for Equation (2.20):

$$\frac{dc(t)}{dt} = D_{eff} \frac{3}{R} \left. \frac{\partial c(r,t)}{\partial r} \right|_{r=R}. \tag{2.23}$$

By correlating (2.21) and (2.23), the link of diffusion and filtration is realized. The initial condition of our problem is in setting methane concentration at the initial moment we set. Let us consider that methane at this moment is equally distributed in the lump:

$$c(r,0) = c_0 = v\rho_0. \qquad (2.24)$$

It is desirable for our problem to measure distance in R units and time as

$$t_d = \frac{R^2}{D_{eff}} \text{ units,} \qquad (2.25)$$

where t_d can be naturally called typical diffusion time. Technically it is useful to solve the above problem via the Laplace transform as to time of the quantities of interest, namely:

$$c(p) = \int_0^\infty c(t) \, e^{-pt} \, dt; \rho(p) = \int_0^\infty \rho(t) e^{-pt} \, dt, \qquad (2.26)$$

where t is dimensionless time in units t_d. Omitting intermediary transformations [10], we find a very interesting correlation between Laplace forms of average concentration and density of methane:

$$\frac{c_0}{p} - c(p) = F(p)\left(\frac{c_0}{p} - v\rho(p)\right), \quad F(p) = \frac{3}{\sqrt{p}}\left(cth\sqrt{p} - \frac{1}{\sqrt{p}}\right) \qquad (2.27)$$

Performing the inverse Laplace transform of Equation (2.27) gives the connection between $c(t)$ and $\rho(t)$ in the form of convolution of two functions:

$$c_0 - c(t) = \int_0^t F(t-\tau)[c_0 - v\rho(\tau)]d\tau, \qquad (2.28)$$

where:

$$F(t) = 6\sum_{n=1}^\infty e^{-\pi^2 n^2 t} \qquad (2.29)$$

is the inverse Laplace transform $F(p)$ from (2.27). As the lump is far smaller than the coal massif (piece of coal), a rough description of gas filtration from the coal massif shows physically infinite small volumes that include many lumps. One can accept that (2.28) works for every spot of massif $\vec{x}$.

Equation (2.28) indicates that methane concentration in lumps at moment t is defined by its density in filtration volume over all previous moments $r < t$. The equation expresses quantitatively and fairly simply the reciprocal influences of diffusion and filtration. The filtration process is dominant while diffusion provides feeding or replenishment of filtration volume. Let us return now to the model for studying methane filtration. Equation (2.14) shows methane content in coal in equilibrium. In a non-equilibrium state, as methane flows on the seam, quantity $c \neq v\rho$ and Darcy's equation is written [11]:

$$\frac{\partial}{\partial t}\left[\gamma_0\rho(\vec{x},t)+c(\vec{x},t)(1-\gamma_0)\left(1-\gamma+\frac{\gamma}{v}\right)\right]=\frac{D_f}{D_{\text{eff}}}\Delta\rho(\vec{x},t). \tag{2.30}$$

Concentration $c(\vec{x},t)$ is connected with density $\rho(\vec{x},t)$ through Equation (2.28) at any $\vec{x}$ of the coal massif. Equation (2.30) is Darcy's equation in dimensionless coordinates of time. Equation (2.28), with (2.25) in mind is true only for spherical lumps. For non-spherical lumps, a value less than R should mean an average typical lump dimension.

To solve concrete problems of methane sorption and desorption, Equation (2.30) must be added to the corresponding boundary conditions in some situations of internal and external gas dynamics. For example, methane flow equals zero on the boundary of a coal seam with bearing strata. On an open coal surface, pressure (and density) may equal to zero. This will be discussed below. Even before solving concrete problems via Equations (2.28) and (2.29), we can draw qualitative conclusions as to methane mass transfer in coal.

Let us consider function $F(t)$ in more detail. At $t \ll 1/\pi^2$, $F(t) \approx 3/\sqrt{\pi t}$, while at $t \gg 1/\pi^2$ this function is exponentially small, namely $F \approx 6e^{-\pi^2 t}$. Consider Equation (1.28) under short time limits:

$$c_0-c(t)\approx\frac{3}{\sqrt{\pi}}\int_0^t\frac{c_0-v\rho(\tau)}{\sqrt{t-\tau}}d\tau=\frac{3v}{\sqrt{\pi}}\int_0^t\frac{\rho_0-\rho(\tau)}{\sqrt{t-\tau}}d\tau. \tag{2.31}$$

We know [11] that in problems on filtration, gas escape under short time limits is proportional to $\sqrt{t}$ and flow speed is proportional $1/\sqrt{t}$, i.e. $\rho_0-\rho(\tau)\approx\rho_0\sqrt{\tau}$. From (2.31) it follows that

$$c_0-c(t)\approx\frac{3v\rho_0}{\sqrt{\pi}}\int_0^t\frac{\sqrt{\tau}}{\sqrt{t-\tau}}d\tau=\frac{3\sqrt{\pi}}{2}v\rho_0 t, \tag{2.32}$$

Methane escapes from the lumps in proportion to t; its escape speed is constant and final while from the filtration volume it escapes much faster and thus allows us to divide methane into fast moving (escaping from filtration volume) and slow (contained in coal lumps). Then when $t \gg 1/\pi^2$, $F(t-\tau)$, considered a function τ, is an extremely "sharp" function. It is exponentially small under all τ except τ values close to t. That's why the integral on the right (1.27) is based on the second theorem of the average of the expression $v[\rho_0 - \rho(t)]\int_0^t F(t-\tau)d\tau$. The integral that belongs to this expression is equal to one with exponential exactness. For this reason, over long time durations, it follows from (2.28) that $c(t) = v\rho(t)$. As this correlation is correct for all $\vec{x}$, Equation (2.29) becomes the standard Darcy's equation during long time durations:

$$\frac{\partial}{\partial t}\left[\left[\gamma_0 + v(1-\gamma_0)\left(1-\gamma+\frac{\gamma}{v}\right)\right]\rho(\vec{x},t)\right] = \frac{D_f}{D_{eff}}\Delta\rho(\vec{x},t) \qquad (2.30')$$

Equation (2.30') shows that over time t_d, methane escapes from the filtration volume synchronously with the escape from lumps, i.e., methane in the lumps with methane in the filtration volume constitutes a single system.

The initial gas release depends on the filtration of methane from the system of open pores. It usually occurs at a time less than diffusion time t_d. It thus appears that all methane escapes from the coal mountain mass during these short times, but in reality, the amount of methane remaining in the coal lumps within the mountain massif may far exceed the amounts that escape. We will show this effect by solving specific problems.

2.2.2 Methane Escaping from Coal into Closed Volume: Role of Backpressure

We now examine coal broken from the mountain mass and contained in a closed vessel with volume V. The total volume of lumps (granules) of coal are designated V_c, and the volume of the left part of the vessel is V_f (Figure 2.1); $V = V_c + V_f$. At the initial moment (when coal is loaded into the vessel), methane density in filtration volume is constant and equal to ρ_0. Within the context of this experiment, ρ_0 is usually less than methane density in the virgin seam as a considerable part of methane escapes from the coal during its transfer from massif to vessel.

According to our model, methane escape occurs as follows. First, gas from the filtration volume flows to volume V_f, which is not filled with coal. Pressure inside the granule drops and that starts the process of diffusion mass transfer of sorbed methane from the lump into filtration volumes. Gas filtration with simultaneous feeding of filtration volume with methane diluted in lumps takes place. Filtration in each granule is described by Equation (2.30). For simplicity, we consider all granules as spheres with radius L so that $L \gg R$. Naturally, Equation (2.30) should be stated in spherical

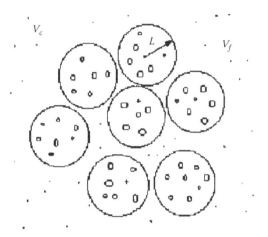

FIGURE 2.1
Coal material contained in closed volume.

coordinates. On the borders of granules $(x = L)$, the following requirement must be observed:

$$\rho\left(\frac{L}{R}, t\right) = n(t).$$

(2.33)

where $n(t)$ is gas density in volume V_f; density in the center $\rho(0,t)$ must be final. Average density $\bar{\rho}(t)$ and concentration $\bar{c}(t)$ in the granule are analogous to Equation (2.22):

$$\bar{c}(t) = \frac{3R^3}{L^3} \int_0^{L/R} x^2 c(x,t)dx, \quad \bar{\rho}(t) = \frac{3R^3}{L^3} \int_0^{L/R} x^2 \rho(x,t)dx.$$

(2.34)

The connection between $\bar{\rho}(t)$, $\bar{c}(t)$ and $n(t)$ is determined by the equation of material balance:

$$n(t)V_f + \left[\gamma_0 \bar{\rho}(t) + (1-\gamma_0)\left(1 - \gamma + \frac{\gamma}{\nu}\right)\bar{c}(t)\right]V_c$$

$$= \left[\gamma_0 \rho_0 + (1-\gamma_0)\left(1 - \gamma + \frac{\gamma}{\nu}\right)c_0\right]V_c$$

(2.35)

In writing (2.35), we assumed that at the initial moment the gas in free volume was not available $[n(0) = 0]$. The connection of $\rho(x,t)$ and $c(x,t)$ was discussed earlier. The formulated problem can be solved with the Laplace transform on times of all unknown values. Two parameters of time dimension, the diffusion time t_d (2.25) and filtration time

$$t_f = \frac{L^2}{D_f}. \tag{2.36}$$

are used to construct the basic dimensionless parameter of the problem

$$a \equiv \sqrt{\frac{t_f}{t_\alpha}} = \frac{L}{R}\sqrt{\frac{D_{eff}}{D_f}}. \tag{2.37}$$

Omitting intermediate calculations, we show the solution in the form of the inverse Laplace transformations of several quantities:

$$\rho_0 - \bar{\rho}(t) = \frac{\rho_0}{2\pi i} \int_{\sigma-i\infty}^{\sigma+i\infty} \frac{gF(y)}{g + \left[\gamma_0 + v(1-\gamma_0)\left(1 - \gamma + \frac{\gamma}{v}\right)F(z)\right]F(y)} \frac{e^{pt}}{p} dp, \tag{2.38}$$

$$c_0 - \bar{c}(t) = \frac{c_0}{2\pi i} \int_{\sigma-i\infty}^{\sigma+i\infty} \frac{gF(y)F(z)}{g + \left[\gamma_0 + v(1-\gamma_0)\left(1 - \gamma + \frac{\gamma}{v}\right)F(z)\right]F(y)} \frac{e^{pt}}{p} dp, \tag{2.39}$$

$$n(t) = \frac{1}{g}\left[\gamma_0(\rho_0 - \bar{\rho}(t)) + (1-\gamma_0)\left(1 - \gamma + \frac{\gamma}{v}\right)(c_0 - \bar{c}(t))\right] \tag{2.40}$$

Here $g = V_f/V_c$, i.e., ratio of free volume to coal massif volume,

$$z \equiv \sqrt{p}, \quad y \equiv a\sqrt{p\left[\gamma_0 + v(1-\gamma_0)\left(1 - \gamma + \frac{\gamma}{v}\right)F(z)\right]}. \tag{2.41}$$

If methane is about escape into open space when $V_f \to \infty$, т.е. $g \to \infty$, in other words, when backpressure is not taken into account, Equations (2.38) and (2.39) are greatly simplified:

$$\rho_0 - \rho(t) = \frac{\rho_0}{2\pi i} \int_{\sigma - i\infty}^{\sigma + i\infty} \frac{F(Y)e^{pt}}{p} \, dp, \tag{2.42}$$

$$c_0 - c(t) = \frac{c_0}{2\pi i} \int_{\sigma - i\infty}^{\sigma + i\infty} \frac{F(Y)F(z)e^{pt}}{p} \, dp, \tag{2.43}$$

$n(t) \to 0$ in this case. Integrals (2.42) and (2.43) cannot be computed in the obvious way. They may be subjected to asymptotic analysis to determine the physical content of methane escape from coal. Let us quote basic results of the analysis for cases $a \ll 1$ $(t_d \gg t_f)$, i.e., long diffusion times, and $a \gg 1$ $(t_d \ll t_f)$, i.e. long filtration times. For $a \ll 1$ (methane escape from a chip; short filtration time), methane escapes from the filtration volume very quickly, according to radical law:

$$\rho_0 - \bar{\rho}(t) \approx \frac{\rho_0}{a} \sqrt{t\gamma_0},$$

or in dimensional form:

$$\rho_0 - \bar{\rho}(t) \approx \rho_0 \frac{\sqrt{t D_f \gamma_0}}{L}, \tag{2.44}$$

The process for lumps is much slower, according to linear law:

$$c_0 - c(t) \approx \frac{3c_0}{a\sqrt{\gamma_0}} t,$$

or in dimensional form

$$c_0 - \bar{c}(t) \approx \frac{3c_0}{\sqrt{\gamma_0}} \frac{t\sqrt{D_f D_{eff}}}{RL}. \tag{2.45}$$

Near the end of this stage in filtration, most methane escapes from the filtration volume, while a small part, about $a\sqrt{\gamma_0}$, escapes from the lumps. Thus we see that under slowed diffusion, methane is clearly fast (contained in filtration volume and leaving it quickly, $\sim L^2/D_f$ and slow (leaving the lumps at diffusion times $\sim R^2/D_{eff}$). Measurements of initial gas release in this case enable the measurement of filtration coefficient but do not yield estimates of coal gas content.

Total quantity $q(t)$ of the released methane (per unit of coal volume) by moment t is calculated in accordance with (2.35), (2.42), and (2.43), using the following formula (when backpressure is absent)

$$q(t) = \frac{\rho_0}{2\pi i} \int\limits_{\sigma - i\infty}^{\sigma + i\infty} \frac{F(y)\left[\gamma_0 + v(1-\gamma_0)\left(1-\gamma+\frac{\gamma}{v}\right)F[z]\right]\exp(pt)dp}{p}. \qquad (2.46)$$

In case $a \ll 1$, at maximum short times, $t \ll \gamma_0 a^2$, the major input into integral is made by $p \gg 1/a^2\gamma_0$. At this limit. it is possible that $F(z) = 0$, but $F(y) \approx 3/y$. Further calculations will lead to formula:

$$q(t) \approx \frac{6\rho_0\sqrt{\gamma_0 t}}{\sqrt{\pi a}}$$

At this stage, methane release from the lumps is fairly small in comparison with methane release from the filtration volume. When $t \gg \gamma_0 a^2$, the major input into the integral is made by $p \ll 1/\gamma_0 a^2$. With it, $y \to 0$ but $F(y) \to 1$. As consequence,

$$q(t) \approx \gamma_0\rho_0 + \frac{\gamma_1\rho_0}{2\pi i} \int\limits_{\sigma - i\infty}^{\sigma + i\infty} \frac{F(z)\exp(pt)dp}{p};$$

$$z \equiv \sqrt{p}, \quad \gamma_1 \equiv v(1-\gamma_0)\left(1-\gamma+\frac{\gamma}{v}\right).$$

According to the compression theorem and taking into account (2.29), we get:

$$q(t) \approx \gamma_0\rho_0 + 6\gamma_1\rho_0 \sum_{n=1}^{\infty} \frac{1-\exp(-\pi^2 n^2 t)}{\pi^2 n^2}.$$

Summand 1 means that at this stage all the methane initially contained in filtration volume has already escaped from coal; summand 2 characterizes the
dynamics of methane escape from lumps into surrounding space through
filtration volume. Asymptotic analysis of the last formula leads to the following results: at intermediate times, $\gamma_0 a^2 \ll t \ll 1/\pi^2$:

$$q(t) \approx \gamma_0 \rho_0 + \frac{6\gamma_1 \rho_0 \sqrt{t}}{\sqrt{\pi}};$$

At the final stage of the process, $t \gg 1/\pi^2$ and $q(t) \approx (\gamma_0 + \gamma_1)\rho_0 - 6\gamma_1\rho_0/\pi^2$
$(-\pi^2 t)$, i.e., approaching equilibrium, it follows the exponential law with
typical time $1/\pi^2$, i.e., in dimension units t_d/π^2. In cases of small size coal
granules (chips), methane escape from coal covers three stages characterized
by the following density values of gas flow from a unit of coal volume (time
is in dimension units):

(1) $\quad t \ll \gamma_0 \dfrac{L^2}{D_f}, \quad j(t) = \dfrac{3\rho_0}{L}\sqrt{\dfrac{\gamma_0 D_f}{\pi t}};$

(2) $\quad \gamma_0 \dfrac{L^2}{D_f} \ll t \ll \dfrac{R^2}{\pi^2 D_{\mathit{eff}}}, \quad j(t) = \dfrac{3\gamma_1 \rho_0}{R}\sqrt{\dfrac{D_{\mathit{eff}}}{\pi t}};$

(3) $\quad t \gg \dfrac{R^2}{\pi^2 D_{\mathit{eff}}}, \quad j(t) = \dfrac{6\gamma_1 \rho_0 D_{\mathit{eff}}}{R^2} \exp\left(-\dfrac{\pi^2 t D_{\mathit{eff}}}{R^2}\right).$

In case $a \gg 1$ for relatively big granules, at time $t \ll a^2$, we substitute $F(y)$ for
$3/y$ and $F(z)$ for 1 and arrive at

$$q(t) \approx \frac{6\rho_0 \sqrt{(\gamma_0 + \gamma_1)t}}{\sqrt{\pi}a}.$$

At earlier t_f, methane escapes at equal rates both from lumps and filtration
volume and thus the speed of methane traveling from lumps to filtration
volume coincides with the speed of filtration outside. It also can be seen from
the asymptotic of function $F(t)$ from (2.29) that travel speed is far longer than

diffusion time. For extremely long time values, we substitute $F(y)$ for unity and similar to the previous operation we get:

$$q(t) \approx (\gamma_0 + \gamma_1)\rho_0 - \frac{6\gamma_1\rho_0}{\pi^2}\exp(-\pi^2 t).$$

As the formula is true only at $t \gg a^2 \gg 1$, this stage reveals almost no gas release because the gas has already escaped. The influence of backpressure on the rate of methane escape from the coal has an impact only under a large coefficient of vessel filling when $g \ll 1$. At this limit, escape time decreases according to law $t \sim g^2$.

2.2.3 Methane Flow from Coal Seam into Worked-Out Space

Physical processes during methane flow from a natural coal seam into a worked-out space do not differ from processes for broken coal although the velocity of methane flow into surrounding spaces may differ considerably. Above all, the filtration coefficient in a broken coal piece may be larger than the coefficient in a coal massif because of mesostructure faults in broken coal.

In broken coal, a typical dimension L of the basic fraction can be singled out. This dimension based on the previous statement is well known and greatly influences the inside gas dynamics of methane. In seam setting, the issue is not the total quantity of escaped methane; it is the density of methane flow from an area unit of exposed coal surface. In general, to solve the practical problems of inner gas dynamics in seams, it is necessary to solve Equation (2.30) taking into account (2.28) and the specific initial and boundary conditions. As a rule, such problems can be solved numerically via algorithms and programs. In some cases, we can obtain estimates of the density of methane flow from a seam.

For this purpose we shall consider formulas (2.42) and (2.43), defining the velocity of methane escape from a sphere of radius L into open space. By simulating the situation with methane in a seam, we perform a transition to large (in a limit of infinite) L. The total quantity of methane escaped from a unit of sphere volume is

$$q = \gamma_0(\rho_0 - \rho(t)) + (1 - \gamma_0)\left(1 - \gamma + \frac{\gamma}{v}\right)(c_0 - c(t)) \qquad (2.47)$$

Let's take into account that under $L \to \infty$:

$$F(y) \approx \frac{3}{a\sqrt{\left[\gamma_0 + v(1 - \gamma_0)\left(1 - \gamma + \frac{\gamma}{v}\right)F(z)\right]p}}, \qquad (2.48)$$

as a is proportional to L. Then:

$$q = \frac{3\rho_0}{2\pi i a} \int\limits_{\sigma-i\infty}^{\sigma+i\infty} \frac{\sqrt{\gamma_0 + v(1-\gamma_0)\left(1-\gamma+\dfrac{\gamma}{v}\right)F(\sqrt{p})}}{p^{3/2}} \exp(pt)dp \qquad (2.49)$$

Multiplying q by the sphere volume $4\pi L^3 / 3$, we get the total methane escaped by moment t (time is in dimensionless units) from the sphere. From (2.49) and definition a, we can see that this quantity is proportional to L^2, i.e., to the area of the sphere surface. Dividing the result by $4\pi L^2$, we calculate the methane quantity escaped by moment t through the unit area of the exposed surface:

$$\frac{q}{4\pi L^2} = \frac{\rho_0 R}{2\pi i}\sqrt{\frac{D_f}{D_{eff}}} \int\limits_{\sigma-i\infty}^{\sigma+i\infty} \frac{\sqrt{\gamma_0 + v(1-\gamma_0)\left(1-\gamma+\dfrac{\gamma}{v}\right)F(\sqrt{p})}}{p^{3/2}} \exp(pt)dp \qquad (2.50)$$

Now if we take the time (dimensional) derivative from both parts (2.50), we will find a relatively simple expression for methane stream density:

$$j(t) = \frac{\rho_0}{2\pi i}\sqrt{\frac{D_f}{t_d}} \int\limits_{\sigma-i\infty}^{\sigma+i\infty} \frac{\sqrt{\gamma_0 + v(1-\gamma_0)\left(1-\gamma+\dfrac{\gamma}{v}\right)F(\sqrt{p})}}{p^{1/2}} \exp(pt)dp \qquad (2.51)$$

The last formula admits simple asymptotic estimates of short ($t \ll 1$) and long ($t \gg 1$) time limits. Under short time limits large p value contributes mainly to the integral (2.51) when $F(\sqrt{p})$ is small and negligible. Then it is easy to calculate the integral (in dimensional time):

$$j(t) \approx \rho_0 \sqrt{\frac{D_f}{\pi}}\sqrt{\frac{\gamma_0}{t}}, t \ll t_d = \frac{R^2}{D_{eff}} \qquad (2.52)$$

In the extreme inverse case, small p contributes mainly to the integral when $F(\sqrt{p}) \approx 1$. The flow in this case is inversely proportional to $\sqrt{t}$ as before but with the other coefficient:

$$j(t) \approx \rho_0 \sqrt{\gamma_0 + v(1-\gamma_0)\left(1-\gamma+\frac{\gamma}{v}\right)}\sqrt{\frac{D_f}{\pi t}} \qquad (2.53)$$

More detailed research of the integral (2.51) shows that the time slice $t \ll t_d$ disintegrates into two intervals: the first is $t \ll \gamma_0^2 t_d / 9\gamma_1^2$ and the second is $\gamma_0^2 t_d / 9\gamma_1^2 \ll t \ll t_d$,

$$\gamma_1 \equiv v(1-\gamma_0)\left(1-\gamma+\frac{\gamma}{v}\right). \qquad (2.54)$$

In the first interval, Equation (1.52) "works." For the second, the equation is

$$j(t) \approx \frac{2\sqrt{2}b\rho_0}{\pi}\sqrt{D_f}\frac{\sqrt{3\gamma_1}}{t_d^{1/4}t^{1/4}}; \qquad (2.55)$$

Here b is a numeral constant of the order of unity. As a rule, $\gamma_1 \approx \gamma$, i.e., γ_1 corresponds to closed porosity. We can check that formula (2.55) "is bound" well with formula (2.52) under $t \sim t_d\gamma_0^2 / 9\gamma_1^2$ and with formula (2.53) under $t \sim t_d$. As the closed porosity is about four to five times the open porosity (typical values: $\gamma_0 = 0.005$, $\gamma = 0.2$), $\gamma_0^2 / 9\gamma_1^2 \approx 1/200$, i.e., the "closing" time of Equation (2.52) is 200 times less than t_d.

Methane escapes from a big coal massif at three stages. During the first stage that does not last long, about $10^{-2}t_d$, the methane flow density decreases according to the inverse radical law. At this stage, methane in the filtration space escapes mainly from the coal mass. Although the stream density decreases with the passage of time at the second stage, it is far weaker according to the law $t^{-1/4}$ because filtration and diffusion take place simultaneously. This stage continues until $t \sim t_d$. At the third and long stage, $t \gg t_d$, the stream density decreases again according to the radical law but slower than at the first stage because now gas escapes from the lumps through the filtration space.

Figure 2.2 depicts the process of the methane escape from the coal massif. The graph of function $N(t) = \int_0^t j(t)dt$ corresponds to the methane quantity escaped during time t from the moment of exposing of the mass and from the unit of area of the exposed coal surface. Theoretically, the methane escape process from the mass continues ad infinitum. In fact the mass is limited; hence methane escapes during a very long but finite time. Some numerical estimates are presented below.

The dimension of a coal microlump is $R \sim 10^{-6}$ m, $D_{eff} \sim 10^{-15}$ m²/sec, and diffusion time is $t_d = R^2 / D_{eff} \sim 10^3 c \sim 20$ min. The filtration coefficient according to (2.8) is 2.5· 10^{-6} m²/sec—nine orders of magnitude greater than the diffusion coefficient; $\rho_0 = 30$ m³/m³ can be considered the gas pressure at 30 atm. It is worthwhile to calculate how many cubic meters of methane will escape from the surface of 1 m² by the moment t_d. $N(t_d) \sim 1$ m³ from the surface of 1 m². Double the amount of methane will escape by the moment $t = 4t_d$. Our example indicates that 2 m³ of methane will escape during 80 minutes from every square meter of the exposed surface.

2.3 Methane Accumulation in Dangerous Coal Lump Regions

2.3.1 Time for Formation of Highly Explosive Methane–Air Mixture

Knowing the methane escape velocity from an open seam and from broken coal permits us to solve the problem of methane accumulations in different coal mine workings, particularly in the dead regions of cul de sacs and bunkers. For the sake of simplicity, we confine our examination to methane concentration dynamics in bunkers.

A coal mine bunker is a container filled periodically with fresh coal and from which coal is periodically removed. That makes it possible to determine the average coal quantity in a bunker and apply the bunker filling coefficient k_y, which is equal to the ratio of the volume taken by the coal substance to the full bunker volume. This value usually varies from 0.5 to 0.7.

Coming from the coal face into the bunker, coal contains a considerable quantity of methane. This methane escapes from the coal into the bunker volume when the coal batch is in the bunker. The methane escape from the coal into the bunker is controlled by such parameters as gas saturation of the working seam, dimensions of coal lumps, diffusion and filtration coefficients, porosity, and solubility. The gas release velocity calculations are well known and covered in detail in the previous section.

Conversely, the interchange of gases between the bunker and the excavation has not been studied thoroughly. In some cases, the bunkers are ventilated forcibly. In other cases, ventilation arises from depressions inside a mine. In both cases, the free bunker space in filled by a methane–air mixture. Concentration of separate components of this mixture (methane, oxygen, nitrogen) can change over time for a number of reasons.

First, methane-containing coal enters a bunker in discrete batches and methane flow from this coal weakens gradually. Second, bunker ventilation may be unequal or even absent if a ventilation cutoff occurs during an emergency situation. Third, to prevent fire, nitrogen may be injected into a bunker. To build a mathematical model describing the changes in concentration of the various components of the mixture, we considered the following principles.

The concentrations of methane and other components in a methane–air mixture are generally heterogeneous and they differ in different areas of a bunker. Nevertheless because of the fast mechanical agitation of gases and their mutual diffusion, the concentration of the components equalizes within seconds according to our estimations. For this reason it is possible to determine average concentrations of certain gases in bunkers.

According to preliminary hydrodynamic calculations, within a fraction of a minute, a regime of interchange of gases is set so that the gas quantity drawn into the bunker is equal to the gas quantity that escapes from it. That makes it possible to apply differential equations that determine the (average) methane concentration in the mixture to analyze the other quantitative characteristics of the process.

Let us examine these considerations in more detail by using the methane example. The same considerations apply to oxygen and nitrogen. An increase in methane concentration takes place during its escape from the coal into the bunker and a decrease occurs when methane is emitted into environment as part of a mixture. Let us designate by $m(t)$, $q(t)$, and $n(t)$ (sec^{-1}) as the quantities of methane, oxygen, and nitrogen, respectively, entering the bunker per unit of volume per unit of time. The sum of these quantities $m(t) + q(t) + n(t)$ is the total gas entering the bunker. Based on these estimates, the same gas amount escapes from the bunker for each unit of time. We assume that the atmosphere entering the bunker is the same and oxygen and nitrogen quantities in the mixture are constant. If we designate the running methane concentration as $c_1(t)$ so the rate of change of this quantity dc_1/dt is estimated by the difference between the velocity of methane entering the bunker $m(t)$ and the velocity of methane escaping from the bunker $[(m(t) + q(t) + n(t))c_1(t)]$:

$$\frac{dc_1}{dt} = m(t) - \left(m(t) + q + n\right)c_1. \tag{2.56}$$

Similar considerations are valid for the oxygen and nitrogen concentrations $c_2(t)$ and $c_3(t)$, respectively:

$$\frac{dc_2}{dt} = q - \left(m(t) + q + n\right)c_2 \tag{2.57}$$

$$\frac{dc_3}{dt} = n - \left(m(t) + q + n\right)c_3 \tag{2.58}$$

Concentrations c_i ($i = 1, 2, 3$) are connected by the correlation:

$$c_1 + c_2 + c_3 = 1. \tag{2.59}$$

It is easy to see the system matching with the correlation (2.59) by summing the system equations. Using the dependencies calculated earlier, we derive the expression for the power of sources of methane entering the bunker in the form:

$$m(t) = \frac{\alpha \rho_0}{\rho_a} \frac{k_3}{l_c} \sqrt{\frac{\gamma_e D_f}{\pi(t + t_{tr})}}, \tag{2.60}$$

where l_c is the radius of coal fractions, (m), P_0 is methane pressure in the seam, P_a is the atmospheric pressure, D_f is the filtration coefficient, t_{tr} is the time of coal transportation to the bunker, $\gamma_e = \gamma_0 + v(1-\gamma_0)(1-\gamma+\gamma/v)$, and a is the geometrical factor that takes into account the form of coal fractions. Oxygen and nitrogen quantities entering the bunker (per unit of bunker volume) can be estimated:

$$q = \frac{Qa}{60V_b}; \; n = \frac{Q(1-a)}{60V_b}. \tag{2.61}$$

Here Q is air consumption (m³/min), V_b is bunker volume (m³), and a is atomic concentration of oxygen in air. Initial conditions of the system [Equations (2.56) through (2.58)] are that no methane is contained in the entering air and oxygen and nitrogen constitute the main parts of the gas mixture:

$$c_1(t)=0, \; c_2(t)=a, \; c_3(t)=1-a. \tag{2.62}$$

The numerical solution of the system under the initial conditions (2.62) allows us to follow the changes in concentration of the components of the methane–air mixture subject to the different characteristics of coal, methane, and ventilation power.

Figure 2.3 illustrates the change in methane concentration in a bunker. Each concentration rise is the result of a charge from a new coal batch. We assume that a bunker with a capacity of 100 m³ is charged every 2 hours by ventilation of 200 m³/min, the radius of coal fractions is 0.01 m, and the coal transportation time to the bunker is 15 minutes. We see that the concentration increase takes place very quickly and the decay begins to decline very slowly. Forced ventilation is desirable for bunkers. The bunker has to be filled less than 50% and coal has to be taken out of the bunker as fast as possible.

Figure 2.2 shows the change in the oxygen content subject to the oscillations of methane concentration. In this case the oscillations of oxygen concentration are small. The large increase of methane concentration at the moment of coal inflow can be accompanied by a relatively small decrease of oxygen concentration, i.e., nitrogen is displaced generally. Research focused on changes of methane concentrations under different ventilation conditions shows that the time for obtaining of the concentration peak does not depend on ventilation capacity. The concentration value maximum reached under these conditions depends on the ventilation value.

Let us consider a situation in which bunker ventilation stops. If the ventilation capacity through the bunker is decreased 10 times (Figure 2.3) leaving only natural leakages, the methane concentration can reach 5% or more during tens of minutes and stay dangerously high over several hours. The conditions for filling a bunker with methane apply to a considerable extent

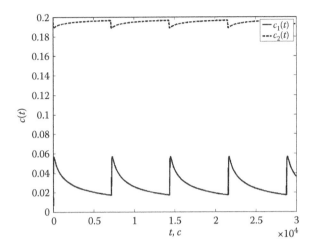

FIGURE 2.2
Change in oxygen content (dotted line) subject to oscillations of methane concentration in a bunker (solid line) during methane escape from coal massif.

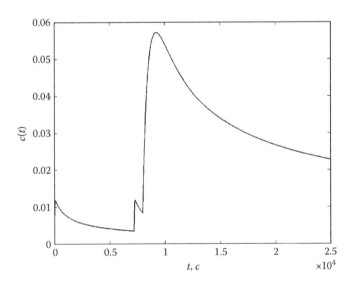

FIGURE 2.3
Variations of methane concentrations when bunker ventilation ceases.

to cul de sacs and other dead spaces. Knowing the condition required to fill dead spaces with methane allows us to predict the appearance of highly explosive methane–air mixtures in these spaces.

2.3.2 Thermodynamic Gibbs Potential for Gas–Coal Massif: Prognosis for Localization of Coal Burst Sections of Seam

Let us consider a coal seam containing pores filled with gas because methane pressure and travel are important safety factors in coal mining. We will apply the model presentations of methane in coal seams developed earlier.

Methane pressure in pores is designated P; outer (rock) pressure on a seam is P_m. To calculate the elastic energy of the coal frame, we start by dividing a seam into spherical areas of radius R_2 containing cavities of radius R_1 filled with methane. We then calculate the frame elastic energy as the sum of elastic energies of all such areas. Applying a known solution of the problem of deformations and tensions on a sphere with a concentric cavity subjected to inner and outer pressures [12], we use the formula for elastic energy of the area under consideration:

$$E_{ynp} = \frac{2\pi}{R_2^3 - R_1^3}\left[\frac{\left(PR_1^3 - P_m R_2^3\right)^2}{3K} + \frac{\left(P_m - P\right)^2 R_1^3 R_2^3}{4G}\right], \qquad (2.63)$$

where K and G are the modules for all-around compression and shift, respectively. We now calculate the ratio:

$$\gamma = R_1^3 / R_2^3 \qquad (2.64)$$

to determine coal porosity. To calculate the density of elastic energy, we use an expression containing parameters defined in laboratory and mine experiments:

$$\varepsilon_{elast}\left(P, P_m\right) = \frac{1}{2(1-\gamma)}\left[\frac{(P_m - \gamma P)^2}{K} + \frac{3\gamma}{4G}(P_m - P)^2\right]. \qquad (2.65)$$

Since ε_{elast} is expressed as pressure, it is an elastic constituent of the thermodynamic Gibbs potential of gas and coal materials. By adding the density of the thermodynamic Gibbs potential of gas in the pores (2.1) to ε_{elast}, we derive the density φ of the Gibbs potential of coal and the gas mass under consideration:

$$\varphi(P; P_m) = \frac{1}{2(1-\gamma)}\left[\frac{(P_m - \gamma P)^2}{K} + \frac{3\gamma}{4G}(P_m - P)^2\right] + \gamma P \ln\frac{P}{P_T}. \tag{2.66}$$

Since the seam is heterogeneous, its elastic modules K and G and porosity γ may differ in different seam sections—the usual case. Furthermore during mining, the outer pressure P_m becomes very heterogeneous. Concentration of tensions caused by the cavity (working) in the massif at maximum (so-called abutment) pressure can be several times higher than the pressure in the virgin massif. In all these cases, the gas pressure redistributes along the seam. To determine the nature of the spatial change of gas pressure at equilibrium, one should minimize the thermodynamic potential of gas and coal material to redistribute gas in the seam. We assume that no gas escapes from or enters the seam, and consider the requirement for keeping the total quantity of methane in the seam. In this case, the thermodynamic equilibrium of the seam gas lies in the homogeneity of its chemical potential (μ):

$$\frac{\partial\varphi}{\partial n} = \gamma\mu = \text{const}, \tag{2.67}$$

where $n = P/T$ is gas density and T is absolute temperature. Methane sorption in the lumps of coal changes methane pressure in accordance with its solubility. Usually the methane solubility in coal does not exceed 10^{-2} and methane sorption during initial calculations of pressures may be ignored.

The application of Equations (2.66) and (2.67) yields an opportunity to localize sections with increased gas pressures and estimate relative changes in those sections. We say for this purpose that the constant on the right side of (2.67) can be derived by considering that under $x \to \infty$ we transfer to the virgin massif. In this way we get the working correlation:

$$\ln\frac{P(x)}{P_\infty} + \frac{1}{1-\gamma}\left[\left(\frac{\gamma}{K} + \frac{3}{4G}\right)P(x) - \left(\frac{1}{K} + \frac{3}{4G}\right)P_m(x)\right]$$

$$= \frac{1}{1-\gamma_\infty}\left[\left(\frac{\gamma_\infty}{K_\infty} + \frac{3}{4G_\infty}\right)P_\infty - \left(\frac{1}{K_\infty} + \frac{3}{4G_\infty}\right)P_{m\infty}\right]. \tag{2.68}$$

The ∞ symbol indicates the section of the seam away from the geological or anthropogenic heterogeneity concentrated near x. When (2.68) is applied, one should bear in mind that elastic modules and porosity may depend on x as well. Let us consider two cases of interest.

Material characteristics are homogeneous; outer pressure is heterogeneous — $G = G_\infty$, $K = K_\infty$, and $\gamma = \gamma_\infty$. While considering the coal material by summands contained in the denominator of the module of all-around compression, K may be neglected in comparison with the summands that contain the shift module G. In coals at medium levels of metamorphism, according to experimental data [13], the shift module G is four to five times less than module K. Under these conditions, considering $\gamma \ll 1$, and using a standard logarithm series expansion we get the following evaluation formula:

$$\frac{P(x)-P_\infty}{P_\infty} = \frac{3}{4G}\left(P_m(x)-P_{m\infty}\right). \tag{2.69}$$

As noted above, during seam extraction, the so-called abutment pressure formed can exceed the rock pressure five to eight times within a few meters of the coal face. Calculations [14] and experiments [15] reveal a wide variation of concentration coefficients of the tensions (two to ten) in the zone of the abutment pressure subject to mining and geological conditions, seam capacity, and geometry of working layouts. Let us specify pre-critical situations. For example, at a depth of ~1000 m the rock pressure ($P_{m\infty}$) is equal to 25 MPa and the abutment pressure $P_m(x) \approx 8P_{m\infty} = 200$ MPa.

Thus, the gas pressure is increased by 15% (~8 atm) at the section of concentration of the abutment pressure in relation to the pressure on the moved seam sections. In fact, the gas pressure increase will be even greater as the elastic deformations can be added by plastic deformation and the effective modules will decrease. The physics of the paradoxical (at first glance) phenomenon of the gas pressure increase on the sections with the high outer pressure lies in the fact that the gas creates the backpressure that decreases the shift of elastic energy of the coal frame.

Even an insignificant increase of gas pressure [8] when a seam is at its maximum state in the zone of abutment pressure may cause the pores to become cracks upon further development and the bursts of coal and gas. Layer-specific tearing occurs in accordance with the concept of Khristianovich [16]. At present, we know of no experimental measurements of the increase of sub-threshold pressure arising from an increase of outer load. We calculated a theoretical value (2.69) of the increase using the concept of homogeneity of the chemical potential subject to the gas influence on the stressed and deformed seam condition.

Outer pressure is homogeneous; coal microstructure is very heterogeneous — Let coordinate x mark the section where elastic modules G and K are considerably fewer than the corresponding modules in the undisturbed section, i.e., $G \ll G_\infty$ and $K \ll K_\infty$. This problem concerns geological seam failures. If the porosity along the seam is constant and the shearing modulus is considerably less than the compression modulus, we can proceed from (2.68) to the following realistic value:

$$P - P_\infty \approx \frac{3P_\infty}{4G}(P_m - P_\infty). \tag{2.70}$$

Since the rock pressure P_m is always more than the methane seam pressure, it follows from (2.70) that methane pressure in the broken seam section ($G \ll G_\infty$) exceeds methane pressure in the unbroken section. The quantity of the effect is defined by the P_∞/G ratio and the rock pressure. For example, if P_∞ = 5 MPa, the shear modulus G = 100 MPa (eight times less than its standard value) and the rock pressure P_m=25 MPa, then $P - P_\infty$ = 0.75 MPa. If seam decay occurs subject to the increase of porosity ($\gamma \gg \gamma_\infty$), we get the following estimation based on analogous reasoning:

$$P - P_\infty \approx \frac{3P_\infty}{4G}\frac{(\gamma - \gamma_\infty)(P_m - P_\infty)}{(1 - \gamma_\infty)(1 - \gamma)}. \tag{2.71}$$

It follows from (2.70) and (2.71) that gas accumulates in sections with low resistance to shear and in sections with heightened porosity, in agreement with mine experiments and observations. The methane pressure accumulations can exceed the average pressure in a seam by 10 to 20% and create conditions prerequisite to dangerous emissions. These techniques allow engineers to make preliminary forecasts of dangerous gas-dynamic phenomena in mines.

2.4 Investigation of Phase State and Desorption Mechanisms of Methane in Coal

Various degrees of metamorphism, heterogeneous porosity, fissuring, and humidity in black coals define a wide spectrum of parameters that influence the kinetics of methane motion but also make comparative analysis of experimental data difficult. As a result, the characteristics of methane distribution, desorption, and phase states in coals remain controversial [17–19].

We believe that filtration and diffusion processes or their superposition may take place in porous materials but this assumption is only theoretical and has not been proven by experimental support. Data concerning gas desorption from massive coal fragments were compiled over a long term and constitute inaccurate measurements that do not permit unambiguous interpretation. Conversely, more accurate measurements on small coal granules show abnormally weak dependence of granule exhaustion rate on size. In this case, the discrepancy between experimental and theoretical results can hardly be explained solely as the result of filtration mechanisms of methane motion in coal.

Progress in understanding methane motion is connected with the development of the concept of block structures of fossil coals [19,20]. For example, Kogan and Krupenia [20] proposed a model of porous coal structure in which the whole mass of coal is subdivided by intercommunicating fissures and macropores into separate structural elements (blocks). Fissures and macropores form the filtration volume that communicates with the external coal surface. The blocks are penetrated by channels of a smaller scale (micropores) the authors assume are the main collectors of methane in coal. Thus, methane desorption bears a two-stage character: at first it diffuses from micropores in blocks and enters filtration channels (fissures and macropores) and then proceeds toward the coal surface.

In studies [19,21] at the Institute for Physics of Mining Processes of the National Academy of Sciences of Ukraine, the idea of block structure of coal has been further developed in a theoretical description of diffusion–filtration methane motion. The physical model of coal containing methane was complemented by the concept of closed pores in blocks. It was assumed that gas desorption from closed pores occurs by solid-state diffusion. Solving the material balance equation in this model determines parameters of a coal–methane system in two cases of methane desorption: from diffusion and filtration.

In summary, the importance of theoretical investigations and their significance is somewhat limited by the adopted model assumptions. Another factor is a noticeable lack of experimental material concerning methane desorption from coal samples that allows comparisons of effects of diffusion and desorption mechanisms. New data, particularly systemic experimental data, will provide additional information about methane distribution and motion in coal.

2.4.1 Methane Phase States in Coal

The studies of methane–coal systems indicate three phase states of methane: free gas in pores and fissures; molecules adsorbed on coal surfaces; and molecules absorbed in coal lumps in the form of a solid solution [22]. Two basic models describe methane–coal systems. One assumes that all the methane is either in a free or adsorbed state [23–25] while coal saturation is provided by a network of small open pores with considerable dispersion of cross-sections. The shortcoming of this model lies in the difficulty of explaining the duration of methane desorption from coal. According to theoretical estimates [26], gas diffusion coefficient in the smallest coal pores should exceed 10^{-6} to 10^{-7} m^2/s. According to [27], this coefficient is considerably smaller (10^{-14} to 10^{-16}) m^2/s and is more typical of solid-state diffusion.

The second model relates the concept of the block structure of coal [21,28,29] based on comparative analysis of methane escape kinetics from coal samples of various sizes. [28] and [30] indicate that experimental data may be explained by assuming that black coal structurally is an aggregate of the smallest formations (microblocks); the free volume between microblocks represents open pores and fissures. The pores communicate with the external surface and provide evacuation channels for the gas after its diffusion from microblocks. In the desorption model, microblocks are the regions of a coal sample not containing open pores and fissures. The size of these regions is assumed to be small compared to the coal granule size.

Both models assume that methane in coal is in both free and adsorbed states but they differ about the character of gas distribution in the coal matrix. In the former case, the main reservoirs of methane are small (Folmer and molecular) pores. In the block model, this role is played by closed pores in microblocks. The block model presents an advantage over other models; it provides more persuasive grounds for interpretation of experimental data. It also explains the low values of methane diffusion coefficients in coals and connects high gas content with the presence of closed pores and inclusions of metastable single-phase formations of a solid solution. For these reasons, the experiments described below involved the block model of coal.

This model allows the general problem of methane distribution to be reduced to a calculation of methane in a free state in open fissures and pores and in an adsorbed state on coal surfaces (including the surfaces of open pores and fissures) and microblocks. It is reasonable to consider that if closed pores are available in microblocks, they contain methane in both free and adsorbed states and the ratio of the phases is the same as in open pores. The amount of free methane in open pores is evaluated according to

$$Q_{op}^{fr} = \frac{P_s \cdot V_{op}}{\psi(T) \cdot P_{atm}} \cdot m_{coal},\qquad(2.72)$$

where V_{op} is the specific volume of open pores and fissures determined by volumetric measurement; P_s is methane pressure in the saturation chamber; $\psi(PT) = P_s \cdot V / P_{atm} \cdot V$ is methane compressibility depending on methane pressure P_s and temperature T; V is the volume of compressed methane; P_{atm} is the normal pressure equal to 760 mm Hg; V_o is the volume of the tested methane at pressure P_{atm} and temperature $T_o = 273$ K; and m_{coal} is coal mass.

Let us use an example of calculating Q_{op}^{fr} for a dry coal sample from the Zasyadko Mine (seam l_1). The initial data are $m_{coal} = 20$ g; $P_s = 3$ MPa; $V_{op} = 0.06$ cm^3 / g $= 6 \cdot 10^{-5}$ m^3/kg; and $T = 298$ K. According to reference data

[31], the quantity of ψ at the specified values of P and T is equal to 1.04. Then:

$$Q_{op}^{fr} = \frac{3 \cdot 10^6 \cdot 6 \cdot 10^{-5}}{1.04 \cdot 10^5} \cdot 20 \cdot 10^{-3} = 0.34 \cdot 10^{-4} \text{ m}^3. \tag{2.73}$$

The amount of methane Q_{op}^{ads} sorbed on the coal surface can be found by comparing the amount of gas escaping from the container with coal after it is saturated with methane or helium. The method is based on the assumption that helium (unlike methane) does not interact with coal, i.e., helium is not adsorbed on a coal surface.

Laboratory studies of coal gas content conducted via methane desorption into a vessel of known volume (volumetric method) involved three stages. The first deals with coal saturation by methane compressed to dozens of atmospheres. The second covers the preliminary release of the compressed gas from the container upon saturation. The third state is collection of methane released by coal into an accumulation vessel (AV). The estimation of quantity Q_{op}^{ads} must be done during the second stage—compressed methane or helium release from the free volume of the container. Note that during the second and third stages, vessels of different volumes and vacuumed in advance are used.

We filled a container of coal granules of 2.0 to 2.5 mm with methane or helium compressed to 30 atm. Upon the closure of the high pressure gas main line, methane pressure in the container decreased as it was sorbed by the coal. In the case of helium saturation, the pressure in the container of coal did not change. In the experiment with methane, the compressed gas was added to the container until equilibrium gas pressure of 30 atm was established (sorption lasted more than 10 days). Before desorption registration, the compressed gas was released from the free volume container into a vessel of known volume. The operation took fewer than 5 sec; after that, helium release into the accumulation vessel was not observed. Conversely, in the methane experiment, methane continued to escape although at far lower velocity than the gas pressure release from the free volume. These results indicate that helium does not interact with coal and the chosen method of estimating Q_{op}^{ads} is well grounded.

The first experiment with helium was conducted under the following conditions: temperature of medium T = 298°C; initial helium pressure in the container with coal = 3 MPa; mass of coal m_{coal} = 20 g; volume of accumulation vessel V = 4804 Cm³; helium pressure in the accumulation vessel at gas release from the free volume of container with coal P_{AV} = (28.13 ± 0.07) kPa; atmosphere pressure P_{atm} = 0.1 MPa.

$$Q_{He} = \frac{273 \cdot P_{AV} \cdot V_{AV}}{298 \cdot P_{atm}} = \frac{0.916 \cdot 28130 \cdot 48.04 \cdot 10^{-4}}{10^5} = (12.4 \pm 0.03) \cdot 10^{-4} m^3. \quad (2.74)$$

For the second (methane) experiment: methane pressure after release of compressed gas from the free volume of the chamber with coal $P_{AV} = 30.39 \pm 0.07$ kPa; the remaining parameters were the same ones used in the first experiment.

$$Q_{CH_4} = \frac{273 \cdot 30390 \cdot 48.04 \cdot 10^{-4}}{298 \cdot 10^5} = (13.4 \pm 0.03) \cdot 10^{-4} m^3. \quad (2.75)$$

We also determined the value of the correction factor k for the different compressibilities of helium and methane. At a compressed methane temperature T = 298 K and 30 atm pressure, methane and helium volume change ratio due to expansion during escape is 1.033 ± 0.005. In these studies of the escapes of compressed gases, it is necessary to consider the influence of the considerable Joule-Thomson effect for methane [32]. Using the obtained data, we found the methane quantity adsorbed on the surfaces of a coal free of moisture:

$$Q_{op}^{ads} = Q_{CH_4} - (1.033 \pm 0.005) \cdot Q_{He} = (0.60 \pm 0.13) \cdot 10^{-4} \ m^3 \quad (2.76)$$

Thus the quantity of adsorbed methane on the open coal surfaces was almost twice the quantity of methane in the free condition. Consider that value Q_{op}^{ads} can be considerably more because of high mobility and the small helium atoms that penetrate into small pores that are inaccessible to methane molecules.

We also used the volumetric method to determine the quantity of gas Q_{mb} in microblocks. After compressed methane escaped from the container of coal, it was possible to observe the slow desorption process of methane escape from the coal microblocks. When methane gathers in an AV, desorption cannot be complete because of pressure in the AV. The presence of gas in coal in the free and adsorbed states may make the volumetric method essential. The gas can be removed easily by cooling the AV used for desorption in liquid nitrogen. Due to low pressure maintained in the vessel (the pressure of saturated methane vapors at T = 77 K = ~1.33 kPa), methane escape will be more complete. After desorption occurs under such conditions, the vessel containing coal is isolated, returned to room temperature, and the pressure is recorded.

We now explain the method of calculating Q_{mb} of methane in coal micro-blocks from the Zasyadko mine. Methane saturation pressure is 3 MPa; coal mass is 20 g; volume of the accumulation vessel $V_{AV} = 12.17 \cdot 10^{-4}$ m³; AV pressure at completion of desorption $P_{AV} = (18.8 \pm 0.07)$ kPa; medium temperature is T = 298 K.

$$Q_{mb} = \frac{T_o \cdot V_{AV} \cdot P_{AV}}{T \cdot P_{atm}} = \frac{273 \cdot 12.17 \cdot 10^{-4} \cdot 18.8 \cdot 10^3}{298 \cdot 10^5} = (2.00 \pm 0.03) \cdot 10^{-4} \text{m}^3. \quad (2.77)$$

Thus, the full amount of methane contained in 20 g of dry coal from Zasyadko (seam l_1) after saturation with methane compressed to 30 atm amounted to

$$\sum Q_{CH_4} = Q_{op}^{fr} + Q_{op}^{ads} + Q_{mb} \approx 2.93 \cdot 10^{-4} m^3, \quad (2.78)$$

which corresponds to gas content of coal equal to 14.6 m³/t.

It is important to note that one-third of methane is lost even before measurement of its content in coal is started. This phenomenon arises from a break of equilibrium in a coal–methane system resulting from the release of compressed gas from the free volume in a container of coal or separation of a coal piece from a gas-containing seam. The result is fast and intensive methane escape from open pores and cracks.

It is logical to ask what quantity of methane remains in pores after depressurization in a saturation chamber and after termination of the first phase of methane escape, i.e., at the moment of desorption registration. If the methane pressure in pores remains high, the method of estimation of the Q_{op}^{ads} of adsorbed methane stated in the first phase should be considered erroneous or requiring more precise definition. This question is all the more important as its solution affects forecasts of pollution at sites where coal is produced and stored.

We studied the dependence of gas flow through coal on gradients of gas pressure. Apparently this dependence may be of use for determining methane pressure in the transporting channels when it is desorbed from coal. Figure 2.4 depicts an appliance for measuring coal permeability under these conditions. We used air and methane as gases and a cylindrical coal sample free of moisture (13 mm diameter; $l_{cyl} = 12$ mm height). Registration of amounts and velocities of gas release was based on changes of pressure in an AV with a volume of $V_{AV}^{stat} = 340$ Cm³ chosen to minimize the influence of the accumulated gas on experiment results. The gas pressure that exceeded the vacuum level was defined with a mercury–oil vacuum meter.

Table 2.1 presents the results from studies of the quasi-stationary motions of methane molecules through a coal sample from seam l_1 of the Zasyadko

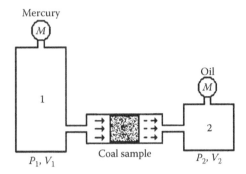

FIGURE 2.4
Appliance for measuring dependence of gas flow through a coal sample on gas pressure. 1. Gas reservoir. 2. Accumulation vessel of known volume. M = pressure meters in vessels 1 and 2.

TABLE 2.1

Influence of Methane Pressure Difference on Velocity of Gas Pressure Change in Accumulation Vessel

P_1	P_2	$\Delta(P^2)$	ΔP_{AV}	Δt	$\left(\overline{\Delta P}/\Delta t\right)$
125	15	15973	59	5400	0.011
161	45	28125	93	3900	0.024
245	30	60937	70	1700	0.041
325	0	105469	114	1500	0.076
453	0	205460	162	1170	0.139
577	0	333594	199	870	0.229

mine using the following parameters: Δt is experiment duration (sec); P_1 and P_2 are gas pressures at sample entrance and exit (mm Hg) in a mercury column; $\Delta(P^2)$ is the difference of these pressures squared; ΔP_{AV} represents changes of gas pressure in the AV during the experiment (mm) in the oil column; and $(\overline{\Delta P}/\Delta t)$ is the velocity of change of pressure (mm of oil column/s). These data were supplemented with the results of a study of air molecule transfer shown in Figure 2.5. The linear character of dependence of $\left(\Delta P_{AV}/\Delta t\right)$ from $\Delta(P^2)$ shows that even in an area of low pressure (lower than atmospheric) viscous flow of gases in the open pores of the black coal occurs and we can apply Darcy's law to describe it. According to [33], under stationary laminar gas flow in an isothermal regime, the molecular release of gas through transportation channels (pores) is described by the following formula:

$$\frac{\partial N}{\partial t} = \frac{\kappa}{\eta} \frac{S(P_1^2 - P_2^2)}{2m_0\beta L}$$

(2.79)

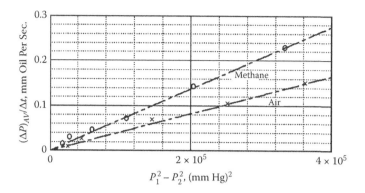

FIGURE 2.5
Dependence of velocity of gas pressure changes in an accumulation vessel on the difference squared of pressures on coal sample ends.

where N is the number of gas molecules, m_o is the mass of one molecule, S is the cut area of a coal sample end, β is the proportionality coefficient between gas pressure and density $\beta = P/\rho$, l_{cyl} is length of a coal sample, P_1 and P_2 are gas pressures on the ends of a coal sample, η is gas viscosity, and κ is coal permeability. $\Delta P_{AV} \, \partial N / \partial t = \left[V_{AV} \cdot (\partial P_{AV} / \partial t) \right] / k_B \cdot T$, then from Equation (2.79) we derive a formula for coal permeability κ:

$$\kappa = \frac{V_{AV} \cdot 2 \cdot l_{cyl}}{k_B \cdot T \cdot S_{cyl} \cdot (P_1^2 - P_2^2)} \cdot \left[\eta \cdot \beta \cdot m_o \cdot \frac{\Delta P_{AV}}{\Delta t} \right]. \tag{2.80}$$

We will now calculate coal permeability based on experimental data from methane: $V^{stat}{}_{AV} = 340 \cdot 10^{-6}$ m³; $l_{cyl} = 12 \cdot 10^{-3}$ m; Boltzmann constant $k_B = 1.38 \cdot 10^{-23}$ J/K; temperature T = 300 K; cylinder end area $S_{cyl} = 1.32 \cdot 10^{-4}$ m²; pressure square drop along sample length $\Delta(P^2) = 105469 \cdot (133.322)^2 = 0.187 \cdot 10^{10}$ Pa²; pressure change rate in SB at this $\Delta(P^2)$ value $(\Delta P_{AV}/\Delta t)^{stat} = 0.076 \cdot 8.7 = 0.66$ Pa/sec; mass of methane molecule $m_o = 26.5 \cdot 10^{-27}$ kg; $\eta = 1.08 \cdot 10^{-5}$ n/m²; and $\beta = 10^5/0.717 = 1.395 \cdot 10^5$ m²/sec.

$$\kappa = \frac{8160 \cdot 10^{-9}}{102.2 \cdot 10^{-17}} \cdot \left[26.35 \cdot 10^{-27} \right] = 2.1 \, 10^{-16} m^2 = 0.21 \text{ mdarcy.} \tag{2.81}$$

Estimation of air permeability yields about the same value as the difference in velocity between air molecule permeability and methane permeability and the differences in viscosity of these gases.

We also compared the methane flow in the above experiment with the flow of gas during desorption from the coal batch. The batch mass was 20 g, size of granules $R_{gr} = 2.0$ to 2.5 mm, and preliminary saturation was with

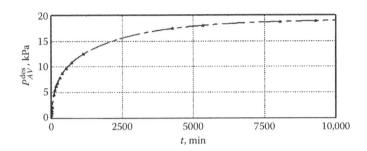

FIGURE 2.6
Methane pressure in accumulation vessel as function of desorption time (symbols depict experimental data points; solid line shows corresponding interpolation function $P_{AV} = a(1 -(1+bt^{-0.5}) + c(1 - \exp(-t/d)))$.

methane compressed to 30 atm. Registration of desorption motion started at 5 seconds—the time necessary to release the pressure of compressed methane in the free volume of the container. During the experiment, we registered the change of gas pressure in the AV as the gas escaped from the coal. Figure 2.6 shows the methane desorption motion from coal into a previously vacuumed vessel with volume $V_{AV}^{des} = 1217$ Cm³. We used a computer interpolation function to determine pressure: $P_{AV} = a(1 - (1 + bt)^{-0.5} + c(1 - \exp(-t/d)))$. More details on choosing interpolation type are in [34].

With data on stationary gas flow, one can estimate the quantity $\Delta(P^2)^{des}$ whose gradient in the lengths of open pores in the coal granules determines the gas flow observed during desorption:

$$\Delta(P^2)^{des} = \left[\frac{\Delta(P^2)}{\partial P_{AV}/\partial t}\right]^{stat} \cdot \frac{V_{AV}^{des}}{V_{AV}^{stat}} \cdot \frac{R_{gr}^2 \cdot S_{cyl}}{3 \cdot V_{coal} \cdot l_{cyl}} \cdot \left[\frac{\partial P_{AV}}{\partial t}\right]^{des}. \tag{2.82}$$

In Equation (2.82), the first square bracket corresponds to the permanent gas flow; the value in the second represents flow by desorption. V_{coal} is the total volume of coal granules in the hinge, in this case, ~15 · 10⁻⁶/ m³. The result for the first square bracket in our experiment was 2.97 · 10⁹ Pa/sec (or 2.227 · 10⁷ mm Hg/sec). As we can see from Figure 2.7, $(\partial P_{AV}/\partial t)^{des}$ of 1.5 Pa/ sec represents maximum gas flow.

Substituting other values into Equation (2.82), we find that $\Delta(P^2)^{des} = (P_1^2 - P_2^2)^{des}$ equals 6.1 · 10⁶ Pa² at the initial time of registration of methane desorption. In a case in which gas escapes from the coal into a pre-degasified vessel, it is possible to consider that $P_2 \approx 0$ at the initial time of desorption. Hence it follows that the methane pressure drop in the open pores does not exceed $\sqrt{6.1 \cdot 10^6} \approx 2.45 \cdot 10^3$ Pa (or 18 mm Hg) at desorption from granules 2.0 to 2.5 mm

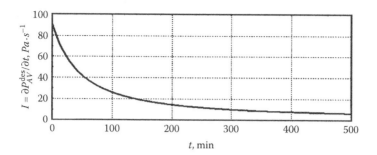

FIGURE 2.7
Methane pressure change rate in accumulation vessel as function of desorption time.

in size. Analogous investigations of desorption on coal in granules 0.2 to 0.25 mm and 9.0 to 10 mm in size showed that the methane pressures in pores are ~2.5 and 31 mm Hg, respectively.

Let us stress that the above calculations were based on data obtained for penetrability κ of the main coal pores, i.e., pores that provided connections between the ends of a comparatively large coal piece. The effect of gas filtration on methane desorption can be connected to small blind pores (with one outlet on the granule surface) and not attributed to permanent flow. Their penetrability is far lower and thus their methane pressure drop during desorption is higher.

The obtained values for differential pressure on the open pores can be used to define the time when the gas pressure in pores decreases from the maximum (after coal saturation) to the value achieved at the beginning of methane desorption from coal microblocks. For this purpose, it is sufficient to determine the time τ_f of the filtration process via the formula cited earlier [28]:

$$\tau_f = 4 \cdot R_{gr}^2 \cdot \eta \cdot \gamma / \kappa \cdot \pi^2 \cdot P_1. \qquad (2.83)$$

Value τ_f varies for different sizes of coal granules and equals 0.048, 0.8, and 7.6 sec for granules with diameters of 0.2, 2.0, and 9.5 mm, respectively. Note that the obtained τ_f values are valid for a particular coal saturated with methane under a certain pressure.

Research reveals how much gas escapes from open pores with large cross sections during the first seconds after unsealing of the equilibrium system coal-methane; and the amount of gas remaining in such pores is negligibly small. Gas escape from the open coal pores during first seconds upon its separation from a seam saturated with gas may exceed 25% of its total content in the coal. This feature must be taken into account during coal extraction and also when taking samples for diagnosing methane subsystems in a coal seam.

The experimental data presented above question the credibility of the known methods of diagnosing methane states in mine seams. In fact, the methods for determining in-seam methane pressure and methane content in coal based upon measuring desorption have one common deficiency: considerable gas loss occurs in the interval between sample removal and seam sealing because of the inevitable appearance of fissures in the sample. The result is distorted information about the initial methane content.

One can avoid these discrepancies and reduce the time for analysis of coal–methane samples by correlating the results of all measurements made under mine conditions with data from the desorption passport (DP) of the coal seam that contains experimentally established numeric or graph data that correlates the intensity of methane release from coal, seam methane pressure, and methane content. To complete a DP, it is necessary to make laboratory measurements of methane desorption kinetics and determine its content in coal. The measurements are made after preliminary saturation with methane in containers under different equilibrium methane pressures P_{sat}. The coal samples of equal mass in granules of 0.2 to 0.25 mm or 1.0 to 1.5 mm (depending on coal type) and natural moisture are used. The samples are obtained by sifting coal chips on sieves.

The first part of a DP establishes dependence of the intensity of methane desorption on the equilibrium gas pressure during coal saturation. Desorption is made into an airtight AV filled with air in which changes of pressure ΔP^{des} are registered during period Δt. Data showing changes of ΔP^{des}

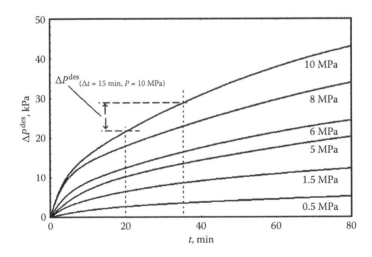

FIGURE 2.8
Change of pressure in accumulation vessel during methane desorption from coal after saturation. Each curve corresponds to a certain equilibrium pressure of saturation.

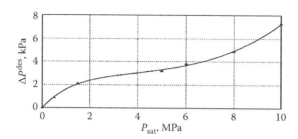

FIGURE 2.9
Change of pressure in accumulation vessel during methane desorption within 20 to 35 min as function of saturation pressure.

during desorption are presented in graph form for several values of P_{sat} that yield a family of curves ΔP^{des} $(t)=F(P_{sat})$ similar to those in Figure 2.8.

The data obtained are then used to derive the dependence of ΔP^{des} on P_{sat} at any time interval. For example, Figure 2.9 depicts the increase of pressure ΔP^{des} (dots) in the AV during 15 min of desorption (desorption interval = 20 to 35 min) based on the methane pressure P_{sat}, saturating the coal. The graph of function $\Delta P^{des}(P_{sat})=a \cdot \left(1-\exp\left(-P_{sat} / b\right)\right)+c(P_{sat})^d$ corresponds to its minimal deviation from experimental dot values. Optimization of parameters a, b, c, and d of the interpolation function is performed by a computer program. A set of numerical values of ΔP^{des} from P_{sat} forms the first segment of the DP. The delay in entering information (20 min in this case) is very important because the first 15 to 20 minutes of mine measurement are usually spent on drilling, sieving, and other preparations. Thus, we can compare measurements obtained under mine conditions with data from the first section of a DP to determine seam methane pressure without taking a coal sample.

The second part of a DP establishes the dependence of the methane in coal on saturation pressure. Having such information about seam methane pressure enables us to estimate the amount of methane in coal. Completing this segment of a DP in a laboratory requires the following operations:

1. Saturation with methane at different pressures (0.5 up to 10 MPa) of several samples of coal chips containing natural moisture
2. Determining the amounts of methane in the samples according to the described methods
3. Optimization of the parameters of the interpolation function $Q_{CH_4} = a(1-\exp(-P_{sat} / b))$

Figure 2.10 illustrates the measurement results (data points) and fitting function. A set of numerical values of function $Q_{CH_4} (P_{sat})$ comprises the second section of the DP.

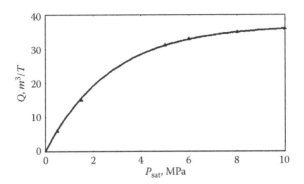

FIGURE 2.10
Dependence of methane amount in coal on equilibrium saturation pressure P_{sat}.

Comparing the results of a quick and simple mine measurement with laboratory data yields information about seam gas pressure and methane content in coal without the need to obtain a sample. The scientific work that led to the DP system served as the basis for constructing the DS-01 desorption tester at the Institute for Physics of Mining Processes of the National Academy of Science of Ukraine. The design of the portable device allows its use in underground worked-out spaces of mines that are unsafe because of the presence of gas and coal dust and also in highly explosive zones under and on the surface.

2.4.2 Kinetics and Mechanisms of Methane Desorption from Coal

Black coal is a dense, fissured, and porous medium with a heterogeneous chemical composition. Its nature makes it difficult to achieve a comparative analysis of the kinetics of gas desorption. For this reason, experimental data on the kinetics of methane escape are rather scarce.

We conducted research on the kinetics of methane escape from black coals via desorption into a vacuumed vessel of known volume. The accuracy of the method depends on the accuracy of measuring the vessel volume, its thermostatic performance, and gas pressure in the vessel. The method is not influenced by air moisture and allows measurement of desorption over a long interval. To exclude variations due to coal moisture content, the experiment was conducted with pre-dried samples. The subsequent computer processing of results yielded parameters of desorption kinetics that reveal mechanisms of methane escape from black coals.

Figure 2.11 illustrates the installation used to study the kinetics of methane desorption from coal. Registration of desorption was done via measuring gas pressure in an AV of known volume. The installation consisted of a metal container containing sample (1). A valve (2) connects the container to the AV (3).

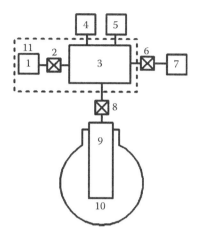

FIGURE 2.11
Low temperature desorption installation.

The pressure in the AV is measured by mercury–oil (4) and reference (5) vacuum meters. The container and accumulation vessel are connected to a vacuum pump (7) via a valve (6). A valve (8) connects the AV with a container (9) that is plunged into a Dewar bottle (10) containing liquid nitrogen. The coal container and AV are confined in a thermal container (11) made of foam plastic.

We used samples of volatile coal from the Korotchenko Mine; fat coal from the Zasyadko Mine; lean coal L_1 from a jammed coal band from the Glubokaya Mine; lean coal L_2 from the Kirov Mine; and anthracite from the Communist Mine. After crushing and sieving, volatile, fat, gaseous, and lean coals were dried at temperatures below 370 K for several hours. Anthracite was dried 30 days or more. The drying quality was checked by analyzing spectra from ^{1}H NMR [35].

The next stage included air removal via vacuum pump from the sample container, after which the container was joined to the tank with compressed methane and the sample remained in the methane atmosphere (6 MPa) for 14 days. Then the coal sample container (9) saturated with methane (valves 2, 6, and 8 were closed) was connected to the pre-vacuumed AV. After these steps, the container (9) was plunged into the liquid nitrogen medium (valve 8 closed).

Gas escape was registered from both the free volume (between coal granules) and from the coal sample. The first stage was compressed gas release from the container into the AV. This is a fast process (less than a minute) and does not provide information we need. Upon the release of the compressed gas and slowing of its emission from the container, the AV pressure was registered. With valves 2 and 6 closed, valve 8 was opened and fast methane condensation occurred in the container (9). Valve 8 was then closed and valve 2 opened to desorb the methane from the coal. The container volume, coal mass, and AV (3) volume were chosen so that methane pressure P_{AV} desorbed from coal into the vessel (3) would not exceed ~100 mm of oil to minimize the influence of abutment pressure in the AV during desorption. The AV and vacuum meter function together as a measuring probe. The more the volume of the AV, the less is its influence on the desorption process. Test measurements proved that additional pumping of the AV increased the methane release from coal no more than 1%. We observed the following conditions:

- Different in sized fractions were made from a single coal lump.
- Methane saturation was done under consistent gas pressure.
- Desorption registration ceased as soon as the change of gas pressure in the AV became less than 1 mm Hg per day.

The number of gas molecules N_{AV} released in the AV of volume V_{AV} at the registration moment was derived according to the well-known gas state equation:

$$N_{AV} = \frac{P_{AV}V_{AV}}{k_B T} \tag{2.84}$$

where k_B is the Boltzmann constant and T is the temperature in Kelvin. The fractional error of determining N_{AV} did not exceed 3%. After desorption, valve 8 was opened, the gas was condensed in the container (9) and the composition of the gas released from the coal was determined via the pressure of the saturated vapors. The admixture of other gases was not registered. To find the methane desorption mechanisms, we estimated the influence of coal sample size on specific time τ_{AV}^{des} of filling the AV with gas while it was desorbed from coal:

$$\tau_{AV}^{des} = \frac{N_{AV}^{max} - N_{AV}(t)}{\partial N_{AV}(t)/\partial t} = \frac{N(t)}{\partial N_{AV}(t)/\partial t}, \tag{2.85}$$

where $\partial N_{AV}(t)/\partial t$ is the velocity of change in the number of gas molecules in the AV (gas flow from coal) and N_{AV}^{max} is maximum number of gas molecules in the AV upon the completion of methane desorption. The last equality in (2.85) is written with the assumption that in the above experiments on desorption the methane in coal $N(t)$ is determined by expression $N(t) = N_{AV}^{max} - N_{AV}(t)$. In this case, for the specific time of methane desorption from coal:

$$\tau^{des} = \frac{N(t)}{\partial N(t)/\partial t} = -\tau_{AV}^{des}. \tag{2.86}$$

Further, we shall be interested only in absolute values of quantities τ^{des} and τ_{AV}^{des}, so while discussing the experimental data we shall use only τ^{des}. Parameters N_{AV}^{max}, $N_{AV}(t)$ and $N(t)$ can be determined by graphical and computer analysis of the curve describing the course of desorption.

The choice of specific time τ^{des} as a basic parameter in the comparative analysis of desorption kinetics was based on two factors. The first is the high sensitivity of time to the sizes of coal granules. The second, no less important, is that quantity τ^{des} does not depend on the number of coal granules in samples.

In analyzing experimental data on methane desorption (or sorption) from black coals, it may be difficult to choose the type of interpolation function describing the gas escape process. In cases where measurement accuracy 7 is low, the function may have the form $N(t) = N_0 \cdot \exp(-t / \tau^{des})$ where $N(t)$ is amount of gas in coal and τ^{des} is the specific time of gas escape, assuming that the specific time of relaxation to equilibrium of the system coal–gas is a constant quantity not depending on the desorption phase $\tau^{des} = \text{const}$.

The filtration theory [35] and research [28, 30] show that this image of desorption kinetics is simplified. In reality, it is impossible to describe gas escape by a single exponential function. Thus, the choice of an interpolation technique grounded in physics is an important element in studies of transient gas release from black coals.

Figure 2.12 presents the typical course of methane desorption from a black coal into a vacuumed vessel of known volume using methane amount N_{AV} (or gas pressure P_{AV}) in the AV as a function of time. Graphs of fitting function $P_{AV} = a \cdot (1 - \exp(-t/\tau^{des}))$ are presented as well. Optimization of parameters a and τ^{des} in the fitting function was done by a special computer program. Cases in which desorption registration ceased after 600, 1,800 and 10,000 minutes of gas release were considered. It is evident that the values of a and τ^{des} depend on experiment duration. That is, the desorption parameters depend on time and the interpolation function in the form of one exponent cannot accurately describe the whole process of methane escape from coal. Our analysis shows that in choosing the type of interpolation, less mean-square error is provided by a combination of the square and exponential dependencies.

$$P_{AV}(t) = a\left(1 - \left(1 + bt\right)^{-0.5}\right) + \left(1 - exp\left(-t/\tau\right)\right). \tag{2.87}$$

Calculated according to Equation (2.87), $\tau_{AV}^{des}(t) = P_{AV}(t)/\left[\partial P_{AV}(t)/\partial t\right]$ changes continuously over time and more accurately describes the desorption process. The disadvantage of this interpolation is also evident: the clarity of physical interpretation of parameter b in the summand of the expression is lost (2.87). That is, we will use an approximation of the sum of two exponents:

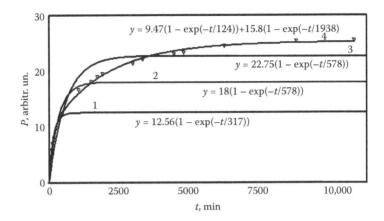

FIGURE 2.12
Desorption registration time influence on calculated parameters of function. $y = a(1 - \exp(-t/b))$. Curve 1 = registration time 600 minutes. Curve 2 = 1,800 minutes. Curve 3 = 10,000 minutes. Curve 4 = possibility of using interpolation function $y = a(1 - \exp(-t / \tau_1^{des})) + c(1 - \exp(-t / \tau_2^{des}))$.

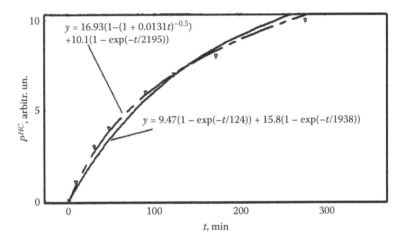

FIGURE 2.13
Comparison of experimental data (points) fitting results for Equations (2.87) and (2.88).

$$P_{AV}(t) = a(1 - \exp(-t/\tau_1^{des})) + b(1 - \exp(-t/\tau_2^{des})), \tag{2.88}$$

This expression, as it follows from Figure 2.12, presents a satisfactory description of the experiment. Parameter $\tau_{AV}^{des}(t)$ in this case also depends on t, but this dependence is weaker than in the case of interpolation based on Equation (2.87). Figure 2.13 shows a fragment of the initial phase of methane escape and illustrates the desorption course with functions of types (2.87) and (2.88).

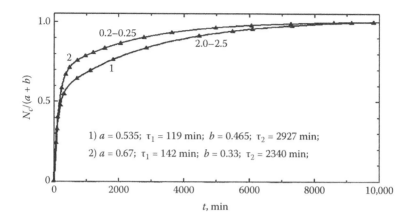

FIGURE 2.14
Pattern of methane desorption from volatile coal. (1) 2.0 to 2.5 mm. (2) 0.2 to 0.25 mm.

Thus, having chosen Equation (2.88), we sacrifice the accuracy of quantitative estimates for the sake of convenience in qualitative analysis. Note that the sum of parameters a and c corresponds to gas pressure in the AV vessel after the end of its release from the coal. If another method of desorption registration (e.g., NMR) is used on methane protons in coal, the interpolation of $a\exp(-t/\tau_1^{des})+c\exp(-t/\tau_2^{des})$ can be applied and parameters a and c are connected with the number of methane molecules in the coal. We note the significant consequence of using interpolation as a sum of several exponential dependences: in this case during the final desorption phase, when $t\gg\tau_1$, the correlation $\tau_{AV}^{des}(t)\approx\tau_2^{des}$ takes place. Figures 2.14 through 2.18 present the results of measurements and the corresponding interpolation curves of methane desorption from volatile, fat, lean$_1$, lean$_2$, and anthracite coals for fractions of 0.2 to 0.25 mm and 2.0 to 2.5 mm. General ideas of the physics of transient gas release from a porous solid material allow us to assume that the first (smaller) value τ_1 in the interpolation function (2.89):

$$Y = a(1-\exp(-t/\tau_1^{des}))+b(1-\exp(-t/\tau_2^{des})) \qquad (2.89)$$

is determined by the gas escape from the near-surface layers of the coal granules. The high initial gradient of gas concentration decreases during desorption, mainly at the expense of a shift of the concentration maximum toward the center of the granule. Upon completion of the transient processes, the second (final) phase of desorption begins when the gradient of methane concentration decreases only from the concentration difference in the center of the granule and near its surface.

The τ_2^{des} represents a specific desorption time at the final stage. This value is of particular interest as it directly depends on the sizes of the coal granules. That is why in further comparing values of τ_2^{des} in samples of different sizes of coal granules (R_1 and R_2) we will use $_1\tau^{des}$ and $_2\tau^{des}$ to describe values in the last decomposition exponent (2.88). We must say that this interpretation of the summands of function y does not fully reflect desorption physics because τ^{des} during laminar gas release is always a continuous function of time t.

The comparative analysis of curves in Figures 2.15 through 2.19 reveals two features. The first is that parameter a in the first summand of Equation (2.88) for coal fractions of smaller size exceeds the similar parameter for the larger fraction. In small fractions, the part of the near-surface layer (and methane in it) in the granule volume exceeds the part in the granules of larger size. It is interesting that in the last summand of function (2.88), during the passage from granules of 0.2 to 0.25 mm to granules of 2.0 to 2.5 mm, the expected change of parameter τ_2^{des} by 100 times was considerably less. This indicates the anomalous dependence of desorption velocity on the sizes of coal granules. Thus, for anthracite, the ratio is $_1\tau^{des}/_2\tau^{des} = \approx 30$; for coals with intermediate degrees of metamorphization (fat, lean$_1$, and lean$_2$), $_1\tau^{des}/_2\tau^{des}$ varies from 2 to 12.

One can better understand the mechanism of methane escape by assuming that each coal granule consists of small (microblock) fractions and the volume between them is the volume of fissures and open pores (similarly to the model in [19], [21], and [30]. Thus, a microblock is a part of coal without fissures and open pores; the surface area of all microblocks in a granule is equal to the surface of all open filtration channels. That is, microblocks can be regarded as results of excision of a homogeneous body by these channels. These formations are very small ($r_{mb} \ll R_{gr}$), distributed over the

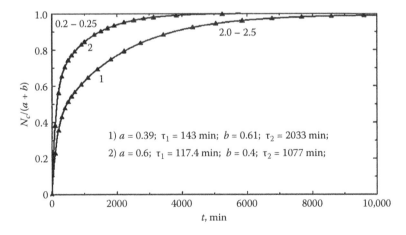

1) $a = 0.39$; $\tau_1 = 143$ min; $b = 0.61$; $\tau_2 = 2033$ min;
2) $a = 0.6$; $\tau_1 = 117.4$ min; $b = 0.4$; $\tau_2 = 1077$ min;

FIGURE 2.15
Pattern of methane desorption from fat coal. (1) 2.0 to 2.5 mm. (2) 0.2 to 0.25 mm

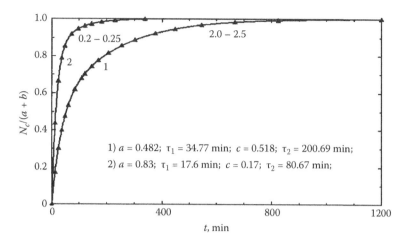

FIGURE 2.16
Pattern of methane desorption from lean$_1$ coal. (1) 2.0 to 2.5 mm. (2) 0.2 to 0.25 mm.

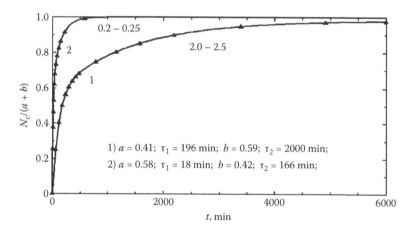

FIGURE 2.17
Pattern of methane desorption from lean$_2$ coal. (1) 2.0 to 2.5 mm. (2) 0.2 to 0.25 mm.

whole volume of granules, and connected to the coal surface through a system of fissures and open pores. Figure 2.19 shows a variant of the block composition of coal.

In this model, methane diffusion occurs from the closed pores of small particles (microblocks) whose average size $\overline{r_{mb}}$ does not depend on the sizes of the coal granules. Since microblocks and closed pores are very small, their large total surface allows diffusion gas flow of sufficient intensity. If one ignores the filtration effect in pores, the model must not show a correlation between granule sizes and the velocity of gas escape. Gas escape is possible if the gas enters pores with large enough openings. The kinetics of methane

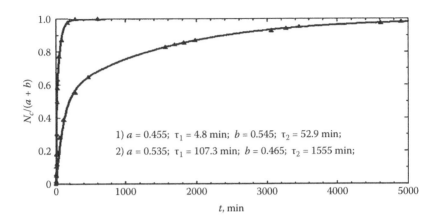

FIGURE 2.18
Pattern of methane desorption from anthracite. (1) 2.0 to 2.5 mm. (2) 0.2 to 0.25 mm.

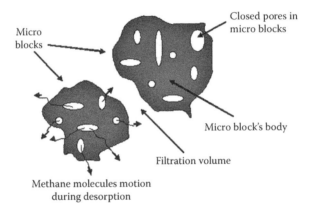

FIGURE 2.19
Coal structure in block model.

desorption from coal at a low level of metamorphization (volatile) is similar in pattern to the described case.

Conversely, in anthracite, the dependence of quantity τ_2^{des} on particle size is the strongest when one of the desorption mechanisms (diffusion or filtration) dominates. However, because anthracites possess developed networks of micropores, the assumption about solid-state diffusion as the main mechanism of gas release should be rejected.

The kinetics of methane desorption from all types of coals can be explained by the assumption that if it is assumed that along with the diffusion from the microblocks, gas filtration from the open pores takes place in these coals and adjusts the velocity of gas passing from the microblocks. To ground this assumption we should consider the process of methane escape from coal in more detail.

We shall assume that coal granules of medium radius R_{gr} consist of spherical formations (microblocks) whose average size r_{mb} is far smaller than R_{gr}. The space between microblocks is the filtration volume where methane from the closed pores through the microblocks body releases by diffusion. This desorption makes an explicit thesis: at a certain moment, a balance of flows of the gas released by diffusion $[\partial N/\partial t]_d$ and filtrated by Poiseuille pores $[\partial N/\partial t]_f$ occurs. The characteristic filtration time is determined by the viscosity η of gas, apparent porosity γ, and the permeability κ of coal. It also depends on the length of the filtration channel and gas pressure drop $P_f(t)$ in the channel [29,33]:

$$\tau^f(t) = R^2/D_f(t) = R^2\eta\gamma/\pi^2\kappa \cdot P_f(t) \tag{2.90}$$

The expression for τ^f assumes that the methane pressure at the end of the filtration channel can be neglected. The τ^f value increases in the initial phase of desorption while pressure P_f decreases. The specific time of the gas escape from microblocks is $\tau^d(t) = r_{mb}^2/D_d$. Most cases include a time function t as a result of the dispersion of microblock size and diffusion coefficient D_d over a coal granule. The gas pressure release P_f in the filtration pores at desorption is followed by a decrease of filtration parameter D_f and weakening of gas flow in the filtration channels, leading to conditions that balance the diffusion and filtration flows that remain until desorption is finished.

Desorption in cases where the dominant process is gas diffusion from microblocks and equality is observed $[\partial N/\partial t]_{des} = [\partial N/\partial t]_d = [\partial N/\partial t]_f$ will be described in detail. Figure 2.20 shows the distribution of methane molecule concentrations in different elements of coal during desorption.

At desorption, the concentration $n_1(t)$ of methane molecules in the closed pores of microblocks exceeds their concentration $n_f(t)$ in the filtration channel near the microblock body. The methane motion from the closed pores into the filtration channels results from gas diffusion through the microblock body. Diffusion in this case is connected with the concentration difference $\Delta C(t) = C_1(t) - C_2(t) = v[n_1(t) - n_f(t)]$ of gas molecules between the opposite walls of the microblock body. The amount of methane in coal $N(t)$ is the sum of molecules in the closed pores $N_1(t)$, filtration channels $N_f(t)$, and solid liquid (microblock body) $N_2(t)$:

$$N(t) = N_f(t) + N_1(t) + N_2(t). \tag{2.91}$$

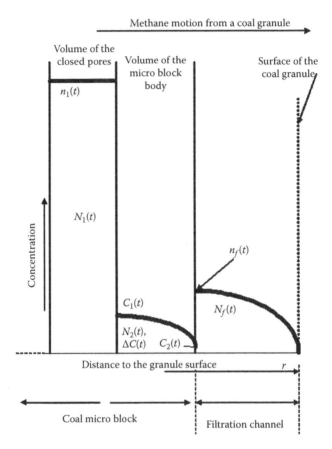

FIGURE 2.20
Pattern of methane concentration distribution in coal structural elements at desorption.

Under these conditions, the balance of flows for the specific desorption time τ^{des} by analogy with (2.85) can be written:

$$\tau^{des}(t) = \frac{N_f(t) + N_1(t) + N_2(t)}{\left[\partial N(t)/\partial t\right]_{des}} = \frac{N_f(t)}{\left[\partial N(t)/\partial t\right]_f} + \frac{N_f(t)}{\left[\partial N(t)/\partial t\right]_f} \cdot \frac{N_1(t)}{N_f(t)}$$

$$+ \frac{N_2(t)}{\left[\partial N(t)/\partial t\right]_d} = \tau^f(t)\left[1 + \frac{N_1(t)}{N_f(t)}\right] + \tau^d(t).$$

(2.92)

The specific desorption time is a linear combination of specific times of filtration and diffusion:

$$\tau^{des}(t) = \tau^f(t)\left[1+U(t)\right]+\tau^d(t) \tag{2.93}$$

Parameter $U(t)$ reflects the correlation of gas amount in the closed and open pores of coal. In materials with $\tau^f(t)(1+U(t)) \ll \tau^d(t)$, the ratio $_1\tau^{des}/_2\tau^{des}$ will be close to the ratio for granules of different sizes (R_1 and R_2). For $\tau^f(t)\left[1+U(t)\right] \gg \tau^d(t)$:

$$\lim_{\frac{\tau^d}{(1+U)\tau^f}\to 0} {_1\tau^{des}}/{_2\tau^{des}} = {_1\tau^f}/{_2\tau^f} \approx \left(R_1/R_2\right)^2 \tag{2.94}$$

Thus, by comparing the course of gas desorption from the granules of different sizes, for example, 10 times larger, we can calculate desorption times that will differ from 1 to 100 times, depending on the type of desorption mechanism prevailing in the given material. For example, the mechanism in anthracite will be $_1\tau^{des}/_2\tau^{des} \approx 30$ and in fat coals it will be $_1\tau^{des}/_2\tau^{des} \approx 2$. In the fat coal case, the role played by the diffusion component of the mechanism of gas escape is considerably greater than the role of diffusion in anthracite.

The clearest influence of diffusion can be observed in volatile coals where the contribution by τ^d is so substantial that the filtration effect is barely noticeable. In anthracite, the dependence τ^{des} on granule size is the strongest and represents clear evidence of the weak influence of the diffusion component of desorption (little τ^d).

Using experimental values of τ^{des}_{AV} for granules of R (large) and r (small) size, we can derive numerical values of the summands of expression (2.93) if we assume that $_1\tau^f/_2\tau^f \approx \left(R_1/R_2\right)^2$, and τ^d does not depend on the sizes of coal granules. Table 2.2 shows the results of calculations of specific times τ^d and filtration components of methane desorption from black coals of different degrees of metamorphism.

Thus, the behavior of desorption curves and the table data indicate that the velocity of the directed motion of methane in small granules of volatile

TABLE 2.2

Values of Filtration and Diffusion Components at Specific Time of Methane Desorption from Black Coals

Coal Type	$\tau^{des}_{0.2}$ (Min)	$\tau^{des}_{0.2}$ (Min)	$(1 + U)\tau\,^f_{0.2}$ (Min)	$(1 + U)\tau\,^f_{2.0}$ (Min)	τ^d (Min)
Volatile	2900	3030	1.3	130	2898
Fat	1075	2940	18.8	1880	1056
Crushed lean$_1$	79	201	8.2	129.7	71.1
Lean$_2$	165	2000	18.5	1850	146.4
Anthracite	78	1625	15.6	1560	61.2

fat and lean coals is determined mainly by their diffusion characteristics; the relaxation time of diffusion decreases from the volatile to lean coals and correspondingly the total desorption time decreases. These coals reveal remarkable values of D_d coefficients that are characteristic of solid-state diffusion.

In anthracite, dependence τ_2^{des} on granule sizes in samples is the strongest and points to the weak influence of the diffusion component of desorption. Analysis of data from Table 2.2 also shows that the specific diffusion time in this case appears to be the least one and the diffusion coefficient $D_d \sim 1/\tau^d(t)$ correspondingly the biggest. At first sight, this result appears unexpected because we know that among coals anthracite has the greatest density and it would be logical to expect the converse result—the lowest value of D_d. Apparently, in anthracite, a considerable admixture of Knudsen diffusion several orders of magnitude higher than the solid-state result occurs. This interpretation of results for anthracite (and possibly for the lean coals) requires consideration of the existing system of small open pores.

Thus, the results of studying methane desorption in black coals show that desorption kinetics cannot be explained by the assumption of a single mechanism of methane escape (diffusion or filtration) through the open pores. The kinetics of methane desorption can be explained as a hybrid mechanism of release that involves a certain type of superposition of the diffusion and filtration processes.

We have not discussed the reasons for the dependence of specific diffusion time $\tau^d(t)$ on time t. This parameter and the specific time $\tau^{des}(t)$ change from coal depletion as the methane escapes, i.e., they depend on the desorption registration time. A simple (but not complete) explanation may be the dispersion of microblocks in the coals, which are heterogeneous materials based on density and composition. Let us analyze gas desorption from coal and try to determine the peculiarities of its kinetics based on dispersion of the coal of microblock size.

During the diffusion, small microblocks deplete more quickly than larger ones. That is why microblocks that are partially or fully free of methane appear during desorption. This indicates that during diffusion the average size of the "active" (methane-containing) microblocks will change. The character of such change [28] approaches the exponential $\exp(-D_d \cdot t/r_{mb}^2)$.

The function of the distribution $N(r_{mb})$ of the active microblocks according to their size r_{mb} and the transformation of this function during desorption are shown in the lower part of Figure 2.21. The upper part illustrates the pattern of change of methane amount Q in microblocks of different sizes over time. In the course of methane escape, the average size of microblocks subjected to diffusion increases. This explains why, during the final phase of gas desorption, the specific time of diffusion achieves the highest value.

A similar effect occurs during decomposition of the desorption curve into two components. Figure 2.22 shows the transformation of gas molecule distribution in small and large microblocks. In the initial state (before desorption), total distribution corresponds to the initial curve ($t_1 = 0$) in Figure 2.22. Over

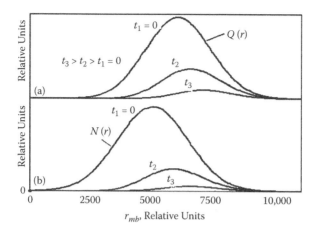

FIGURE 2.21
Representation of methane content $Q(r)$ change in microblocks (a) and number $N(r)$ of active microblocks (b) according to their size during desorption.

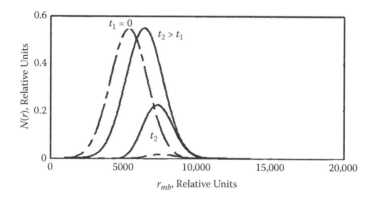

FIGURE 2.22
Pattern of transformation of size distribution functions of small and large active microblocks during desorption. Continuous line = large blocks. Dotted line = small blocks.

time, the ratio of number of small and large active microblocks decreases and as a result methane diffusion during the final phase of desorption occurs chiefly from large microblocks. This effect is described by the first component of Equation (2.88).

Thus, we can state that in Equation (2.88) the constituent characterized by larger τ describes the kinetics of gas desorption during its diffusion from a group of microblocks of larger size. Parameter τ^d in Equation (2.93) has a mean value $\tau^d = r_{mb}^2 / D^d$, as determined by the mean value squared of size r_{mb} in the group of large microblocks. Although large blocks are fewer than small ones, their total volume is probably large because they carry considerable methane in coal.

The analysis indicates that another "fast" constituent of the interpolation function must be connected with the escape of methane from smaller blocks of coal. Artificial partitioning of the microblocks into two groups of mean values $\overline{(r_{mb})_1}$ and $\overline{(r_{mb})_2}$ leads to the same partitioning of τ^d into two components. Thus, the dispersion of the coal microblock sizes predetermines the choice of the interpolation functions for several constituents. Each one describes the gas escape from a certain group of coal microblocks. For interpolation in the form of the sum of two functions, the microblocks are divided into two groups that differ based on the mean value of their sizes and the corresponding time of methane escape. Apparently, the more constituents used for interpolation, the more accurate will be the description of the desorption.

2.4.3 Transformation of Methane Desorption Mechanism: Three Stages of Desorption

In studying the kinetics of methane desorption, it is acceptable to consider the initial and final phases the most informative ones. According to the physics of transient processes, the kinetics of desorption in the initial phase must be defined to a considerable extent by the initial parameters of the equilibrium gas condition of the coal-methane system. Gas filtration in open coal pores is a distinctive peculiarity of the initial phase when the pressure in the open pores decreases from the initial (equilibrium) value in megapascals to kilopascals. The velocity and short duration of the process often led to complete desorption before registration could be attempted.

Conversely, the final phase is characterized by longer duration and weak intensity of gas release. It is assumed that by the final stage the transient processes have ended and desorption kinetics are fully determined by the current content of gas in the coal and the sizes and structures of coal lumps. The absence (or weak influence) of the transient processes aids in discovering the mechanisms of methane desorption from coal. Based on the block model of coal, the gas released through diffusion from the blocks (the smallest elements of coal without open pores) passes into the open pores and fissures and on its way to the surface experiences the effect of filtration. The assumption about flow balance in this phase proved exceptionally productive and helped explain the abnormal kinetics of methane desorption from the coal lumps of small sizes [28].

Against the background of some progress in explaining the mechanisms of mass transfer in the two phases described above, the third (transient) phase has not been thoroughly studied. Until recently, no experimental evidence indicated the duration of the phase. One reason for the lack of study may be the absence of proven methodology. A second may relate to the specific character of the black coal structure. Its porosity, heterogeneity, and chemical composition distort desorption characteristics. Taking measurements means dealing with "washed-away" average values of kinetic parameters. Furthermore, the specific character of coal composition makes it difficult to analyze the course of desorption in the transient phase. A distinctive peculiarity of the transient phase must be condition: $I^{filtr} \neq I^{diff}$.

These difficulties can be avoided by studying the dependence of gas flow I^{des} from coal starting from the initial methane pressure P^{sat} in a coal–methane system. By assuming the availability of two nonequal and weakly connected gas flows (diffusion flow from the microblocks and filtration flow from the open pores), we can determine the complex dependence I^{des} on P^{sat}. The method is based on the different sensitivities of the two flows to the quantity P^{sat}. Following the principles of transient process physics, determining the complex correlation between I^{des} and P^{sat} is most likely during the initial stage of methane desorption from coals with small sections of open pores such as types T and A.

Below we present the results of studying the correlation of methane desorption flow from coal and the initial equilibrium pressure of coal saturated with methane. The laboratory experiments to assess the velocity of methane escape were carried out by gas desorption in a vessel of known volume (the volumetric method). The experiments involved three stages. The first was coal saturation by methane compressed to some tens of atmospheres. The second stage was the preliminary release of compressed gas from the container of coal after saturation. The third stage was the accumulation of methane released from the coal into a vacuumed vessel of known volume. We used pieces from a single lean coal lump (volatile-matter escape from coal is 12 to 18%). After the coal was crushed and sieved, granules of 1.0 to 1.5 mm were picked out and buttons of 20 g were formed. The samples were then dried, poured into the containers under high pressure, and saturated with compressed methane for 14 days. Before the desorption registration, the compressed gas was released from the free volume of the container into a larger vessel. After about 5 sec, the methane escape was in progress but its velocity was several orders of magnitude lower. Immediately after the methane emission slowed, its flow was directed into another vacuumed AV of known volume. The course of desorption in the initial state was registered during the first 120 minutes. The optimization of the interpolation parameters of desorption at each state was performed by an appropriate computer program.

Earlier in this chapter we presented a set of curves that reflected the change of methane pressure P^{des} in the AV during the first 120 minutes of desorption (Figure 2.9). The numbers in the figure designate the values of methane pressure P^{sat} under which the coal was saturated.

Note that as pressure P^{sat} grows, the velocity of gas release does not reach saturation and continues to increase. This contradicts the generally accepted ideas about the direct proportionality of desorption flow and gas content in coal. We know that the methane content of coal with a rise of P^{sat} tends to reach its maximal value (methane intensity) and it would be logical to expect such a change for the flow of desorbed gas.

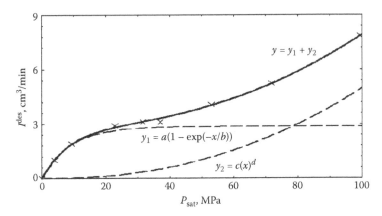

FIGURE 2.23
Pattern of gas flow change I^{des} during desorption depending on initial equilibrium methane pressure.

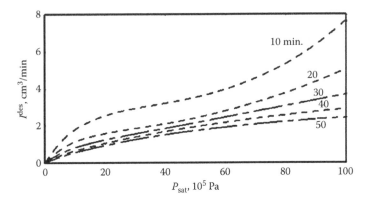

FIGURE 2.24
Transformation of dependence of gas flow on value of P^{sat} in transient phase.

The details of the dependence of intensity of methane emission I^{des} on P^{sat} appear more explicit if all the values for gas flow are taken at the same time of desorption. The plus symbols (+) in Figure 2.23 designate values of I^{des} after 10 minutes of methane desorption from samples saturated with methane at pressures of 4, 9.3, 22.6, 31.3, 33.2, 53, 72, and 100 atm. The dependence of gas flow I^{des} on P^{sat} is easily interpolated by the sum of exponential and power functions $y = a(1 - \exp(-x/b)) + c(x)^d$. The computer optimization of interpolation parameters gives values of $a = 2.85$, $b = 8.9$, $c = 0.00013$, and $d = 2.3$. Figure 2.23 illustrates this function as a continuous line. The dotted line represents the exponential and power components. It is clear that under the low pressure of saturation, the dependence of I^{des} on P^{sat} has an exponential nature. Further, with the increase of P^{sat}, the influence of the power dependence of gas flow on the value of P^{sat} becomes more distinct.

The pattern of dependence of methane release intensity on the value of equilibrium pressure can be explained if we assume that under low values of P^{sat} the main accumulators of methane are closed pores in the lumps of coal. Due to the growth of P^{sat} and approaching maximal gas content of coal, the sorption ability of the closed pores gradually decreases while the adsorption in the open pores of large sections and coal fissures continues. In the framework of such an assumption about the early stage of the reverse process, the unfinished initial phase of gas filtration by transportation channels (open pores) will be noticeable. One should expect it to occur under comparatively high initial gas pressures in pores. Based on the theory of viscous laminar gas flow during filtration, the intensity of the gas flow depends on the pressure difference in the pores as $\Delta(P^2)$. This type of dependence occurred at high values of P^{sat}.

Figure 2.24 shows the transformation of the pattern of the dependence of I^{des} on P^{sat} during desorption. In the early stage (10 and 20 min after start), the unfinished filtration processes of the initial phase are rather strong based on the influence of the squared contribution of the interpolation function. With methane release, this contribution weakens and becomes almost invisible after 50 min of desorption. We can also observe the dependence of I^{des} on P^{sat} to acquire the form typical for the gas release when the leading process of desorption is diffusion from coal blocks. As stated earlier [28], it is possible only under conditions of balance of diffusion and filtration flows that indicate the final phase of desorption.

The experiment indicates that the duration of the transient phase is considerably longer than expected, taking into account the fact that it is connected with the admixture of unfinished filtration processes in the initial phase. The long duration can be easily explained by the residual manifestations of the initial phase characterized by small difference of pressure in the open –pores. Thus for $\Delta P = 1$ Pa, pore length of $0.7 \cdot 10^{-3}$ m (in coal granules of 1.0 to 1.5 mm), methane viscosity $\eta = 1.08^{-5}$ nm^2/sec, permeability $\kappa = 2 \cdot 10^{-16}/$ m^2 [36], and coal porosity $\gamma = 0.06$, the specific time τ_f of the filtration process will be $\tau_f = 4 \cdot R_{gr}^2 \cdot \eta \cdot \gamma / \cdot \pi^2 \cdot P = 648$ sec or ~10 min.

The results show that between the initial and final phases of methane desorption from coal is a lasting transient phase characterized by a shift of the leading role of mechanisms of methane escape from filtration to diffusion. For researchers dealing with practical applications of scientific works, the important fact is the breach of proportionality between the methane content in coal and the intensity of its release during separation of coal lumps from a seam.

2.5 Conclusions

For objective reasons, the description of the coal–methane system is a very difficult physical problem. Solving it requires a correct choice of a structural model of coal and a mechanism of methane motion in the coal matrix. The theory of methane motion in coal set forth in the first section of the chapter is based on two assumptions: the diffusion–filtration nature of gas molecule transmission and the availability of closed pores in coal lumps. Solving the equation of material balance made it possible to detect three phases of methane escape from the coal massif that all display different kinetics. The research outcomes have practical significance. It is now possible to describe methane release from a seam and also estimate the time of formation of highly explosive methane–air mixtures in cul de sacs and bunkers.

The second section of the chapter discusses the most important outcomes that complement the theoretical concepts. The dependence of methane desorption kinetics on granule sizes and other coal structure parameters has been studied. The results of the experiments can be explained only within the framework of a hybrid mechanism of methane emission—a concrete type of superposition of diffusion and filtration processes that reveals the dependence of methane escape time on the sizes of coal granules.

The phase state and the quantity of methane in the structural components of coal were determined. The quantity of methane adsorbed on the surface of coal may amount to 25% of the total content; methane emission after coal separation from a seam has a burst nature. The intermediate (transient) desorption phase was experimentally proven. In this phase, the leading mechanism of methane escape shifts from filtration to diffusion.

References

1. Landau L.D. and Lifshitz E.M. 1978. *Statistical Physics. Vol. 2: Theory of Condensed State*. Moscow: Nauka.
2. Lifshitz E.M. and Pitayevskii L.P. 1979. *Physical Kinetics*. Moscow: Nauka.
3. Van Krevelen D.W. 1993. *Coal Typology, Chemistry, Physics, Constitution*, 3rd ed. Amsterdam: Elsevier.
4. Kovaleva I.B. 1979. Binding Energy of Methane to Coal in Coal Beds. Abstract of thesis. Moscow: Institute for Problems of the Complex Reserve Development of Academy of Sciences of the USSR.
5. Ettinger I.L. and Kovaleva I.B. 1979. Swelling stress and the free energy of the gas–coal system. *Dokl. akad. nauk. SSSR*. 244, 659–663.
6. Alexeev A.D., Sinolitskii V.V., Vasilenko T.A. et al. 1992. Closed pores in fossil coals. *Fiz. tekhn. probl. razrab. polezn. iskop.* 2, 99–106.
7. Alexeev A.D., Vasilenko T.A., and Kirillov A.K. 2008. Fractal analysis of the hierarchical structure of fossil coal surface. *Fiz. tekhn. probl. razrab. polezn. iskop.* 3, 14–24.
8. Khodot V.V. 1961. Sudden Outbursts of Coal and Methane. Moscow: Gorgostekhizdat.
9. Alexeev A.D., Vasilenko T.A., and Ulyanova E.R. 2004. Methane phase states in fossil coals. *Solid State Commun.* 130, 669–673.

10. Feldman E.P., Vasilenko T.A., and Kalugina N.A. 2006. *Fiz. tekhn. vysok. davl.* 16, 99–114.
11. Alexeev A.D., Vasilenko T.A., Gumennuk K.V. et al. 2007. Diffusion–filtration model of methane escape from a coal seam. *Z. tekhnich. fiz.* 4, 65–74.
12. Landau L.D. and Lifshitz E.M. 2003. *Theoretical Physics. Vol 7: Theory of Elasticity.* Moscow: Fizmatgiz.
13. Starikov G.P. 2001. Peculiarities of coal deformation and destruction under three-dimensional compression. In *Geotechnology at the Turn of the XXI Century,* 4, 81–87.
14. Petuhov I.M. and Linkov A.M. 1983. *Mechanics of Rock Bumps and Outbursts.* Moscow: Nedra.
15. Kolesov O.A., Vainshtein L.A., and Kolchin G.I. 1991. Judgment on the mode of deformation of the bottom hole massif section by the parameters of seismoacoustic signal. In *Proceedings of 24th International Conference of the Research Institute for Safety of Mining Works,* Donetsk, pp. 391–401.
16. Khristianovich S.A. 1953. On the outburst wave. *Izvestia akad. nauk.* SSSR 12, 1679–1688.
17. Zheltov Y.P. and Zolotarev P.P. 1962. On gas filtration in fissured rocks. *Prikl. mehan. tekhnich. fiz.* 5, 135–139.
18. Petrosian A.E. 1975. *Gas Emission in Coal Mines.* Moscow: Nauka.
19. Alexeev A.D., Feldman E.P., Vasilenko T.A. et al. 2004. Methane transport in coal due to joint filtration and diffusion. *Fiz. tekhn. vysok. davl.* 14, 107–118.
20. Kogan G.L. and Krupenia V.G. 1972. Methane motion in fossil coals. In *Physico-Chemistry of Gas Dynamics Phenomena in Mines.* Moscow: Nauka, pp. 84–94.
21. Alexeev A., Feldman F., and Vasilenko T. 2007. Methane desorption from a coal-bed. *Fuel* 86, 2574–2580.
22. Alexeev A.D., Airuni A.T., Vasyuchkov Y.F. et al. June 30, 1994. The property of coal organic substance to form metastable single-phase solid-solution-type systems with gases. Discovery Diploma 9, Application A-016-M. Reg. 16.
23. Ivanov B.M., Feit G.N., and Yanovskaya M.F. 1979. *Mechanical and Physico-Chemical Properties of Coals in Outburst-Hazardous Beds.* Moscow: Nauka.
24. *A guide to determination of adsorbed and free methane content in fossil coals.* 1977. Makeevka Donbass Research Institute.
25. Zheltov Y.P. and Zolotarev P.P. 1962. On gas filtration in fissured rocks. *Prikl. meh. tekn. fiz.* 5, 135–139.
26. Nikolaev N.I. 1980. *Diffusion in Membranes.* Moscow: Himiya.
27. Vasilkovskii V.A. 2007. Estimation of diffusion coefficient and diffusion characteristic time under methane desorption from black coal. *Proc. Int. Miners' Forum,* Dnipropetrovsk: National Mining University, pp. 100–106.
28. Vasilkovskii V.A., Molchanov A.N., and Kalugina N.A. 2006. Methane phase states in coal and mechanisms of its desorption. *Fiz. tekhn. probl. gornogo proizv.* 9, 62–70.
29. Barenblatt G.I., Zheltov Y.P., and Kochina I.N. 1960. On the main concepts of the theory of filtration of homogeneous liquids in fissured rocks. *Prikl. mat. mekh.* 24, 852–864.
30. Alexeev A.D., Vasilkovskii V.A., and Kalugina N.A. 2005. Kinetics and mechanisms of methane desorption from coal. *Fiz. tekhn. probl. gorn. proizv.* 8, 9–21.
31. *Chemist's Reference Book.* 1952. Moscow, pp. 572–579.

32. Kikoin I.K., Ed. 1976. *Tables of Physical Quantities*. Moscow, Atomizdat, pp. 67–75.
33. Leibenzon L.S. 1947. *Motion of Natural Liquids and Gases in a Porous Medium*. Moscow: Ogiz.
34. Vasilkovskii V.A. and Ulyanova E.V. 2006. Some aspects of interpretation of methane desorption kinetics from black coal. *Fiz. tekhn. probl. gorn. proizv.* 9, 56–61.
35. Alexeev A.D., Zaidenvarg V.U., Sinolitskii V.U. et al. 1992. *Radiophysics in the Mining Industry*. Moscow: Nedrap.
36. Alexeev A.D., Vasilkovskii V.A., and Shahzko Y.V. 2007. On methane distribution in coal. *Fiz. tekhn. probl. gorn. proizv.* 10, 29–38.

3

Nuclear Magnetic Resonance Studies of Coal and Rocks

Nuclear magnetic resonance (NMR) is widely used for scientific research and qualitative and quantitative analysis during manufacturing. Recent technical progress has greatly improved NMR spectrometers and developed methods to extend the sphere of its applications. Since its invention, NMR technology proceeded in two directions: (1) high resolution NMR that can analyze lines from millionths to thousandths of oersted (Oe) in width and (2) low resolution NMR for analyzing lines of tenths of oersteds and larger. Spectrometers implementing these methods include various types of technologies and instrumentation. Their only similarity is functioning is based on NMR principles.

To analyze the structures of coals and rocks, their metamorphism, porosity levels, and phase state of water and methane, as well as diffusion and filtration processes, we used stationary and impulse NMR spectrometers of both high and low resolution.

3.1 Experimental Techniques

3.1.1 High Resolution Spectrometers

Modern spectrometers of high resolution (produced by Bruker and Varian) were developed mainly to study isotropic liquid samples, but they can also work with solid samples in special stands. They are characterized by high sensitivity and wide temperature range (~5 to 600 K). Research under pressure up to 350 atm is also possible.

High resolution means that under a given external field the nuclei of the same atoms absorb the energy of the applied high resolution field at different frequencies because different chemical environment nuclei are shielded from the applied field in different ways. The distance between resonance frequencies of the examined nucleus and reference compound is called chemical shift. It is represented as a dimensionless parameter in parts per million (ppm):

$$\delta = \frac{H_i - H_{ref}}{H_{ref}} \cdot 10^6 \ [ppm],$$

where H_i and H_{ref} are the resonant field of the analyzed i-th nucleus and the reference, respectively. In complex compounds, the same atoms, for example carbon atoms in coal, can be surrounded by different electrons and thus yield several absorption lines instead of one. However, the necessary line resolution cannot be achieved by direct methods such as increasing data accessing time. Additional problems appear during registration of coal spectra on isotope ^{13}C because the percentage of this isotope in the natural environment is small and because the spectrum widens due to strong dipole ^{13}C and ^1H interaction. Coal spectra were determined by the combined use of multi-impulse patterns and sample rotation at a "magic" angle.

In Bruker spectrometers, proton-amplified NMR is used. Magnetization of the proton system is transferred to the ^{13}C-system. This leads to improved sensitivity and reduced experiment time due to shorter times of T_1 protons. In addition, the dipole ^{13}C and ^1H interaction is suppressed by turning on the decoupling field, making it possible to separate aromatic and non-aromatic coal components. The effect connected with residual line width (70 to 100 ppm) and caused by the anisotropy of chemical shift is eliminated by fast rotation of the sample at the angle of 54°44' (the "magic" angle) to the axis of the magnetic field. Thus, combining proton amplification and sample rotation yields well-resolved coal spectra.

To achieve high resolution, small amounts of samples are usually analyzed. To achieve higher sensitivity, a large number of signal accumulations is needed. Thus hours and even days are required to achieve a well-resolved spectrum. This limits applications to analyses of kinetic processes. Unfortunately, high resolution NMR proved ineffective for analyzing high speed (transient) processes of methane desorption whereas Bruker stopped producing wide line spectrometers.

3.1.2 Wide Line Spectrometers

Powerful permanent magnets make expensive cryogenic magnetic systems (such as with NMR of high resolution) unnecessary. Spectrometers of wide lines can be equipped with supplementary devices in the magnet bore. For example high pressure chambers, helium and liquid nitrogen cryostats, and other equipment allow NMR studies that would have been impossible for high resolution spectrometers. Scientists in Russia and Ukraine have long carried out NMR research and now manufacture low resolution spectrometers.

Relaxometers are produced in Kazan, Russia. Using the high impulse gradient of a magnetic field, they can measure diffusion and self-diffusion coefficients in liquids and solids at 10^{-8} to 10^{-16} m^2/sec. The Institute for Physics of Mining Processes of the National Academy of Sciences of Ukraine created a spectrometer that examines samples under high hydrostatic pressure up to 1000 MPa in a high pressure cylinder piston type chamber, samples at gas pressure up to 25 MPa in a special gas container, and samples up to 150 MPa with a cryocompressor. It is possible with a continuous-flow liquid nitrogen cryostat to conduct temperature research at 90 to 300 K and achieve long-term temperature stability within 0.5 K accuracy.

In NMR spectrometers of wide lines, permanent magnets with field intensity at 4 kOe and homogeneity $\sim10^{-6}$ Oe/cm are used. The sensitivity of the spectrometer reception path is ~1 mV with a signal-to-noise ratio ~1. Lines from 0.01 Oe wide can be resolved. Figure 3.1 shows a spectrometer flowchart.

The process of spectrum registration is carried out via computer monitoring with special software. NMR wide lines techniques led to the development of devices for quantitative express analysis of methane-rich material in coal mines. Modern wide line NMR spectrometers used for physical research utilize equipment whose total cost is dozens of times lower than the cost of a high resolution spectrometer. This fact and the ability to use the equipment for temperature and baric research promotes its development, improvement, and distribution for research purposes.

The physical principle of NMR is that nuclei of many atoms have magnetic moments that precess in a magnetic field at a frequency ω_0. The materials under our analysis—water, methane, and coal material—contain atoms of hydrogen, oxygen, and carbon that exhibit magnetic moments. If a substance with precessing magnetic moments is placed in a magnetic field of intensity H_0 and gliding frequency ω_n, resonance will occur at $\omega_n = \omega_0$. Under resonance, the nuclei transfer to the next energy level via energy absorption registered in the form of an absorption spectrum.

Figure 3.2 illustrates the ^1H NMR spectrum (PMR = proton magnetic resonance) of a fluid-rich coal sample. The spectrum consists of two lines. The Lorentz line (5) is a narrow spectrum component (Equation 3.1). The Gaussian line (3) is a wide component (Equation 3.2) whose parameters (width, intensity, etc.) contain information about the amount of hydrogen in the sample structure, retained fluid, and phase states of hydrogen-containing fluids.

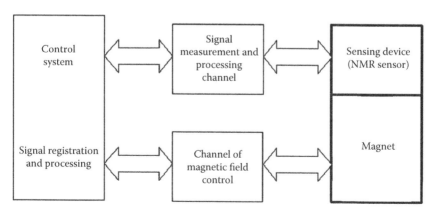

FIGURE 3.1
Flowchart of NMR spectrometer.

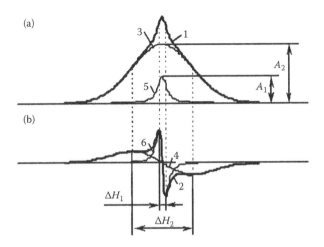

FIGURE 3.2
Absorption line of ^1H NMR of fossil coal containing fluid (a), and derivative from absorption line (b). 1. Absorption spectrum. 2. First derivative of absorption spectrum. 3. Wide component of absorption. 4. Its first derivative. 5. Narrow component of absorption spectrum. 6. Its first derivative.

$$f(H) = A_n \cdot \frac{\Delta H_n^2}{(H - H_m)^2 + \Delta H_n^2} - k \qquad (3.1)$$

$$f(H) = A_w \cdot e^{\left(-2\frac{(H-H_m)^2}{\Delta H_w^2}\right)} - k \qquad (3.2)$$

Here: A_n is the amplitude of the narrow component of the prototypic absorption line; $\Delta H_n = \Delta H_1$ is the width of the narrow component of the prototypic absorption line at height $\sqrt{2}A_n/2$; H is the magnetic field increment, H_m is the modulation amplitude; A_w is the amplitude of the wide component of the prototypic absorption line; $\Delta H_w = \Delta H_2$ is the width of the wide component of the prototypic absorption line at height $A_w/2$; and k is the instrument parameter that determines the position of the zero line on the coordinate axis.

To register the PMR spectra of moisture and methane-rich coals we use a continuous autodyne spectrometer and modulation of spectra registration. Initially, the first derivative of the absorption line is registered (Figure 3.2b). It represents the sum of the derivatives of the absorption line components. To interpolate the experimental ^1H NMR spectra, Equation (3.3) is used:

$$\frac{df(H)}{dH} = A_w \left[e^{\left(-2\frac{(H+H_m)^2}{\Delta H_w^2}\right)} - e^{\left(-2\frac{(H-H_m)^2}{\Delta H_w^2}\right)} \right]$$

$$+ A_n \Delta H_n^2 \left(\frac{1}{(H + H_m)^2 + \Delta H_n^2} - \frac{1}{(H - H_m)^2 + \Delta H_n^2} \right) \qquad (3.3)$$

It presents the superposition of the first derivatives of the Lorentz and Gaussian components. Absorption spectra contain important information. The line width ΔH characterizes atomic mobility. The wider the line, the less mobile the atom. Thus, it is possible to characterize the phase state of the material under analysis. In our case for methane, the phase state is gas, both adsorbed and, as we found, dissolved in the internal coal structure. The area under the adsorption curve is proportional to the number of resonant nuclei. NMR is a high precision quantitative technique.

3.1.3 High Pressure Technique

The practicability of studying coal material under high pressure is conditioned by the fact that fossil coal under natural conditions is always under pressure and stress from tectonic processes within the massif. Coal production process changes in its deformation mode. Donets Basin (Donbass) mines are as deep as 1500 m where geostatic pressure is 45.5 MPa. As a mine is worked, the voids can concentrate pressures up to 8 times. That is why loads up to 300 MPa in a rock massif are common.

As an organic mass turns into coal, it goes through a plasticity stage under certain pressure and temperature conditions during the coalification process. It is logical to assume that during coalification the coal pores are filled with methane (formed during the restructuring of the organic mass). Then the pore size will decrease to the point when interstitial pressure plus matrix resistance do not equalize the pressure of the overlying material, that is, methane pressure in closed pores equals γH minus plastic matrix strength during coalification. The higher the coal strength, the less interstitial pressure it has.

It is incorrect to equate methane pressure measured in spurs in coal to methane pressure in closed pores. During a spur drill, the gas from the peripheral spur areas in transport channels and pores "gets out" via filtration over tens of seconds. After hermetization, the pressure in a spur is created by methane from the closed pores via solid-state diffusion. This process is very slow. For anthracites, for example, methane takes 3 months to "get out" of coal particles as big as 3 mm. Measurements in a spur last for half a month on average; this is why maximum measured pressure at depth was 10 MPa. The pressure in closed pores is much higher.

We used ^1H NMR together with uniform hydrostatic compression of samples. The spectrometer receiving circuit with the analyzed sample was in a channel of a high pressure chamber and linked to the autodyne input through a special electric lead-in. Figure 3.3 shows a double high pressure chamber with autofrettaged external layer [1].

The choice of perfluorooctane C_8F_{18} as a pressure medium in a high pressure chamber was based on a number of factors. First, it allows easy hydrostatic compression of a coal sample without introducing "outside" hydrogen. Thus, all possible changes in the spectrum will reflect only processes in

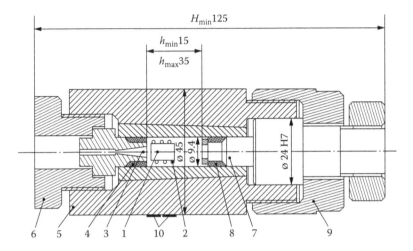

FIGURE 3.3
High-pressure chamber for NMR studies of coal. 1. Test sample. 2. Autodyne receiving circuit.
3. Obdurator. 4. Sealing rings. 5. Case. 6. Screw nut. 7. Plunger. 8. Sealing rings. 9. Screw nut.
10. Tensiomanometer.

the sample. Second, sample preparation is easier because the sample does
not have to be isolated from the working fluid with hermetic shells. Finally,
despite its volatility, perfluorooctane is compressed easily in a chamber
(much easier than benzene and spirit mixtures that are known to provide the
best hydrostatic conditions in high pressure chambers) and produces high
compression hydrostatics over a whole range of working pressures in a high
pressure chamber.

In the first experiments using NMR under pressure [2], a question arose
about using the phase states of fluorine-containing fluids under pressure
to assess the value of the transferred pressure to the test sample by these
fluids. In the experiment, pressures from 1 to 1000 MPa were used at temperatures from 77 to 295 K. The phase changes between liquid and solid
states in perfluorooctane, perfluoroheptane, perfluorochemical ether, and
perfluoraminetetrafuran were studied. Because the fluids all contained fluo-
rine, NMR analysis was carried out with ^{19}F. Under normal pressure, the
temperature of phase change is between 143 K (perfluoroheptane) and 183
K (perfluoroheptane) for all liquids. The NMR line width in a solid state is
11.3 to 13.0 Oe. At indoor temperature, phase changes occur above 450 MPa.
Perfluoroheptane does not solidify even at 10^3 MPa. The simultaneous influence of high pressure and temperature results in a shift of phase change to
higher temperatures.

The same author designed an installation for carrying out x-ray and NMR
experiments at gas pressures between 0.1 and 300 MPa and temperatures
from 0 to 200°C [3]. One difficulty was consolidating dynamic connection
seals for gas to ensure laboratory safety. Several structures of high pressure

chambers for x-ray structural analysis of solids are known. One common component is an external gas pressure generator (compressor, intensifier). The systems differ in whether they handle gas void compression via mercury, siphon, pressure generation by thermal compression, or the kinetic energy of a "flying bullet."

We suggested two-level technology for handling high gas pressure. The installation is presented in Figure 3.4. The functional units include a cylinder with methane (1), intensifier (2) measuring cell (3), high pressure cutoff valve in intensifier case (4), high pressure tubing (5), high pressure manometer (6), cooling bath (7), thermal stabilization system (8), low pressure cutoff valves (9), lock nut on intensifier plunger (10), and x-ray diffractometer or spectrometer NMR (11).

Pressure generation starts when all gas channels of high pressure chamber components cooled in the bath (7) to –70°C are filled with methane (20 to 30 MPa) from the gas cylinder (1). After fixing t at –70°C, the system is disconnected from the cylinder and low pressure channels by cutoff valves 4 and 9. Chamber temperature is brought to the required level (0 to 200°C) by the thermal stabilization system (8). As a result of thermal compression, the pressure in the chamber is increased according to Amagat's law $P = PV/P_0V_0$; P_0V_0 denotes the initial conditions.

If the experiment requires pressure higher than the corresponding value resulting from thermal compression, the increase is achieved by the

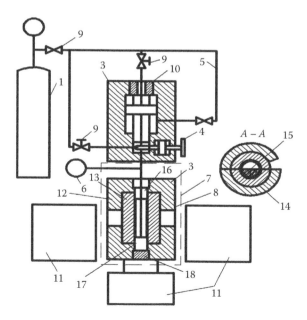

FIGURE 3.4
X-ray structural analysis and NMR installation for coal research under high methane pressure.

intensifier (2). Shifting the plunger compresses the void gas down to the necessary value with the help of low pressure gas. After that, the plunger position in the channel is fixed by the cutoff valve (9) or the lock nut (10). Then the chamber is placed on the goniometer table of the diffractometer or between the pole terminals of the NMR spectrometer, thermostatted, and measured.

The measuring cell characteristics depend on the technique (x-ray structural analysis or NMR). A cell for x-ray structural analysis has two base layers: the external band (12) of alloyed steel (X18H9T, 5XHM, 40XH) and the internal layer (13) of beryllium. The band has a double sector slit (35 degrees, 1 mm height) for x-ray input and output. Layers 12 and 13 are mated with tightness on the conical surface $2\alpha = 2$ degrees. The diameter of the cell-working channel is 10 mm; the average value of the external diameter of the beryllium layer and its height is 20 mm, and the external diameter of the band is 50 mm.

The holder (14) of the coal sample (15) is in semi-cylindrical form. Edge locks on sealing plugs 16 and 17 provide spatial orientation of the sample plane relative to the x-ray axis. The upper sealing plug (16) is latched in the axial direction by the clamp; the lower plus (17) is latched by the nut (18). Axial thrust is grounded from gas pressure on the chamber band. Sealing of the plugs is carried out by a set of bronze, fluoroplastic, and rubber gaskets. The face seals of the nipples mating with the plugs is carried out according to the cone–cone conjugation scheme and the slope angle of the cone generators differs by 1 or 2 degrees.

The mechanical features of beryllium define the ultimate pressure in the chamber. Estimates show that under the guaranteed level of yield point at operating voltage $\sigma_s \leq 260$ MPa and isotropy of the material strength, the minimum calculated working stress to make the inner layer plastic is at $P_{ss} \leq 165$ MPa and collapsing pressure $-P_b \leq 290$ MPa. The dependences $P = f(t)$ are shown in Figure 3.5. The maximum mean square error of approximation for a curve with initial pressure ($t = -70°C$) $P_H = 10$ MPa makes $S = 2.3$ MPa. The approximation equation in this case is

$$P = 26.02 + 0.19t \tag{3.4}$$

For initial pressures 20 and 30 MPa, final pressure with error $S = 1$ to 1.5 MPa can be calculated from the proportions (3.5) and (3.6), respectively:

$$P = 34.56 + 0.21t \tag{3.5}$$

$$P = 43.03 + 0.21t \tag{3.6}$$

During our studies of changes in coal under pressure [4], as noted above, perfluorooctane was used for loading coal samples. Dissolution of coal in perfluorooctane was not noted over a period of several days. Perfluorooctane

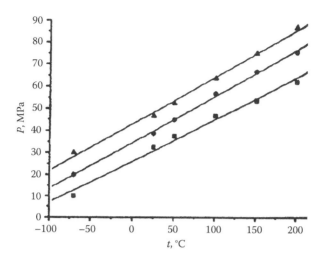

FIGURE 3.5
Dependence of final pressure P on temperature t and initial pressure in chamber at 70°C.

purity was controlled. No water or impurity containing hydrogen appeared in any state. No changes in the ^1H NMR spectrum of coal were noted when the sample was placed in the receiving circuit of the autodyne in a test tube with perfluorooctane.

NMR spectra were registered by autodyne spectrometer according to Robinson's scheme [5].

We noted an unusually strong dependence of the spectrum intensity on the value of the applied hydrostatic pressure and the sign of increment load. Figure 3.6 presents the spectra transformation. Under increments of pressure from 0 to the maximum (1000 MPa), the amplitude gradually decreased. Under a negative increment of load (decompression) beginning at 500 MPa, we see a dramatic increase in the intensity of the narrow component of the full NMR signal. The narrow line intensity under complete decompression is far larger than the intensity of the same line before the pressure increment.

Detailed mathematical processing of spectra [6] showed that the changes in coal structure begin during the load increment. At 700 MPa, we see a relative increase of the narrow line intensity. Under high pressures, the increasing density of perfluorine octane and the analyzed material disguise the noted effect. Under gradual unloading of the high pressure chamber, the effect becomes evident.

The appearance in the spectrum of a narrow line corresponds, in our opinion, to the appearance of mobile protons in the sample structure. This may be connected with the mechanical and chemical reactions stimulated by pressure treatment [7]. Coal chemistry studies [8] show that the redistribution of oxygen and hydrogen is possible in the organic coal mass through

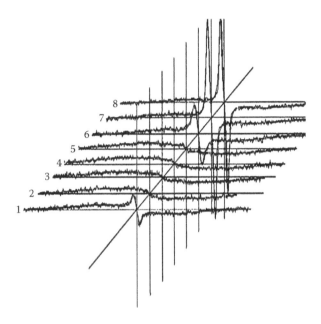

FIGURE 3.6
Sequence of ^1H NMR spectra in coke coal during one measurement cycle. Loading;
1. $P = 0$ MPa. 2. $P = 700$ MPa. 3. $P = 1000$ MPa. Unloading: 4. $P = 700$ MPa. 5. $P = 500$ MPa.
6. $P = 300$ MPa. 7. $P = 100$ MPa. 8. $P = 0$ MPa.

the influences of various factors. In our case, the factor was high pressure.
The reorientation of an internal coal structure by high pressure is possible
and under certain conditions leads to valence bond breakage accompanied
by isolation of separate radicals or their association in a new structure. In
our experiments, we are most likely to find a new hydrogenous material or a
significant increase of existing materials under high pressure based on their
isolation from a rigidly bound state in a mobile form. This is indicated by
small line width compared with the line width of a liquid or gas. The effect
of the working fluid on proton presence carried out immediately after high
pressure chamber disassembling showed a negative result. The fluid was
still clear, proving its effectiveness for this type of experiment.

In summary, the NMR method proved effective, accurate, and universal
for studying fossil coals and their physical processes. High resolution NMR
made it possible to determine the distribution of coal material structural
elements and their quantitative changes under temperature, pressure, and
other factors. Despite rapid development of methods and equipment for
high resolution NMR, the wide line NMR method is still relevant. New
computerized NMR wide line spectrometers equipped with the necessary software are characterized by high sensitivity and resolving capacity. They can successfully analyze porous structures of fossil coals and
reveal qualitative and kinetic characteristics of processes of fluid sorption
in fossil coals.

3.2 Coal Structure Research

Despite the development of several models (Chapter 1), all the differences in coal qualities evident among the ranks and within a single rank taken from different seams are not explained. For that reason, the focus is on carbon and oxygen distribution based on functional groups and the orders of structural elements and links between them.

The organic mass of coal contains sets of molecules of various chemical compositions interconnected by multiple bonds, mainly electron donor–acceptor (EDA) interactions. EDA interactions in organic masses of coal are created by irregularities of electron density distributions in macromolecules. This is due to the presence in coal of various functional groups with heteroatom participation (O, N, S) and carbon atoms with various hybridizations of valence electrons. Some structural elements have separate zones that reveal electron donor and acceptor qualities. The character of these zones depends on chemical structures of groups, their surroundings, level of macromolecule aromaticity, and aromatic structure. The same functional groups can be both donors and acceptors of electrons, resulting in more EDA bonds in coal.

NMR research allows more accurate separation of coals of similar structures. At the same geological age, the structural groups of petrographic and elemental compositions of coals can differ.

3.2.1 Determining Structural Components of Coals through ^1H NMR Data

Richards and Van Krevelen conducted the first NMR examinations of fossil coals [9,10]. In NMR spectra, the ^1H of solid structural hydrogen of dry coal is represented by a wide component with line width ΔH_2 and a second moment M^2 that depends on the distance between hydrogen atoms in a structure. The distance between hydrogen atoms in aromatic rings is larger than the distance in aliphatic chains. That is why it was possible to define the correlation between aliphatic and aromatic hydrogen in coals by measuring M^2. While registering NMR spectra in the zone of normal temperature ($T = 296$ K), it is almost impossible to resolve this line with respect to nuclei H_{ar}, H_{al}, and H_{OH}. However, Van Krevelen [10] showed that the values of the second moments of M^2 spectra, registered at temperatures below 110 K, depend on the rate of coal metamorphism. It is convenient to use the following relation to calculate the second moments [11]:

$$M^2_{exper} = c^2 \sum_n f_n n^3 \left/ \left(3 \sum_n f_n n^3 \right) \right.,$$

(3.7)

where c is the splitting scale in a magnetic field; f_n is an ordinate of the n-th point; and n is splitting in the horizontal coordinate of a spectrum. Richards [9] developed a formula to calculate relation P_A of hydrogen quantity in aromatic rings A to hydrogen quantity in aliphatic groups b depending on values of M^2:

$$P_a = \frac{a}{b} = \frac{M_{al}^2 - M_{exper}^2}{M_{exper}^2 - M_{ar}^2} \qquad (3.8)$$

where M_{exper}^2 denotes the second moment of NMR spectra from hydrogen atoms contained in the coal structure; and $M_{al}^2 = 174.2 \times 10^3$ and $M_{arom}^2 = 61.4 \times 10^3$ (A/m)^2 indicate the second moments from protons of aliphatic and aromatic groups, respectively, determined by modeling connections. As M_{exper}^2 for fossil coals of Donbass varies from 147.8×10^3 to 81.2×10^3 (A/m)^2 (with decrease of coal-volatile matter ranging from $V^G = 39$ to 5%) [12], according to Richards the amount of aromatic carbon increases from 0.75 to 0.98. That is, with the growth of metamorphism, coal becomes more ordered and the content of the side groups (aliphatics) decreases.

Since value M_{exper}^2 varies almost linearly from 147.8×10^3 (A/m)^2 in coals with carbon content C = 75% to 81.2×10^3 (A/m)^2 in coals with C = 95%, an increase of C on average by 1% results in a decrease of M_{exper}^2 to 3.2×10^3 (A/m)^2. The dependence result can be used to assess C percent: C = 75 + $(147.8 \times 10^3 \times M_{exper}^2)/3.2 \times 10^3$. Accuracy of practical calculations of value M_{exper}^2 means 5%, absolute C value has been calculated with an error within $\pm1\%$ [13].

As metamorphism grows, the atomic fraction H_{al} of methyl and methylene hydrogen decreases and the H_{ar} share of aromatic hydrogen increases accordingly. Aggregate content of organic H_0 is expressed by the sum of aromatic H_{ar}, aliphatic H_{al}, and hydroxyl H_{OH} of hydrogen. Calculation of hydrogen distribution is carried out by means of the modified equation [9]:

$$M_{exper}^2 = \frac{a_1}{a_1 + a_2 + a_3} M_{ar}^2 + \frac{a_2}{a_1 + a_2 + a_3} M_{al}^2 + \frac{a_3}{a_1 + a_2 + a_3} M_{OH}^2 , \qquad (3.9)$$

where $a_1 + a_2 + a_3 = 1$ indicate fractions of atoms H_{ar}, H_{al} and H_{OH}, respectively. $H_0 = H_{ar} + H_{al} + H_{OH}$, then $H_{ar} = A_1 H_0$; $H_{al} = A_2 H_0$; and $H_{OH} = A_3 H_0$. M_{Ar}^2 and M_{Al}^2 were obtained from modeling aromatic and aliphatic structures. Value M_{OH}^2 varies little from M_{Ar}^2 [13].

3.2.2 Application of ^{13}C NMR for Analyzing Coal Structures

Improvements of NMR spectroscopy of high resolution in solids (such as the NMR spectrometers with strong magnetic fields of Varian and Bruker) provide better resolution of spectral lines [14]. Japanese researchers [15] analyzed brown coal from Australia containing 66.1% C, 5.3% H, and 27.7% O (elementary composition determined relative to combustible part of coal). They used a Bruker CXP-300 spectrometer with resonance frequency ~75 MHz; rotor rotational frequency was 4 KHz. The authors achieved better resolution of spectral lines than they had in previous works. ^{13}C NMR revealed a number of absorbing maxima of ten types of carbon bonds. (Figure 3.7) [15].

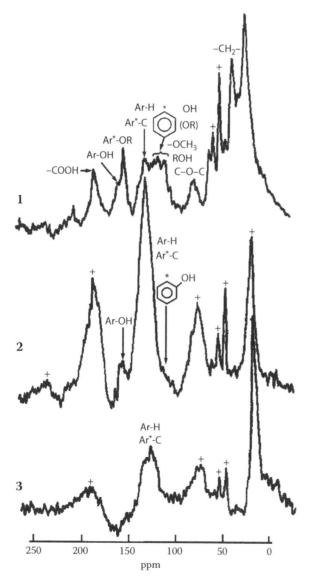

FIGURE 3.7

Spectra of Yallourn coals. 1. Fossil coal. 2. Coal annealed at 773 K. 3. Coal annealed at 1073 K. (Adapted from Richards R.E., Yorke R.W. *J. Chem. Soc.* 6, 2489–2497, 1960.)

In the spectra, maximum of absorption from carboxyl groups (COOH), aromatic carbons bound to oxygen (Ar-OH), aromatic carbons bound to alkyne ethers (Ar-O$_r$), hydroaromatics, ether, methylene (CH$_2$), and methyl (CH$_3$) groups have been resolved. After thermal treatment at 500°C, the coal spectrum changes drastically. The resonances conditioned by methyl and methylene groups disappear completely in aliphatic zones. In aromatic zones, the intensity of the resonance line grows dramatically. It can be explained by the growth of an aromatic carbon fraction during a pyrolysis reaction. Maximum of absorption from phenol carbon remains invariable. Obtaining such detailed descriptions of coal structures and the results of treatments is possible only by using high resolution fields.

In another study [16], several coal ranks from New Zealand containing 62 to 92% carbon, 4.7 to 3.9% hydrogen, and 38.3 to 3.9% oxygen were examined with a Varian XL-200 NMR spectrometer. The authors obtained ^{13}C NMR spectra for coals exhibiting several absorption maximums, thus proving the conclusions in [15]. ^{13}C NMR spectra give more detailed information about coal structure than simple separation into aliphatics and aromatics.

3.2.3 Research of Donets Basin Coals

^{13}C NMR-spectra show alterations of coal structures at different degrees of metamorphism and also alterations in structures caused by outbursts. Using a specialized Bruker CXP-200 spectrometer, ^{13}C NMR spectra of Donets Basin fossil coals of different degrees of metamorphism were taken and differences in molecular compositions of coals before and after outbursts were examined [17]. Measurements were carried out at indoor temperature. Rotor speed was 200 MHz. A standard method of cross-polarization with sample rotation at the "magic angle" was used. A mushroom-shaped rotor was made of deuterated plastic. Rotation frequency was 3.5 KHz; the 90-degree polarizing pulse was 12×10^{-6} sec; timing of polarization transfer was 2×10^{-3} sec. The phase of the first polarizing pulse changed by 180 degrees at each scanning with a simultaneous change of free induction decay sign. Therefore, the aggregate signal lost some coherent interference. The amount of accumulation was 512; scanning time was 1 sec.

Much research has been conducted on fat coal (C = 84.8%), coking coal (C = 88.5%), lean coal (C = 89.7%), and anthracite (C = 92.2%). The spectra consist of four maximums. According to the value of chemical shifts (comparing with modeling connections), several functional groups were singled out [17].

The biggest maximum of absorption can be found in the zone of 126 ppm due to the presence of heteroaromatic and aromatic carbons. Absorption in the zone of 72 ppm relates to carbon linked by bridges to oxygen, nitrogen, and sulfur heteroatoms. Absorption in the 18.20 ppm zone is caused by the availability of carbon atoms belonging to methyl groups. The line of absorption at 30 ppm in spectra of fat coals is due to an increase of the number of carbon atoms of methyl and methylene bridge groups that are bound to

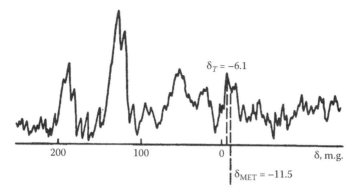

FIGURE 3.8
^{13}C NMR spectrum of methane-rich lean coal. (Adapted from Alexeev A.D., Serebrova N.N., and Ulyanova E.V. 1989. *Dokl. akad. nauk.* USSR B 9, 25–28.)

nitrogen atoms and haloids. Absorption at 176 to 200 ppm arises from carbonyl groups of aldehydes and amides.

Figure 3.8 [18] is a ^{13}C NMR spectrum of methane rich lean coal. Along with four maximums of coal organics, we can see a maximum of the methane sorbed. A chemical shift of methane sorbed varies from −4.2 (volatile coal) to −9.3 ppm (anthracite). A chemical shift of free methane equals −11.5 ppm. For the first molecular layer, chemical shift varies from −7 to +5 ppm [19].

Samples taken from the zones of coal layers (pillars) that presented no danger of methane outburst and from outburst layers exhibited considerably different NMR spectra. The spectra of the coals thrown by outbursts [17] showed a decrease of line intensity in the zone of 127 ppm, demonstrating the detachment of light carbons bound to aromatic parts of the coal. A substantial decrease of the maximum 20 and 73 ppm intensity suggests a decrease in a number of aliphatic chains caused by destruction of many bonds during the outburst. Natural coal has typical polymer features [20]. For example, polymer deformations lead to restructuring of molecular structure accompanied by formation of gaseous substances. Drawing an analogy of polymers and coal destruction, the authors [18] conclude that coal structure destruction during outbursts is accompanied by methane formation. This explains why, during outbursts, large amounts of methane that exceed levels in efficient methane-bearing coal seams are released. This assumption was proved later [21,22].

3.3 Phase States of Water and Methane in Coal

In fossil coal the role of fluid collection is performed by transport channels and pores (open and closed) that contain water and methane in different phase states. Gas and water are by-products of coal formation (coalification) as original plant cells lose hydrogen and oxygen in the form of water, methane, and other kinds of gases. Methane dominates in coal extracted from the Donets Basin. Pressure in overlying rocks and release of gaseous products created a specific dispersive porous-fractured structure of fossil coals of seven orders from commensurable sizes of methane and water molecules (diameter = –0.414 nm) to fractures with gaps up to several millimeters. Pores differ from fractures in that they are characterized by similar sizes in three dimensions. In a single fracture, sizes can differ by several orders of magnitude.

One of the main problems of extracting coal is estimating outburst dangers of coal seams. Efforts to control sudden coal outbursts and gas explosions are discussed in Chapter 5. This problem relates to the amounts and phase states of moisture in different surface-active agents; moisture is forced naturally or artificially in a coal massif. Thus, analysis of water behavior in coals at different stages of metamorphism is of great importance for national and world economies.

The efficiency of watering as a way to control coal and gas outbursts does not depend on the total amount of water forced into a seam. It depends on the water sorbed, i.e., physically linked to the coal surface. Its effect is shown both in wedging and weakening of intermolecular bonds of coal mass particles in microcracks. Sorption forces of moisture interaction with solids can change coal mass into a plastic state at maximum moistening, thus excluding dangerous outbursts. That is why the moisture sorbed W_{sorb} value of coal is one of the main indices of an outburst-dangerous coal seam.

To determine the W_{sorb} level necessary to change a brittle fracture into a plastic one, several experiments were conducted on fractures of different coal ranks with different water contents. Experiments involved triaxial compression on unequal components (see Chapter 4) using moisture-containing cubic samples (edge length = 50 mm). The conditions for coal outburst arise from adjustment of principal stresses σ_1, σ_2, and σ_3. No outburst occurs when moisture content reaches 2 to 3% (depending on coal rank) and there is no point to further increasing moisture content. Extensive mine testing carried out in different geological conditions confirmed that a specific amount of moisture is sufficient to reduce outburst danger. Furthermore, at such moisture content, gas is displaced 50 to 80% from porous spaces whose size constitutes 4 to 6%. Gas left in coal does not constitute an outburst danger because it is partly dissolved in water or partly enclosed in dead-end cracks by adsorbed moisture.

NMR is an efficient method for studying adsorbing interactions and water phase states in solids. It has high resolving power and tests samples without destroying their structures. It also allows us to clearly differentiate physical adsorption and chemical adsorption—a difficult task with other methods especially when chemical adsorption is irreversible.

3.3.1 Water–Coal Adsorption

Physical adsorption is an increase of substance concentration (water in this case) on the surface of a solid (surface of a pore in coal) as a result of attraction and its retention at the interphase boundary by Van der Waals forces. Physically adsorbed molecules keep their chemical identities and adsorption is reversible. The nature of surface greatly influences the character of intermolecular interactions and parameters of the boundary phase.

The basic thermodynamic works by Gibbs [23] led to investigation of the fluid structure near the interphase boundary. Water has a special place among fluids due to its unique properties; that is why the question of water structure at the interphase boundary has always interested researchers [24]. Significant research indicates that structural and dynamic water properties near the interphase boundary are significantly different from the properties of volumetric fluids [25,26]. Note that the formation of an interphase layer of adsorbed water is not a unique characteristic. Interphase layers form in all fluids [27]. However, polymer characters of a water structure and availability of a cooperative net of hydrogen bonds are reflected in interphase properties. Studies of intermolecular interaction character and surface influence on interphase properties is also of great importance.

Some research assumed that interphase layers are lengthy—10^2 to 10^3 nm [28]; that estimate was later shown to be overstated. Some researchers assumed that water interphase layer thickness d_{in} did not exceed ~10 to 100 nm [29]. Subsequent works showed that d_{in} barely exceeds 1 to 2 nm [30]. One peculiarity of the water interphase layer is water molecules orientation ordering near the interphase boundary, which causes, in particular, surface potential jump [24].

In assumptions about the character of adsorbed molecule movements, scientists who played great roles include E. Rydil [31], J. De Boer [32], and their disciples [33]. De Boer [32] chose two ultimate cases from different models of monomolecular adsorptive layers: localized and non-localized adsorption. In a model of localized adsorption, the molecules adsorbed are strongly bound to adsorption centers; they have no translational degrees of freedom and do not oscillate perpendicularly or parallel to the surface. The localized character of adsorption does not signify that adsorptive molecules remain motionless. The presence of an adsorptive balance between the gas and adsorptive layer shows that molecular motion occurs both perpendicularly and parallel to the surface.

Adsorptive molecule migration along the surface occurs in discrete steps from one adsorptive center to another. The connection of molecular residence time in one section τ_C and activation energy V_0 for discrete steps along the surface is shown in Frenkel's equation [34]:

$$\tau_c = \tau_0 \cdot e^{V_0/RT},$$

(3.10)

where τ_0 is the constant connected to adsorptive molecular vibration. V_0 is a difference between adsorption molecular heat in the active center of the surface and in its section between two active centers. That is why V_0 is significantly lower than adsorption heat Q.

During physical adsorption on hydrophilic (polar) surfaces, V_0 usually constitutes a third to a half of $Q \geq 40$ kJ/mole [35]. It is much bigger than heat motion energy $RT \sim 2.4$ kJ/mole (at $T = 290$ to 295 K). Therefore, time of delay τ exceeds vibration period τ_0, which has order 10^{-13} sec, but it is less than adsorption time t. Research examples [32,36] show that during physical adsorption on polar adsorbents, the ratio t/τ_C is ~5000. When V_0 is less than the energy of thermal motion, the molecules adsorbed move freely along the surface in both directions. Adsorptive layers consisting of such molecules behave as two-dimensional gases. In this case, adsorption is described as a non-localized model. According to De Boer, during ideal non-localized adsorption, the adsorbed molecules retain rotational and two translational degrees of freedom.

The surface of a polar adsorbent, as a rule, is energetically non-homogeneous [37]. Non-homogeneity molecular arrangements in different adsorptive centers are no longer equiprobable and the role of combinative energy in overall entropy of adsorption has decreased. Therefore, no models can cover all varieties of adsorptive systems.

In analyzing adsorption entropy in real systems, it is necessary to consider that during adsorption both partial and complete loss of translational mobility can occur. At the same time, the number of rotational degrees of freedom decreases because the surface forms a potential barrier limiting molecular rotation. Based on these results, it is necessary to note that water molecules on any hydrophilic surface are adsorbed locally and interact with polar groups.

Scientists later became more interested in studying water adsorption on the surfaces of hydrophobic materials [38,39]. Most hydrophobic adsorbents have hydrophilic adsorption centers on their surfaces. Their active centers are different π bonds, single electronic pores, cations of small radii, oxygen-containing groups, and CH, CH_2, and CH_3 radicals. Water molecules are adsorbed first in the hydrophilic centers. This results in localized water adsorption on carbon and little movement to coal surfaces.

However, the literature lacks data about adsorptive water molecule states on hydrophobic adsorbents with different concentration and hydrophilic center polarities. To a certain extent, the lack arises from the experimental difficulties of defining quantities of hydrophilic groups on the surfaces of hydrophobic adsorbents independent of methods of measuring water adsorption. Another difficulty surrounds the intentional formation of specific adsorption centers with different polarities on the surfaces of such adsorbents.

Wide line and impulse NMR methods have special places in interphase water structure and dynamics research. The first NMR experiments studying adsorptive moisture on the surfaces of solids were carried out on activated carbon, cellulose, silica gel, and synthetic zeolites. However, the results of the experiments are not totally useful for describing adsorptive processes in coals. One reason is that coals have different adsorptive centers (π bonds, single electronic pairs, branched aliphatics, cations of small radii). Moreover, zeolites and silica gels have porous systems that have similar radii. Coal contains systems of micropores (0.4 nm) and macropores ($1 \cdot 10^5$ nm). For these reasons, researching water phase state in coals and defining water quantity are complex tasks.

Studies of the dependence [39–41] of water sorption on degree of coal metamorphism defined the quantity of moisture sorbed as 0.3 to 0.5%, accurate within 0.1%. The experiments tested artificially moistened coals because the adsorbed H_2O state in both artificially and naturally moistened coals is identical. Coals with different degrees of volatility (5.8, 11, 21, and 42%) and moisture content (2.4, 0.77, 1.23, and 1.67%) were examined. First the coal was crushed to 0.1 to 1 mm fractions. Since the micropore radius is 1 nm and its volume in coals with high metamorphism is about 80% of free internal volume (with low metamorphism, 50 to 60%), the crushing did not affect the natural micropores system that mainly defines the adsorptive surface and it did not change the absorptive properties.

In the NMR spectra of the coal–water system (Figure 3.2) two lines of the hydrogen nuclei with different mobilities are resolved. The nuclei that are strongly fixed related to neighboring nuclei (hydrogen within the organic coal mass) yielded an extensive line (ΔH_2 = 5.5 to 6 Oe). Water in coal is in a physically associated state. As compared to free water (line width smaller than 0.1 Oe), it is less mobile because molecules in the first monolayer seem to be attached to the coal surface and vibrate around the adsorptive centers. As soaking intensity increases, molecule mobility grows. The line width of physically adsorbed water is ΔH_1 = 0.1 to 0.5 Oe. For coals with natural moisture, the values of NMR spectra line width were 2.4%, 0.21 Oe; 1.67%, 0.36 Oe; 1.23%, 0.35 Oe; and 0.77%, 0.45 Oe. All coal samples with natural moisture were heated to 115°C for 0.5 to 0.7 hours and vacuumed at the same time. The heating continued until evaporation controlled by NMR spectra was complete. The narrow line indicating moisture availability is missing on the vacuumed coal spectrum.

The vacuumed samples were saturated with water vapor and held in vapor for 20, 30, 40, 60, and 90 min. After 90 min, the moisture of all coal ranks reached the original level. An increase of saturation time up to 4 hours and more did not raise coal moisture in comparison with the original amount under natural conditions. Adsorption of water vapor by coals at a steady temperature normally does not exceed the value of natural vapor. Temperatures are steady in mine conditions and provide good evidence of coal saturation in natural conditions. Coal cannot absorb more water even if moistening time increases. To increase the quantity of moisture adsorbed, it is necessary to choose appropriate surface-active substances. Anthracite is an exception because its moisture reached 3.2% after 2 hours of saturation.

TABLE 3.1

Dependence of Spectrum Line Width on Moisture

Moisture (%)	Line Width (Oe)			
	Anthracite	Lean	Coking	Volatile
0.7 to 0.8	0.23	0.41	0.45	0.49
0.95	0.23	0.38 to 0.4	—	0.45
1.1	0.22	0.38	—	0.39
1.2	0.22	0.35	—	0.36
1.7	0.21	—	—	0.36
2.7	0.19	—	—	—
3.2	0.19	—	—	—

Note: Natural moisture (%): anthracite 2.4, lean 1.23, coking 0.77, volatile 1.67.
Volatility outlet (%): anthracite 5.8; lean 11; coking 21; volatile 42.

In Table 3.1, the widths of NMR spectra narrow lines for moistened coals are shown for anthracite, lean, coking, and volatile ranks. As moisture increases, the line width of NMR spectra for all coal ranks decrease. A decrease of line width is caused by molecule mobility and is good evidence of a weakening connection to the adsorbent surface. Line narrowing occurs when new adsorption layers appear. However, line narrowing occurs differently in various coal ranks. Thus, the change of anthracite sample line width by 0.04 Oe corresponds to a moisture content change from 0.7 to 3.2%. For the volatile sample, the change of line width is four times that when moisture changes only from 0.7 to 1.7%. The change is caused by different porosities. Coal porosity is high at the early stage of metamorphism. Pores have significant sizes (up to 500 nm). As metamorphism continues, lamellae numbers in packets and oriented parallel to stratification increase; the structure becomes more compact and porosity decreases. In studies of volatile coal (low metamorphism) and lean coal (much higher metamorphism), we noticed an obvious line width reduction with an increase of water content. One can suppose that coal moisture increases due to new sorption layers, which proves that relaxation time increases in impulse NMR.

Moistened coal adsorption at low temperatures was studied. Wide line NMR can show the influences of porous structures and active surface centers on activation energy V_0 and correlation time τ_C more easily than measurements of entropy in wide interval T [42], dielectric losses at different frequencies, and nuclear quadrupole resonance (NQR) frequency [43].

Using the Bloembergen-Purcell-Pound approximate theory [44] and experimental data on low temperature NMR line broadening of sorbed water information on τ_C, correlation time change based on water molecule mobility was obtained. According to Fedin [45] τ_C, molecule correlation frequency of line width ΔH and ΔH_0 at a given temperature and rigid lattice temperature, respectively, are connected by the following relation:

$$(\Delta H)^2 = \frac{2(\Delta H_0)^2}{\pi} \mathrm{arctg} \frac{\alpha\gamma\Delta H}{\omega_c}, \tag{3.11}$$

where ΔH_0 is line width of the frozen lattice; $\alpha = 8\ln2$; $\gamma = 4258 \text{ c}^{-1}\text{Oe}^{-1}$; $\omega_C = 2\pi/\tau_C$ reorientation frequency. Hence, the formula, defining correlation time τ_C can be given as

$$\tau_c = \frac{2\pi}{\alpha\gamma\Delta H} tg \left[\frac{\pi}{2} \left(\frac{\Delta H}{\Delta H_0} \right)^2 \right]. \tag{3.12}$$

The low temperature dependence of line width and second moments was determined for several coal ranks [19,46] (Figures 3.9 and 3.10). The characteristic peculiarity of low temperature spectra of NMR coals is the difference of temperatures during the phase transformation of adsorbed water from a liquid to a crystal-like state for coals of different ranks [19] (Figure 3.11).

The transformation is not abrupt. It occurs gradually as the molecules that are less closely connected with the surface are the first ones frozen. As Bloch showed [47], the two or three monolayers nearest the surface maintain their mobility even at very low temperatures at which rotatory mobility is frozen. For anthracites, the transformation starts at 183 K; for coals with $V^G = 26\%$ at 158 K; and for coals with $V^G = 39\%$ at 233 K. According to Uo and Fedin

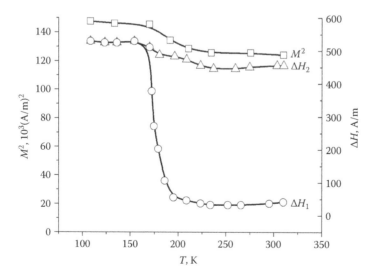

FIGURE 3.9
Dependence of line widths ΔH_1 and ΔH_2 and second moments M^2 of NMR spectra on temperature in anthracite coals ($V^G = 5\%$). (From Alexeev A.D., Krivitskaya B.V., and Pestryakov B.V. *Chemistry of Solid Fuel.* 2, 94, 1977.)

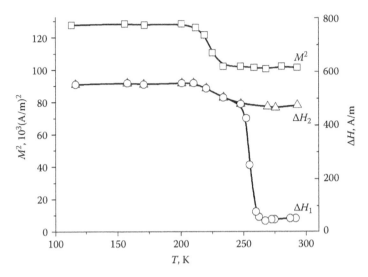

FIGURE 3.10
Dependence of line widths ΔH_1 and ΔH_2 and second moments M^2 of NMR spectra on temperature in coking coals ($V^G = 26\%$). (From Alexeev A.D., Krivitskaya B.V., and Pestryakov B.V. *Chemistry of Solid Fuel.* 2, 94, 1977).

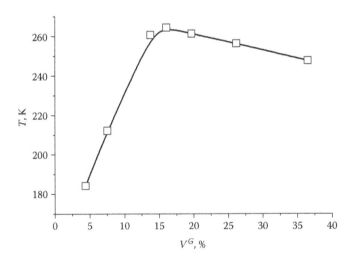

FIGURE 3.11
Dependence of temperature T of liquid–crystalline phase transition of adsorbed water on volatile gas outlet V^G in coals. (Adapted from Alexeev A.D., Zaidenvarg V.E., Sinolitskiy V.V. et al. 1992. *Radiofizika v ugolnoy promyshlennosti.* Moscow: Nedra.)

[45], the potential barrier of stagnated motion U_0 is calculated according to a simple formula:

$$U_0 = 155.4 \, T \, (K) \, J/mole, \tag{3.13}$$

where T is the phase transition temperature. The maximum value U_0 is 40.74 kJ/mole for coals ejecting volatile matter ($V^G = 39\%$); the minimum is 28.43 kJ/mole for anthracites ($V^G = 5\%$). Thus, the connection of water molecule sorption centers is strongest for coals of medium rates of metamorphism. As follows from Equation (3.12), the correlation time τ_C for coals with $V^G = 39\%$ is 2.8×10^{-5} at 233 K for coals with $V^G = 26\%$—the same correlation time for water molecules at 253 K; for anthracites with $V^G = 5\%$, the correlation time is 3.1×10^{-5} at 173 K.

Adsorption in coals of low and medium rates of metamorphism according to the classification of de Boer is localized. Adsorption of water in anthracites is closer to non-localized. At low temperatures, NMR reveals a sharp division of water and methane content in different phase states in fossil coals. However, one difficulty is that under natural conditions water and methane in coal both contribute to the narrow line of a spectrum. The intensity of the narrow line bears information about the general quantity of sorbed water and methane molecules. As long as the mechanisms of water and methane sorption by coal are different, the two fluids will react differently when the temperature drops (Figure 3.12).

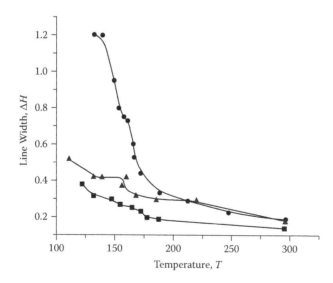

FIGURE 3.12

Dependence of line width ΔH of sorbed fluid in PMR spectra on temperature. ■ = Methane. ● = Water. ▲ = Methane–water mixture.

For sorbed water the phase change occurs at 243 to 173 K, depending on coal rank. However, the phase transition temperature of methane adsorbed by anthracite for example, is lower than 103 K. Thus, in a water–methane mixture adsorbed on coals, the phase transition for sorbed water will occur much earlier than for methane. This phenomenon allows division of water and methane adsorbed by coal.

Sorbed water (from a water–methane mixture) transforms into a crystal-like state at the same temperature as the water in the damped sample. Sorbed methane remains in the mobile state (at $T = 110$ K). The potential barrier that slows molecule motion is $U_0 = 6.6$ kCal/mole for water adsorbed on anthracite.

3.3.2 Adsorption by Impulse NMR

A successful implementation of impulse NMR in studies of adsorbed water is based on the fact that the correlation time of molecular processes in interphase boundary water is, as a rule, greater than correlation time for volumetric water ($\tau_C \cong 3 \times 10^{-12}$ sec) and falls within the interval of maximum response of NMR impulses (10^{-11} to 10^{-6} sec). The experimental data on spin–lattice T_1 and spin–spin T_2 relaxation times demonstrate the dynamic properties of interphase water and methods to determine correlation times. The NMR relaxation method also shows potential for coal porosity studies because T_1 and T_2 depend on the sizes, volumes, and areas of pore surfaces and the extents of their filling [48].

Studying pore structures of coals is difficult because the pore diameter sizes vary greatly from micropores to macropores [49]. Most studies focused on hydrogen and results are difficult to interpret in relation to T_1 and T_2 values for coal cores [48,50,51]. Therefore, silica gels were used as model sorbents.

The time of spin–spin relaxation T_2 of water adsorbed in silica gels with pore sizes of 0.9, 0.25, 1.4, and 6.5 nm in volatile, lean, and anthracite coals was studied [52–54]. The T_2 value of water protons was measured at $T = 298$ K at 20 MHz frequency. The double pulse method (90 to 180 degrees τ) was used. The minimum delay τ was $100 \cdot 10^{-3}$ sec; when changing the relaxation time (less than 160×10^{-3} sec), τ decreased to 60×10^{-3} sec. To measure T_2 while the magnetic field was slowly changed, 1H NMR spectra were given for different delay values τ between 90 and 180 impulse degrees. The dependence of line magnitude NMR on delay τ corresponds to exponential law:

$$A = A_0 \exp(-2\tau/T_2) \tag{3.14}$$

from which time T_2 was defined. A model of calculation of the monolayer thickness h of sorbed water has been suggested [53]. The sorbed water is divided into a surface layer (monolayer) that closely interacts with the surface of the solid body and the volumetric part [55–59]. The results suggest a short-term interaction between the pore surfaces and the molecules of the monolayer liquid and also between the monolayer and the volumetric part in the sample. Thus the speed of the spin–spin relaxation was given as

$$\frac{1}{T_2} = \frac{v}{v_s}\frac{1}{T_s} + \frac{v-v_s}{v}\frac{1}{T_b}, \tag{3.15}$$

where T_s and T_b are the times of the crossed relaxation (index 2 was omitted for convenience) for the surface and volumetric parts of the liquid, thus v is the liquid volume in the pore and v_s is the liquid volume in the surface layer of the pore. It is suggested that $T_s \ll T_b$ and layer h should be as thick as the distance between the molecules, based on the assumption that the interaction between the molecules of the liquid and the cell walls is a nuclear dipole–dipole interaction whose value is proportional to r^6. The simple reductions result in the following:

$$T_2 = \frac{v}{v_s}\frac{T_s}{1 + \dfrac{v-v_s}{v}\dfrac{T_s}{T_b}}. \tag{3.16}$$

As $T_s \ll T_b$:

$$T_2 = \frac{v}{v_s}T_s \tag{3.17}$$

If the liquid contains pores of the same size and they are homogeneously filled with liquid, the induced nuclear magnetization will decrease exponentially when the time of crossed relaxation T_2 equals:

$$T_2 = \frac{V}{V_s}T_s \tag{3.18}$$

where V is liquid volume in the sample and V_s is volume sample in the surface layers of all the pores. Equation (3.18) is true only for $V \geq V_s$. If local values of the relation T_s/T_b differ because of the heterogeneous distribution of liquid over the sample, large differences in pore sizes, or the condition of a fast molecule interaction is not fulfilled, the T_2 values will be different for different parts of the sample. In this case, nuclear magnetization decreases non-exponentially, as it is a sum of exponential functions with different T_2 values. In this case, Equation (3.18) shows an average value T_2. If we gradually reduce the moisture content in the sample to volume $V = V_S$ (V_S = volume of first monolayer) and assume that $V_S = hS$ (S = total area of pore surface in sample), Equation (3.18) becomes:

$$T_2 = \frac{V}{S}\frac{T_s}{h} \tag{3.19}$$

It has been suggested that the area of the monolayer identically equals the total area of the pore surface, i.e., during the process of drying no dry pores appear.

According to the Equations (3.18) and (3.19) the time of relaxation is proportional to the volume of water in the sample. In order to check the theoretical functions we showed the experimental function of the relaxation time T_2 depending on the relative content of water in the sample, i.e., on the value V/V_0, where V is current volume of water, V_0 is volume of water when the pores are fully filled, thus V_0 is pore volume.

In Figure 3.13 [52], function T_2 depends on the relative content of water in silica gels with 0.9 nm pore radius. The experimental data are approximated with two straight lines. A straight line serves a theoretical function for large volumes (3.18) because it passes through the origin of coordinates. The change of T_2 noted with a small quantity of liquid in a sample arises from the reduction of the number of water molecules in the first monolayer; in this case the interaction between the molecules of water changes, and the correlation time changes [48].

The experimental data from a little watering of the sample were optionally approximated with a solid line. The joint matches the border of the relative monolayer. All the approximations were made to achieve the smallest mean root square error. As shown in Figure 3.13, there is agreement between experimental data and theoretical function (3.18).

Based on figure data and the theoretical function (3.18), we determined the time of cross (spin–spin) relaxation for the most completely filled monolayer ($T_2 = 200 \times 10^{-6}$ sec) and the volume of water V_S that matches the complete filling in of the surface monolayer: $V_S = 0.31\ V_0$ or $V_S = 6.6 \times 10^{-5}$ m^3/kg, which is that of the layer thickness $h = 0.103$ nm.

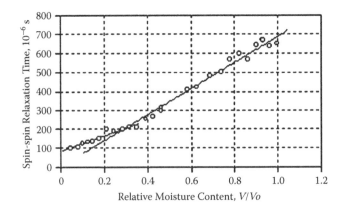

FIGURE 3.13
Dependence of spin–spin relaxation time on relative moisture content of silica gel with pore radius of 0.9 nm.

The dependence of the time of the spin-spin relaxation T_2 on the relative water content of a silica gel with a pore radius of 1.4 nm is shown in Figure 3.14. This dependence is similar to the previous one. The above description of the dependence T_2 for silica gel with 0.9 nm pores is also applicable. One can observe a good agreement of experimental data with theoretical function (3.18), although in the zone of water content that matches the border between the first and second monolayers, the experimental points are below the theoretical line. From the experimental data, and Equation (3.18), we see that $V_S = 0.11V_0$, i.e., $V_S = 4.2 \times 10^{-5}$ m^3/kg, and $h = 0.084$ nm. From the function, we can calculate the spin–spin relaxation for the filled monolayer at $T_2 = 170 \times 10^{-6}$ sec

The thickness of the first monolayer of water is approximately three times smaller than the sizes of the water molecules. Such a big difference cannot be explained by ignoring the roughness of the pore surface when h is determined. To confirm the difference, we used nuclear magnetic relaxation to calculate $h = 0.3$ nm for the water adsorbed in the pore glass with a pore radius of 1.75 nm while the time of spin–spin relaxation for the first monolayer was $T_2 = 1.1 \times 10^{-3}$ sec [48]. The underestimated h values can be explained by the fact that the water molecules in the surface layer are adsorbed around the hydroxyl groups and not spread over the whole geometric surface [60,61]. Small values of T_S caused by a closer connection of water molecules with the surface hydroxyls of silica gels do not contradict such s hypothesis [56].

We recalculated the volume of monolayer per surface unit, keeping in mind that every water molecule is connected with one OH group [60]. For each square nanometer of surface, we found 3.5 and 2.8 hydroxyl groups for silica gels with pore radii of 0.9 nm and 1.4 nm, respectively.

This method was used to research wet samples of volatile ($V^G = 39.0\%$), lean ($V^G = 9.0\%$), and anthracite ($V^G = 5.0\%$) coal ranks. The volatile coals ($V^G = 39.0\%$),

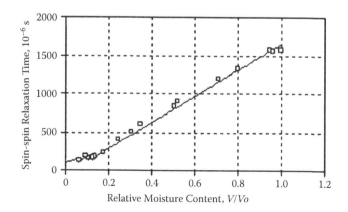

FIGURE 3.14
Dependence of spin–spin relaxation time on relative moisture content of silica gel with pore radius of 1.4 nm.

whose water content can be as much as 13% exhibited the best sorption ability. The relaxation time T_2 of water protons of known water content was determined according to the stagnation of nuclear magnetism with that water content using an adjustable curve consisting of the sum of two exponents:

$$A = a\exp(-2\tau / T_{2c}) + b\exp(-2\tau / T_2), \qquad (3.20)$$

where T_{2c} is relaxation time for coal protons, A is constant; b and T_2 are variables; and τ is the delay time between 90 and 180 degree impulses. Value b linearly depends on water content. The adjustment was made to ensure the smallest mean root square error.

The dependence of relaxation time T_2 on the water content for volatile coals with natural water content of 13.6% (Figure 3.15) is also non-linear and can be approximated with two straight lines. The change of the slope of the straight lines when $W = 7.5\%$ shows that at a lower concentration there is water in the sample in the form of a monolayer. The time of spin–spin relaxation for the completely filled monolayer of water is $T_2 = 120 \times 10^{-6}$ sec and the adsorbing container of the monolayer is $\lambda = 2.31$ mole/kg. The time of relaxation measured for a completely dried sample was $T_{2c} = 20 \times 10^{-6}$ sec. These data are the same as those found using a sorption isotherm for volatile coals [62].

For anthracite with $V^G = 9.0\%$, the spin–spin relaxation T_2 for each percentage of water content was determined using an adjustable curve. Instead of constant T_{2c} (25×10^{-6} sec for the sample), we used a variable constant T^*_{2c} because the experimental curve of the stagnation of nuclear magnetism could not be decomposed accurately to two exponents for all water content when $T_{2c} = const$. Table 3.2 shows the T^*_{2c} and T_2 results based on water content.

The effect of non-conservation of constant T_{2c} in the first exponent of Equation (3.20) that describes the damping of nuclear magnetism from

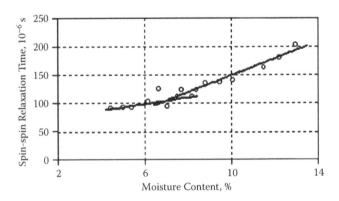

FIGURE 3.15
Dependence of spin–spin relaxation time on percentage of water content in volatile coal. (From Alexeev A.D., Troitskii G.A., Ulyanova E.V., Zavrazhin V. V. *Physicotechnical Problems of Mining Production* (Alexeev A.D., Ed). Donetsk: Donetsk Inst. for Phys. & Eng. of NAS of Ukraine, 1999 p. 3.)

TABLE 3.2

Dependence of Spin–Spin Relaxation Times for Protons of Water and Coal

W (%)	5.80	5.74	4.60	4.35	4.07	2.94	2.65	2.05
T_2, 10^{-6}	482	447	384	360	325	203	233	190
T^*_{2c}, 10^{-6}	36.0	35.5	34.0	31.5	30.0	28.5	27.5	27.5

coal protons can be explained. Coal contains both open and closed pores. In anthracites, the volume of closed pores is the same or greater than the volume of open pores [62–64]. The open pores are connected with the outer surface and this allows different liquids and gases to penetrate quite quickly into coal substance and leave it just as quickly. The closed porosity of the fossil fuels is not connected by transport channels with the outer surface. The evacuation of the water molecules is carried out only by means of solid-state diffusion—a process of considerable duration.

If we assume that some sorbed water is in closed pores, in NMR there must be influence of water protons in the closed pores due to the lack of fast molecule exchange between that water and the water in open pores. Closed pore sizes differ considerably—from holes containing a few molecules to large cavities in which water in volumetric condition may cover pore surfaces with several layers. Thus, the relaxation times of water protons in different pores will differ considerably and range from a time close to T_{2c} to the time approaching T_2 for water protons in open pores. The closed pores that were totally filled lose moisture during drying, although it happens with a slight delay in time compared to the loss of water in the open pores. Under this condition, the amplitudes of signals from water in the closed pores decrease and the relaxation time of the signals decreases accordingly. Time T^*_{2c} in the first exponent of Equation (3.20) by which the non-exponential function is approximated (sum of exponents from coal protons and water protons in closed pores) will reduce and approach T_{2c}.

The fully filled monolayer for lean coal ($V^G = 9.0\%$) corresponds to moisture content of 2.2% (with high moisture content T_2 linearly depends on moisture content). The size of monolayer is $\lambda = 0.55$ mole/kg. The relaxation time T_2 for the fully filled monolayer is $T_2 = 180 \times 10^{-6}$ sec. The dependence of spin–spin relaxation time T_2 on the moisture content is shown in Figure 3.16.

The data for anthracite ($V^G = 5.0\%$) were determined with a continuous method. Using the received values of line width we determined T_2 relying on the Lorentz form of the NMR line. The dependence has a typical format (Figure 3.17) [53]: linear dependence with high moisture content and excursion from linear dependence $W = 0.9\%$, which corresponds to the full filling of the monolayer. Time T_2 for the fully filled monolayer is 80×10^{-6} s, the size of the monolayer is $\lambda = 0.28$ mole/kg.

The results received while researching silica gels made it possible to determine the sizes of nominal monolayers in fossil coals. NMR spectrographic studies were performed to study the structures of fossil coals and the

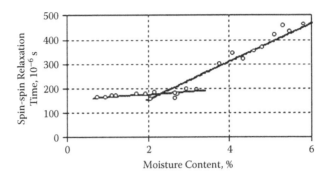

FIGURE 3.16
Dependence of spin–spin relaxation time on moisture content in lean coal. (From Alexeev A.D., Troitskii G.A., Ulyanova E.V., Zavrazhin V. V. *Physicotechnical Problems of Mining Production* (Alexeev A.D., Ed). Donetsk: Donetsk Inst. for Phys. & Eng. of NAS of Ukraine, 1999 p. 4)

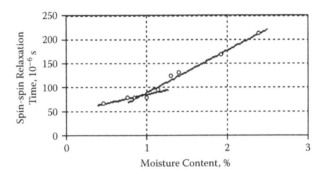

FIGURE 3.17
Dependence of spin–spin relaxation time on relative moisture content for anthracite. (From Alexeev A.D., Troitskii G.A., Ulyanova E.V., Zavrazhin V. V. *Physicotechnical Problems of Mining Production* (Alexeev A.D., Ed). Donetsk: Donetsk Inst. for Phys. & Eng. of NAS of Ukraine, 1999 p. 5)

processes of methane and water desorption from the host rocks [65–66]. To measure T_2, we used the double-pulse (90 to 180 degrees τ) spin-echo method of Carr and Purcell [67]. The time of relaxation T_2 was determined as the geometrically average quantity between surface and volumetric speeds [52]. The formula for the speed of cross relaxation is

$$\frac{1}{T_2} = (1 - \frac{\lambda S}{V})\frac{1}{T_b} + \frac{\lambda S}{V}\frac{1}{T_s} \tag{3.21}$$

where T_b and T_s are the times of relaxation for the volumetric and surface parts of the liquid, respectively; S and V represent the local area of surface and the content of the liquid in the pore; and λ is the length or dimension of interaction of the adsorbate molecules with the surface that is responsible for the increase of relaxation speed.

Proportion $\rho = \lambda/T_s$ (surface relaxivity) characterizes the intensity of the interactions of the molecules of the liquid with the surface of the sorbent and penetration of the surface. In fact, $\rho \to \infty$ corresponds to an ideally penetrating surface and a higher value of ρ means a higher penetrating ability of sorbent [68]. When pore diameters differ, one can calculate an average value of the relaxation time according to the relaxation curve [69]:

$$\rho \frac{1}{T_2} = \sum_{j=1}^{N_p} \rho \frac{1}{T_{2j}} \frac{n_j}{N} = \sum_{j=1}^{N_p} \frac{S_j}{V_j} \frac{n_j}{N} = \overline{\left(\frac{S}{V}\right)} \qquad (3.22)$$

In Equation (3.22), to determine the number of pores, n_j/N is relative share of pores of surface area S_j and volume V_j. As molecules are sorbed in the atmosphere saturated with the vapors of the liquid over the whole pore surface, we can consider S a constant. When changing the moisture content of a sample by continuous drying at high temperature, we can determine ρS according to the measured T_2. Note that ρ does not depend on pore size.

In fact, if we implement the regime of moisture content in one layer, using Equation (3.21), we can consider surface S a constant value. Thus, the change of liquid volume V in the pores during desorption will be connected with T_2 linear dependence [70]:

$$T_2(w) = \left(\frac{m_0}{S\rho\rho_w}\right)w, \qquad (3.23)$$

where m_0 is the mass of the dry sample of sorbent and ρ_w is the density of sorbate. The formula in brackets is the tangent of dependence angle $T_2 = T_2(w)$ and allows calculation of yielding of the pore surface S and ρ. We used the density of saturated vapors of water $\rho_w = 3 \times 10^{-2}$ kg/m³ at 303 K. To approximate short time relaxation [68] with spin-echo spectroscopy of porous media:

$$\rho \gg \sqrt{D_0/t}, \qquad (3.24)$$

where D_0 is the coefficient of self-diffusion of water in the volume and t is measuring time. The magnetization curve of the medium response to the impulse change of the interaction in small pore volumes can be approximated with the proportion:

$$\frac{M(t)}{M_0} = 1 - 2\frac{S}{V_p}\left(\frac{D_0 t}{\pi}\right)^{1/2} + O(t). \qquad (3.25)$$

If

$$\rho \ll \sqrt{D_0/t}, \qquad (3.26)$$

the time dependence of the magnetization curve is

$$\frac{M(t)}{M_0} = 1 - \frac{S}{V_p}\rho t + O(t^{3/2}). \tag{3.27}$$

We know the shared pore distribution, surface area of sorbents, and ρ from examinations by NMR, sorbate, and small angle x-ray and neutron scattering methods [68,71,72]. The ρ values of the researched sorbents were <100 μm/sec, which corresponds to the approximation in Equation (3.26). With the simultaneous determination of spin–spin T_2 and spin–lattice T_1 relaxation times, it may be possible to calculate the diffusion mobility of water in pores and on the surface areas of pores S/V_p along with ρ [69] from the experimental data on relaxation in gradient magnetic fields of water protons. We assume that the duration of transverse (spin–spin) relaxation T_2 for water protons is connected with pore space and the gradient of magnetic field G_0 yielding a spin-echo [73] in accordance with

$$\frac{1}{T_2} \approx \rho\frac{S}{V_p} + \frac{2}{3}\gamma^2\tau^2 DG_0^2, \tag{3.28}$$

where γ is the gyromagnetic relation for a proton, D is a coefficient of self-diffusion, and τ is the distance between an impulse and an echo signal. When the induction of the magnetic field $B = 0.4\ T$ and the constant gradient $G_0 = 0.14\ T/m$, the second term is considerably less than the first and may be disregarded. We then have

$$\frac{1}{T_2} \approx \rho\frac{S}{V_p}. \tag{3.29}$$

It is also necessary to consider that the relaxation of a full signal of NMR depends also on the sorbate molecules connected with the surface and on the molecules filling the pore volumes. That is why T_2 is connected to the volume T_b and the surface T_s periodically transverse relaxation by Equation (3.21) or (3.30):

$$\frac{1}{T_2} = \frac{1}{T_s} + \frac{1}{T_b}. \tag{3.30}$$

As the duration of liquid relaxation while volume filling of pores is considerably longer than spin relaxation duration near the surface ($T_b \gg T_s$), $T_s \approx T_2$ will be correct.

The additional information about structural non-uniformities may be obtained using the theory of fractals [74]. The analysis of two-dimensional descriptions of porous material structures with atomic force [75] and scanning microscopy [76] shows fractal dimensionality d_f of the surface hierarchical structure expressed in the interval $1 < d_f < 2$. Using two-dimensional descriptions of the surfaces of mono-fractal domain linear scales of hierarchical structure, the fractal dimensionality of a porous space surface D can be determined. Assume the scaling distribution of pores according to sizes $f(r) \sim r^{-B}$ where f is probability density, $B = D - 1$, and equation $D = d_f + 1$ [72] in the domain of mono-fractals of linear scales of hierarchical structure is fulfilled.

If the lower limit ℓ and upper limit ξ of the fractality domain are known, the minimum scale L_{cg} where spins relax in pore spaces is connected with the duration of relaxation [74] by

$$T_{2,short} = L_{cg}^2 / D_0, \tag{3.31}$$

where

$$L_{cg} = \ell(\Lambda/\ell)^{1/(D-1)}, \tag{3.32}$$

where $\Lambda = D_0/\rho$ and D_0 is a coefficient of water self-diffusion. Some of the magnetization that follows the law of fast relaxation will be determined from $f = (L_{cg}/\xi)^{2-D}$. Using Equations (3.31) and (3.32) to determine the least measured value T_2 while saturation of the sorbent changes, we can determine the thickness of layer λ that which corresponds to the traditional approximation of a monolayer sorption. The thickness of a monolayer can be found using the inflexion point on curves $T_2(w)$ (Figure 3.18).

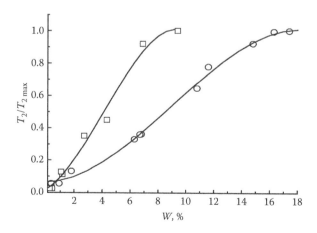

FIGURE 3.18
Dependence of duration of spin–spin relaxation of water protons in porous space of coal on humidity. Squares = anthracite. Circles = volatile coal.

Using descriptions of deposit coal structures, the characteristic scales of fractality and fractal dimensionality of surface hierarchical structures were determined via atomic force microscopy [75]. For the samples of volatile coal and anthracite under consideration, the ranges of linear sizes of 14 to 195 nm and 28 to 389 nm, respectively, may be regarded as mono-fractal where fractal dimensionality $d_f \cong$ const. The characteristic sizes $L_{cg} \approx 439$ and 650 nm for volatile coal and anthracite samples correspond to the $T_{2,short} = 88$ μsec and 213 μsec. Using Equation (3.29), $\rho = 2.6 \times 10^{-3}$ m/sec and 7.3×10^{-4} m/sec. These results are consistent with Equation (3.21) and allow approximation (3.22) of curve magnetization M(t). From (3.21), we calculate the upper limit of time interval t_c:

$$t_c < \left(\frac{D_0}{\rho^2} \right) \tag{3.33}$$

where the approximation of fast relaxation is carried out. For anthracite, $t_c = 3.75 \times 10^{-3}$ sec; for volatile coal, $t_c = 3.05 \times 10^{-4}$ sec. From the coefficient when $t^{1/2}$ in Equation (3.25) is the known value of the coefficient of water self-diffusion D_0, we can obtain the relation of pore surface to the volume of liquid filling the pores (S/V). This relation coincides with the relation (S/V)$_p$ for the pores if sorbate saturation is present in all the pores. The data for (S/V) are obtained by the method of the least squares while taking a logarithm of dependence (3.25).

We know from graph $T_2(w)$ (3.23) that $S\rho$ values are 0.587 m³/sec and 2.27 m³/sec for volatile coal and anthracite, respectively. The obtained values of (S/V) allow us to calculate the specific surfaces of pores when the maximum dampness of the samples S/V = 2.2×10^4 m²/m³ for volatile coal and 3.8×10^4 m²/m³ for anthracite.

Using these data, the specific surface values for maximum dampness are obtained: 221m²/g and 69m²/g for volatile coal and anthracite, respectively. Good consonance with these values is obtained for volatile coal when the specific surfaces from the dependence graph $T_2(w)$ = 185 m²/g are calculated. For anthracite, $S/m_0 = 1.54 \times 10^3$ m²/g and S/V = 4.9×10^5 m²/m³ were obtained. The latter perfectly accords with the results of measuring [77] Australian coal. The surfaces of open pores can be measured by other methods; closed porosity can be measured by non-invasive methods like NMR.

Measuring lateral relaxation duration of proton T_1 spins needed to calculate diffusion coefficient was carried out by progressive saturation [73] simultaneously with measurement of spin–spin relaxation duration T_2. For calibration of connections of the coefficient of diffusion D with relaxation duration T_1, the duration of water relaxation in free space $T_{1,w} = 1.8$ sec and the coefficient of water self-diffusion coincided well with the studies of other authors [78]. We kept in mind that spin relaxation of water protons occurs because of intramolecular interaction. The duration of relaxation τ_C is connected with the coefficient D_r by:

$$M_2 = \frac{3}{10} \frac{1}{T_1 \cdot \tau_c} = 7.4 \cdot 10^8, \qquad (3.34)$$

$$\tau_c = \frac{4\pi a^2 \eta}{3kT}, \qquad (3.35)$$

where η is viscosity, a is molecular radius, and k is Boltzmann's constant. For a diatomic molecule [79]:

$$M_2 = \frac{9}{20} \left(\frac{\mu_0}{4\pi} \right)^2 \frac{\hbar^2 \gamma^4}{r^6}, \qquad (3.36)$$

where μ_0 is a magnetic constant, γ is gyromagnetic relation for a proton, $\hbar$ is Planck's constant divided into 2π, and r is the distance between the nearest protons in a molecule. To calculate coefficient of diffusion D:

$$D_r = \left(\frac{9}{5M_2} \right) \cdot T_1 \qquad (3.37)$$

Substituting M_2 from (3.34) we obtain the relationship between the diffusion coefficient of water and the longitudinal relaxation time T_1: $D_r = 0.25 \times 10^{-8} \times T_1$.

The values of a self-diffusion coefficient for anthracite exceed the values for volatile coal (Figure 3.19). Extrapolation for big values of pore radius gives

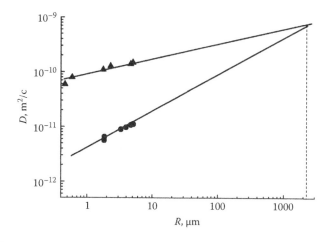

FIGURE 3.19
Coefficients of water self-diffusion in anthracite (▲) and volatile coal (●) depending on pore radius.

the cross point of straight lines that corresponds to the volume coefficient of water self-diffusion D_0.

Water can form conglomerates in pores by means of hydrogen bridges. The average number of molecules in pores is less than in volumetric liquid. Analyzing this mechanism yields the surface water layers λ for volatile coal (220 nm) and for anthracite (155 nm) that considerably exceed the sizes of water molecules (0.18 nm). This means that forming of a surface layer during absorption on a coal pore surface is determined by a cluster-type mechanism, that is, when water molecules form conglomerates with binding energy that exceeds the binding energy of H_2O with active centers of coal pore surfaces [80], the volume filling in the pores by concentrated vapor of adsorbate is realized.

In addition to analyzing minimum scales of porous space structures L_c, NMR measurements of spin-echo allowed the calculation of maximum scale L_{int} [74]. In a diffusion-controlled regime, relaxation duration T_2 is connected with the inner scales of pores L_{int} by:

$$T_{2,long} \approx L_{int}^2 / (2\pi^2 D_0). \tag{3.38}$$

For the measured maximum values, we obtain 5.1 μm and 9.6 μm for volatile coal (665 μsec) and anthracite (2.31). The latest values characterize the maximum pore sizes filled by water. It is interesting that we can find the radii of pores $R = 3 \, \rho T_2$ assuming their spherical form. For volatile coal we have $R = 5.1$ μm, for anthracite $R = 7.2$ μm; these correspond well with the values obtained from fractality measurements. According to the spin-echo results, volatile coal has more ability to absorb water than anthracite. This is indicated also by the greater dampness found in saturated coal samples (17.4% for volatile coal and 9.4% for anthracite). The values obtained for speed of surface relaxation ρ for coal considerably exceed the values for porous glass but can be compared with the results for limestone. The surface layers of molecules of sorbed water considerably exceed the thickness of the monolayer and correspond to the model of the cluster-type mechanism of water layer formation in porous coal spaces.

3.3.3 Water in Rocks

The binding power of adsorbed water in rock pore surfaces is determined by the same method used for coal. Water molecule binding power may be seen in the width of NMR spectrum lines, the potential barrier that hampers molecule rotation, and the time of correlation indicating degree of molecule mobility. Adsorption centers change liquid structures in layers adjoining solid surfaces based on volume phase. At the same time, H bindings are deformed and the deformation spreads a significant distance from the substrate. Structurally transformed layers exhibit oscillations of density that depend on constants of molecular forces and also on damping liquid capabilities.

Active centers on the surface force water to the first layers that dramatically differ from volume water structures. This thin layer of liquid is a boundary phase 8 molecular sizes thick. Between the boundary phase and free liquid lies a transition zone with characteristics similar to those of volumetric water. This zone (fused layer) serves as a bridge between sorbate-associated water and volumetric water; water viscosity is higher here than in volume water [81]. The fused layer has a disorganized structure due to breaking of intra-molecular hydrogen bonds. The boundary and transition layer spread is 7 to 10 nm and determined by solid substrate impact. For sorbate liquid it is characteristic of phase transition to solid condition.

The NMR spectra of naturally damp argillite and limestone (outburst and non-outburst) were studied. The samples of argillite were chosen from the Kalinina Mine (Artemugol) one year after an outcrop resulting from exploratory mining. Argillite is a gray, fine-grained, disorganized material containing clayey basalt cement. The lime samples (outburst and non-outburst) were from Petrovskaya-Glubokaya (Donetskugol), and Krasnoarmeiskaya Kapitalnaya (Krasnoarmeiskugol). The limestone had middle graininess, clayey quartz cement, and disorganized structures.

Comparing the values of line widths and second moments with the known values for the specific types of water, we determined the character of water binding in mine deposit structures. For the molecules of chemically associated water, the second moment changed in a range of 1 to 8 Oe^2. Crystallized water revealed line widths of 10 to 12 Oe^2 and the second moment values were 21 to 36 Oe^2. For adsorbed and capillary water (physically associated), considerable molecule mobility led to line constriction of 0.1 to 0.3 Oe. More constricted lines (<0.1 Oe) are peculiar to free water.

The investigations showed chemically and physically associated water in different percent ratios in argillite. The percent ratio of chemically associated water drops immediately after the face is uncovered and was unchanged after 1 to 1.5 months. A decrease of chemically associated water in crystal lattices of some minerals containing H^+ and OH^- groups proves the crystal lattice transformation and partial destruction of these minerals. Transformation begins as soon as the face is uncovered. We suggest that the reason for crystal lattice transformation is to release tension influencing mine rocks under natural conditions.

Conversely, the percent ratio of physically associated water increases. This is explained by the capabilities of minerals in argillite to adsorb water molecules on their surfaces through electrostatic gravitation. Adsorbed water relates to the mobility phase. For that reason, it concentrates on the surface when it makes contact with newly exposed mine deposits in damp mine atmospheres, and then penetrates through the cracks until saturation. The results (Table 3.3) indicate that argillite saturation by physically associated water occurs evenly. Total saturation of the mineral layer (width of 10 to 12 cm) was achieved within 5 months. The contents of adsorbed water increased in 0.6% per month. During this time, the strength of the argillite decreased 30%, and this decrease must be considered when calculating mine working stability.

TABLE 3.3

Dependence of Water Content (%) on Time of
Saturation in Argillite

Month	Chemically Associated Water	Physically Associated Water
0	4.8	2.4
1	4.2	3.0
4	4.1	5.3
5	4.1	5.4
6	4.1	5.4

For argillite, the most active centers of adsorption are surface hydroxyl groups of kaolinite and mica. At minimal filling, water molecules form hydrogen bonds with surface OH^- groups and with each other. After considerable filling with water, coordination binding may appear.

Recording of NMR spectra at different temperatures revealed the availability of physically and chemically associated moisture in outburst and non-outburst sandstones. Sandstones lack OH^- groups; adsorption is realized through coordination binding determined by the availability of silicon atoms that are coordination-unsaturated on the surfaces of mineral particles. They are capable of forming donor–acceptor associations between free α-orbits and unshared pairs of oxygen electrons of water molecules. Increasing temperature up to 240°C leads to considerable constriction of the line determined by increasing moisture molecule mobility connected with weakening or breaking of molecule links with adsorbing centers of argillite surfaces.

Sandstone, argillite, clay, and marl were used to study narrow line changes. The minerals were ground to fractions of 0.2 to 0.5 mm (diameters controlled with sieves). All the samples of natural dampness were heated to 160°C for 6 h. Heating was carried out until the water evaporated completely, controlled according to NMR spectra and dry samples were saturated with water vapor. Table 3.4 presents the results.

We can determine the types of links in a water–mineral system by the width of the ΔH line because ΔH increases are caused by molecule mobility. Studies of zeolite show that narrowing of spectra lines of water adsorbed on zeolite arises from the appearance of new layers of sorption. After forming the first adsorbing layers, a layer of loosely bound water appears, as shown on the NMR spectra of moistened minerals. At a small percent ratio of water (<2%), the spectra consist only of two components: one wide and one narrow (Figure 3.20). Sorbent water near the surface yields the wide line. The faintly associated component yields the narrow line. The widths of both lines decrease as water concentration in the mineral increases. This proves multi-layered adsorption. The faintly associated water then becomes free.

Figure 3.21 graphs the dependence of NMR line width on dampness at indoor temperature. The dependence of ΔH on moisture W is non-linear. When

TABLE 3.4

Dependence of NMR Line Width on Mineral Dampness

Argillite		Clay		Limestone		Marl	
W %	ΔH, **Oe**	W %	ΔH, **Oe**	W %	ΔH, **Oe**	W %	ΔH, **Oe**
–	1.15	–	1.9	–	1.5	–	–
1.9	–	1.6	–	1.7	–	2	1
	0.28	–	0.4		0.19	–	–
2.3	1.1	2.2	1.37	2.2	1	2.3	0.88
2.9	0.96	2.8	0.98	2.8	0.74	2.9	0.72
3.7	0.84	3.3	0.96	4.4	0.91	3	0.6
5	0.84	5.7	0.9	5.5	0.9	–	–
–	–	6	0.9	–	–	–	–

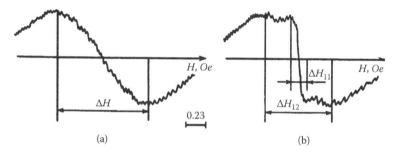

(a)   (b)

FIGURE 3.20

NMR spectra of water protons in clay minerals obtained under the assumption of their spherical form from the equation $R = 3\rho T_2$. (a) Dampness 0.75%. (b) Dampness 2%.

filling of pore spaces of mining minerals is small, the widest line and consequently the most powerful binding between water molecules and substrate is observed in clay; the smallest line was produced by marl. The dependence of ΔH on dampness in clay and argillite is similar in character which can be explained by the highly metamorphic nature of argillite where reconstruction occurred. As dampness increases, line width stabilizes. This occurs in clay when dampness is 5.7% because porous volume is high (up to 30%). In argillite (porous volume of 2%), this kind of stabilization is observed when dampness is 3.7%. After marl was exposed to dampness for 6 h, water content did not exceed 3% and did not stabilize. Porous volume is smaller in marl than in argillite and thus its dampness is lower. When dampness is increased via simple moistening, mobility of H_2O molecules rises and the temperature of a phase transit to a solid condition is 0°C, which proves the volumetric base.

A study of adsorbed water in minerals at low temperatures was conducted (Figure 3.22). The line width of NMR spectra based on water adsorbed by clay and argillite is shown. It is obvious that adsorbed water transforms into a solid condition at considerably lower temperatures than free water. In argillite, the transition starts at –73°C; in clay, the transition is smoother and starts

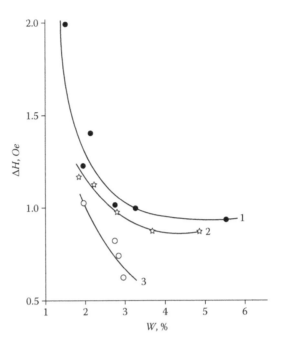

FIGURE 3.21
Dependence of line width of NMR spectra on water amount in minerals at indoor temperature.
1. Clay. 2. Argillite. 3. Marl.

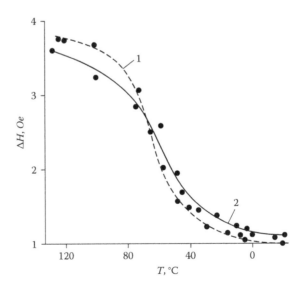

FIGURE 3.22
Temperature dependence of line width of NMR spectra on water in clay (1) and in argillite (2).

at –57°C. The potential barrier that hampers molecule rotation determined by Equation (3.13) for water in argillite is 31.08 kJ/mole and for water in clay is 33.6 kJ/mole (T = temperature in Kelvin). At this temperature, noticeable narrowing of the NMR spectrum line starts.

The time of correlation τ_c characterizing the degree of water molecule mobility on the surfaces of sorbents was calculated by connecting the line width of NMR spectra at given temperatures using Equation (3.12). On the surfaces of sorbents at the initial section of the temperature dependence curve ΔH (from indoor temperature through the moment when abrupt deceleration of H_2O molecules starts), the time of correlation increases from $1.9 \cdot 10^{-5}$ to $2.8 \cdot 10^{-5}$ sec in argillite and from $3.4 \cdot 10^{-5}$ to $5.4 \cdot 10^{-5}$ sec in clay. At –70°C, the time of correlation of water molecules in clay is $11.1 \cdot 10^{-5}$ sec and in argillite is $6.6 \cdot 10^{-5}$ sec. Thus, water molecules in argillite have more mobility than in clay at this temperature. This also proves that the energy of adsorbing centers of clay is powerful. In addition, greater defection of intra-molecular liquid bindings determines the temperature of a phase transit to a solid condition.

3.3.4 Methane in Coal

Applying wide line NMR methods produced wide-ranging information about methane conduct in coal [82–87], the ability of deposit coal to create a metastable condition with methane in the form of a solid solution [88], and the phase state of methane in fossil coal.

NMR spectra of coal–methane samples [82] look like the spectra of damp-saturated coal. However, a comparison of narrow line intensities of NMR spectra of coal–methane and coal–water systems showed that the values of occluded methane intensities (I_{1M}) are considerably larger than the values of adsorbed water intensities (I_{1B}) having the same percent ratio of occluded substances because occluded methane mobility is larger than occluded water mobility (Figure 3.23).

To achieve better accuracy in calculating dependence (Figure 3.23) correlations to wide line intensity I_2 instead of intensities I_{1M} and I_{1B} are used. The intensity of a wide component of NMR spectra I_2 is used as a datum mark because its value remains constant within one grade of coal [82,85]. The standard dependence of the occlusion of methane and water can reveal the sorbate in a coal under investigation. The fluids in coals have different structural forms. The concentration of fluids and localization conditions vary. Thus, contributions of different groups of hydrogen atoms vary among coals and the variation of change of line forms. The narrow component of a coal saturated with fluid reflects the extent and conditions of fluid localization and thus determine the phase states of fluids in coal.

A proton magnetic resonance (PMR) spectrum of coal saturated with fluid has two main well-solved components (Figure 3.2a) of different widths. Hydrogen atoms fixed in coal organic mass contribute to the wide component ($\Delta H_2 \approx 6$ Oe). Hydrogen atoms of fluid absorbed by coal form the narrow

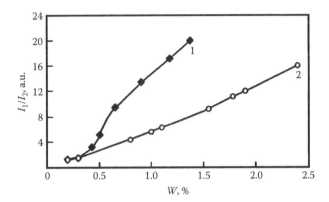

FIGURE 3.23
Dependence of intensity relation I_1/I_2 of NMR spectra lines on percent ratio of methane (1) and water (2) in anthracite.

component of the spectrum. As a rule they have high mobility. Some are almost in free condition in large pores and cracks; most are occluded on pores and crack surfaces. Hydrogen also fills in pores much smaller than the length of free passage of a methane or water molecule that restricts their mobility. The narrow width component of the spectrum ranges from $\Delta H_1 = 0.1$ to 0.5 Oe, sufficient for reliable recording.

In studies of sorbent processes in coal, the narrow component of the PMR spectrum, its parameters, and transformation in desorption are of interest [83,85,87]. While investigating desorption processes [83,84], two phases of methane of different mobilities were discovered [88]. Dried and degassed coal samples were cut into cylinders with diameters of 8 mm and lengths of 20 mm to obtain the best radiofrequency contour from NMR spectroscopy (radiofrequency contour is enclosed in a container of beryl bronze placed into a magnet gap of the spectrometer). Methane was pumped into the container at a pressure of 10.0 MPa. The sample in the methane atmosphere was kept for 10 days until it was completely saturated. Methane desorption was studied from the moment of pressure release. Initially, free methane located inside macropores and crack exits. Within hours, methane occluded in pores leaves. Desorption of a solid coal–methane solution lasts tens of hours (Chapter 2).

Figure 3.24 shows a narrow spectrum line of coal saturated by methane as well as its resolution into components. Based on research on narrow line components of transformations during methane desorption, we can draw conclusions about methane quantities at different phases and its redistribution between phases. A method to differentiate methane quantity and its phase state in coal via NMR is based on reconstruction of the function that determines the kinetics of gas emission [89]. Some methane in coal is practically in a free state. The diffusion of this methane ends just after carbon sample reactivation. These molecules are contained in pores of large diameter

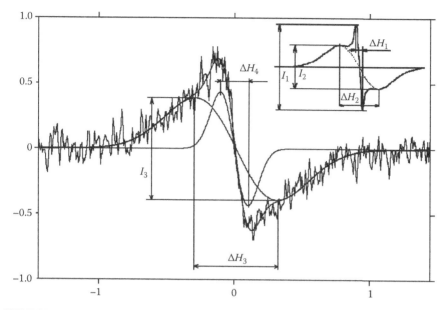

FIGURE 3.24
NMR spectrum of carbon–methane system. Inset shows resolution of narrow line of general
NMR spectrum into components.

and in surface-connected cleavages. This part of methane forms the narrowest line with $\Delta H_4 \sim$0.1 to 0.2 Oe (Figure 3.24).

Another part of methane, penetrated with saturated carbon substance and localized in the closed porosity, molecular pores, and intercrystalline spaces as a solid solution, shows considerably less mobility than in the gas phase while exceeding the mobility of the chemically bound organic compounds in carbon. This produces a well resolvable component in the spectrum ($\Delta H_3 \sim$0.6 to 1.03 Oe). At sufficient amplitude, these components are reliably divided for subsequent mathematical processing. Because the free and sorbed methane is localized and constant in the volumes of pores and cleavages, at each moment a certain quantity of methane leaves by filtration and is replaced as a result of solid-state diffusion. Therefore, it is possible to analyze the number of nuclei of hydrogen defining the area of a narrow line component with ΔH_4, a constant [4]. Subtraction from the total area of a narrow line of the area components with ΔH_4 (Figure 3.24 inset) yields the area of the line formed by nuclei of hydrogen in a firm carbon–methane solution (ΔH_3) and the amplitude I_3 and width of line ΔH_3 for methane in a firm solution. The dependence on time for coking coal ($V^G = 31.0\%$) is shown in Figure 3.25 [87].

Eventually the general intensity I_3 of a line decreases; thus its width ΔH_3 varies slightly. It follows that even after tens of hours from the start of decontamination, the methane remains in a kind of firm solution. The dependence of amplitude of signal I_3 on time is based on established ideas about the course of desorption of methane from coal.

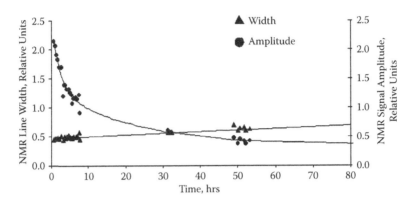

FIGURE 3.25
Dependence of width and amplitude of carbon–methane solution line on time. (From Alexeev A.D.., Shatalova G.E., Ulyanova E.V., Molchanov A.N., Pismenova N.E., Levchenko G.G. *Fizika i tekhnika vysokikh davleniy.* 13, 4, 100–106, 2003.)

We can estimate the content of a fluid and the parities of its various forms (phases) for coal from a given layer. It is possible to consider parameters of a wide line invariable (constant) and the area under a wide component of a spectrum (S_w) based only on the quantity of the sample [19]. We can use S_w from a spectrum of the dry decontaminated sample as a normalizing factor. For one sample of the series, it is necessary to define masses in the dried condition (m_d) and take corresponding spectra in the presence of a fluid (m_{fl}), then use the spectra to define the line areas.

The mass of a fluid in the sample $\Delta m = (m_{fl} - m_d)$; $\Delta m = K \cdot m_d \cdot (S_n/S_w)$, where S_n is the area under a narrow component of a spectrum and S_w is the area under a wide component. The factor for all samples (^1H NMR spectra of coals from the given layer) is $K = (\Delta m / m_d) \cdot (S_w/S_n)$. After calculating the factor, we can estimate quantity of a fluid in any sample of the series by the results of processing of spectra, that is, $\Delta m_i = K \cdot m_d \cdot (S_{ni}/S_{wi})$. Fluid in the investigated samples in more than one state (phase) constitutes the narrow part of the spectrum and consists of two well resolved components. It is possible to estimate the amount of fluid in each state. Assume that $S_n = S_{1n} + S_{2n}$ is the area of a narrow line of a spectrum and $\Delta m_{i1} = K \cdot m_d (S_{1di}/S_{wi})$, $\Delta m_{i2} = K \cdot m_d \cdot (S_{2ni}/S_{wi})$ where Δm_{i1} is mass of fluid in state 1 and Δm_{i2} is mass of fluid in state 2.

Studies of filtration and methane diffusion coefficients in coal were made on a stationary spectrometer of wide lines and by spin-echo impulse. NMR relaxation is the return of magnetization to the initial state after a Boltzmann distribution of population-oriented magnetic moments broken by the impact of a short pulse of electromagnetic radiation of Larmor frequency. Two main types characterize NMR relaxation processes: the dispersion of the scattered energy (T_1 – relaxation) and the loss of phase coherence in a group of spins, caused by local changes in the Larmor frequency (T_2 – relaxation).

T_1 spin–lattice (longitudinal) relaxation depends on the energy exchange between spin system and its environment characterized by the relaxation of the spin system to a thermal equilibrium state, at a certain temperature and magnetic field after excitation of Zeeman energy levels by radiofrequency pulse. The T_2 spin–spin (transverse) relaxation does not require energy transition; it depends on local (or intermolecular) changes in the magnetic field expressed in the change of Larmor frequencies of individual spins. The greater dispersion of Larmor frequency, the faster the decrease of phase coherence of spins precessing in a magnetic field.

The diffusion coefficient estimate was obtained from measuring T_2 during registration of the spin-echo signal from the nuclei of hydrogen, comprising the gas in the pore volume with micrometer and nanometer scale dimensions. The mobility of methane is determined by the interactions of molecules with pore surfaces. Closed porosity, like the diameters of gas molecules, significantly alters the relaxation times and filtration coefficients of self-diffusion of methane. A pulse spectrometer registers the integral signals of hydrogen nuclei in the coal and methane molecules in pore systems with complex size distributions. The dependence of the spin-echo signal on time is

$$I(\tau) = a\exp\left(-\frac{2\tau}{T_{2c}}\right) + b\exp\left(-\frac{2\tau}{T_{2m}}\right)\exp\left[-\frac{D}{3}(G_0\gamma)^2(2\tau)^3\right] \qquad (3.39)$$

The first component of T_{2c} is identified by the signal from the spins of hydrogen atoms in the structure of coal. The exponential factor of the second term with a characteristic relaxation time T_{2T} corresponds to the gas molecules sorbed on pore surfaces. The coefficient with cofactor D describes diffusion in the pore volume. The first component in a spin-echo signal allows us to calculate the change in methane concentration in the sample relative to the number of hydrogen atoms of carbon compounds in relative units. Indeed, it makes it possible to determine the value of $A(t) = b(t/a(t))$ at any time t of record of the magnetization relaxation signal after exposure of the radiofrequency pulse on the magnetic moments of ^1H nuclei.

In our measurements of T_2 by the two-pulse (90 to 180 degrees τ) spin-echo, the method proposed by Carr and Purcell was used [67]. The spin-echo spectrometer at the Institute for Physics of Mining Processes has a constant magnetic field gradient $G_0 = 0.143\ T/m$, resonant frequency $f = \gamma H = 20\ MHz$, and initial value of the delay between pulses $T_0 = 70\ msec$. Dried samples of coking, volatile, and anthracite coals of the Donets Basin (size fractions of 2 to 2.5 mm) were saturated with methane under the pressure of 9 MPa for a month. Relaxation curves were recorded for a few days to trace the desorption of methane from the samples. Losses of mass in the samples and volumes of methane released from samples in closed containers at indoor temperature were registered. Anthracite measurements were made on an autodyne NMR spectrometer of wide lines according to the previously described procedure.

Since the line width ΔH on a stationary NMR spectrometer and the T_2 spin–spin relaxation on a spin-echo spectrometer are connected by relation $1/T_2 = \gamma \Delta H$, $\gamma - 1/T_2 = \gamma \Delta H$ (γ = gyromagnetic ratio for the proton), we could compare the results. The maximum difference was 0.3%. Measurements of the mobility of methane molecules in pores of methane-saturated volatile coal samples were made. Parameter T_{2c} in the exponential dependence was determined at the final stage of desorption of methane and in further processing as $3.4 \cdot 10^{-5}$ sec. A satisfactory approximation of the relaxation of the spin-echo signal was obtained by assuming that the exponential cofactor containing T_2 is not in Equation (3.39). Figure 3.26 shows changes over time of the ratio of b to amplitude A in the NMR signals for the protons of coal substance [the ratio of amplitudes A = b/a in Equation (3.39)]. Changes over time of the ratio indicate changes in the number of resonating hydrogen atoms in a molecule of methane. It was possible to approximate the number by power (scaling) dependence of the form:

$$A = a_0 \cdot t^{-a_1} \tag{3.40}$$

where $A_1 = 1.186$ with correlation coefficient equal to 0.98 (Figure 3.26).

During methane release from pore spaces, the effective sizes of pores that contain most of the gas change. In the first phase of methane desorption, the filtration coefficient $D = 2.94 \cdot 10^{-6}$ m^2/sec. For the longest time interval (nine measurements), D was $(1.90 \cdot 0.41) \cdot 10^{-6}$ m^2/sec. In the restricted geometry of a pore space, transfer of molecules occurs in the regime of Knudsen diffusion if the pore size is less than the length of free path of the molecule. The formula for methane is

$$D_k = \frac{d}{3}\sqrt{\frac{8RT}{\pi M}} \tag{3.41}$$

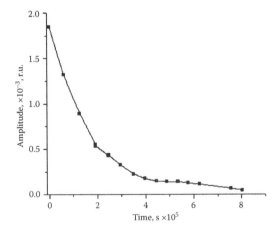

FIGURE 3.26

Changes over time of radio of amplitude b to amplitude a in NMR signals for protons.

where d is pore diameter. From Equation (3.41) we can determine the characteristic pore size at each stage of desorption of gas from a sample. The average speed of methane molecules in the circuit set of a spin-echo at 40°C is 643 m/sec, which yields diffusion coefficient pore diameters $d = 13.7$ and 8.9 nm. Note that the volatile coal has a high porosity ≈10% of the volume. This is mainly open porosity. The largest pores are clearly visible on the surface of an image from a scanning electron microscope (Figure 3.27) at the Institute for Physics and Engineering of the National Academy of Sciences of Ukraine in Donetsk.

At the final stage of desorption (Chapter 2), the basic processes are the output of methane from the blocks in the open porosity by solid-state diffusion and mass transfer to the transport channels via filtration. By analyzing the rate reduction of methane over time by amplitude of a spin-echo signal, we measured the effective solid-state diffusion coefficient D_{eff} for the first time. For anthracite at 313 K $D_{eff} = 3.8 \cdot 10^{-11} m^2/sec$.

Activation energy E is one of the most important parameters for characterizing the coal–methane state to determine the rate of the thermally activated process at a given temperature. Since its value depends on temperature, E can be determined by measuring the diffusion rates at different temperatures. The low temperature method for determining activation energy by recording the temperature dependence of NMR spectra of gas-saturated coal samples allows us to calculate the activation energy of molecular motion (a potential barrier for diffusion) [45] by Equation (3.13).

To calculate activation energy, it is necessary to know the diffusion coefficients at various temperatures. D_{eff} can be defined in two ways: (1) from the tangent of an angle of slope of the curve of decrease of the NMR signal amplitude over time and (2) the desorption characteristics of methane from coals of different fractions. For reliability, we measured the diffusion coefficients by both methods. Anthracite was selected for analysis because it has

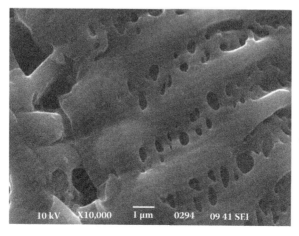

FIGURE 3.27
Image of surface of volatile coal with characteristic pore sizes.

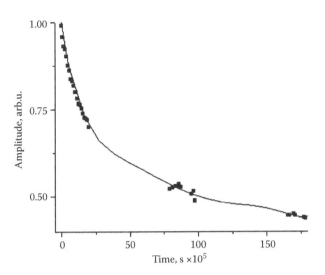

FIGURE 3.28
Changing amplitude of narrow line.

the most developed system of closed pores; thus desorption lasts for several days, sometimes weeks. Figure 3.28 shows changes over time and signal amplitude on a stationary spectrometer at 293 K.

The experiments were repeated on a fraction with particle sizes an order less than 0.25 to 0.5 mm (particle size of the previous group was 2.0 to 2.5 mm). Gas-saturated samples of all fractions were examined sequentially by NMR spectrometry. We recorded the narrow and wide lines of PMR spectra to determine the number of resonating nuclei of hydrogen protons in the structure of coal and methane. Since the proton content of a coal substance is constant, the results of the measurements were determined by the initial relative concentrations of the hydrogen protons of methane in the sample of each fraction by the formula:

$$W_{initial} = \frac{\Delta S_n}{\Delta S_w} \tag{3.42}$$

where ΔS_n, ΔS_w are areas of the narrow and wide lines of the NMR spectrum, respectively. After baseline determination of methane content in coal samples from the two fractions, the samples were kept for 6 h, after which we determined the residual amount of methane $W(t)$ by Equation (3.31). The diffusion coefficient was defined as the average number of protons of methane in three samples for each fraction. The diffusion coefficient of methane was calculated by

$$D = \frac{R_2^2 \ln Q_1 - R_1^2 \ln Q_2}{6t \ln \dfrac{Q_1}{Q_2}}, \tag{3.43}$$

where R_1 and R_2 are sizes of coal fractions (m); t is desorption time (sec); and Q_1 and Q_2 are the relative changes in methane quantity during the desorption corresponding to each fraction:

$$Q_{1,2} = \frac{W_{cp.initial} - W_{cp}.(t)}{W_{cp.initial}}.$$

Calculations by two methods yielded $D_{eff} = 1.75 \cdot 10^{-11}$ m²/sec in close agreement. Since we obtained diffusion coefficients obtained at different temperatures for a single anthracite sample, we can determine activation energy:

$$E = 2,38R\frac{T_1 T_2}{T_1 - T_2}\lg\frac{D_1}{D_2}, \qquad (3.44)$$

where R is a universal gas constant. Based on T_1 = 293 K (stationary NMR) and T_2 = 313 K (spin-echo), E = 29.55 kJ/mol. For comparison, we note that the activation energy of the structural adjustment chains of anthracite coal substance at the change of the energy potential of the molecular structure is 250 kJ /mole. With NMR, we were able to determine the coefficient of Knudsen diffusion in the transport channels and solid-state diffusion coefficient in the organic matter. Application of x-ray analysis with NMR (on the same coals) showed changes in the structure of methane-saturated coal compared with dried coal [87]: increased ordering of the carbon structures of methane-saturated coal and modified x-ray parameters. Gradually, with desorption of methane, the x-ray parameters returned to the originals. This confirms that methane is introduced into a coal structure and changes it and these changes are reversible.

3.4 Phase State of Binding in Rocks

Binding samples of coals and rocks to analyze their behavior at true triaxial compression unit (TTCU) loading is used to prevent sudden outbursts of coal and gas and improve the sustainability of mine workings.

3.4.1 Polymer Compositions in Coals

Physical and chemical measures can prevent sudden outbursts of coal and natural gas during the development of workings, in stopes on cragged outburst-dangerous coal seams characterized by low strength and large differences in strengths of individual coal patches in a seam, and in zones of geological faults, and surrounding areas. Impact is measured after the injection of a polymer solution into a coal seam that transforms into a solid state over time under the influence of a curing agent that alters the mechanical and filtration properties of the massif [90].

The injection of liquid polymer causes a piston displacement of free methane and partial desorption of solution on the surface of the coal. After saturation of the treated area, the reservoir solution hardens, exhibits low gas permeability, and overrides micropore and filtration volumes. On contact, a coal–solid polymer adhesive bond forms and holds the polymer in the occupied volume.

Intentionally changing the natural properties of coals in an array by physical and chemical methods can minimize hazardous conditions:

- Gas dynamics: increasing the plasticity and coal particle bonds and reducing the rate of filtration and diffusion of gas
- Gas emission: blocking methane in the porous spaces of coal
- Dust and dust emissions: use the higher plasticity and adhesion of polymers for control.

For physical and mechanical effects on coal strata, a 24% aqueous solution of urea formaldehyde (carbamide) resin (KM-2) was used along with a hardener of 20% NH_4Cl. One condition for qualitative saturation with KM-2 is the correct choice of the transition time of the solution from liquid to solid. This time depends on the amount of hardener, pore volume filling speed, and quality of the KM-2.

Monitoring the status and dynamics of the permeability of the KM-2 polymer in the seam was carried out by NMR because the technique reveals full information about the phase state of the polymer. The studies were conducted at different concentrations of hardener and under various thermodynamic conditions. The NMR spectrum of liquid urea formaldehyde resin solution is a narrow line of width $\Delta H_1 = 0.22$ Oe (Figure 3.29).

During curing (polymerization), line broadening ΔH_1 to 0.4 Oe and the emergence of a broad line ($\Delta H_3 = 12.1$ Oe) with small intensity were observed. Figure 3.30 shows the NMR spectra of solid polymer (a) and of coal–solid polymer systems (b). The spectrum of the polymer consists of two components with a line width $\Delta H_1 = 0.4$ Oe and $\Delta H_3 = 12.1$ Oe. The spectrum of coal represented by one line was $\Delta H_2 = 5.5$ Oe.

Polymerization times may differ based on the amount of curing agent added to the aqueous KM-2. In NMR spectra, the start (after increasing the width of the narrow ΔH_1 line) and the end of polymerization (ΔH_1 reaching 0.4 Oe and appearance of a wide line) can be determined. During polymerization, resin KM-2 is a solid and its low strength is easy to destroy. It is therefore important to know when polymerization is finished. NMR spectra precisely define the beginning and end of polymerization. Figure 3.31 shows the dependence of polymerization time on the amount of hardener. This serves as a basis to determine the amount of curing agent required for any specific time of polymerization.

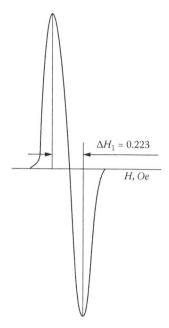

FIGURE 3.29
NMR spectrum of resin of 24% aqueous solution of urea–formaldehyde.

When analyzing the temperature dependence of the line width of NMR spectra of liquid and solid polymers, specific patterns appear (Figures 3.32 and 3.33). For a solid polymer (at $T = 24°C$) $\Delta H_1 = 0.4$ Oe and $\Delta H_3 = 12.1$ Oe (Figure 3.32). At 180 to 190°C, the polymer is destroyed. As the temperature decreases, the narrow component of the spectrum broadens to $\Delta H_1 = 5$ Oe and disappears after $T = -100°C$. The broad component ΔH_3 reaches 15 Oe.

Changes to the liquid polymer are somewhat different from changes to the solid. At $T = 24°C$, the spectrum consists of only one narrow component at $\Delta H_1 = 0.22$ Oe. At $-10°C$, a wide component $\Delta H_2 = 13$ Oe broadens with decreasing temperature to 16 Oe. The narrow component of the spectrum, up to 5 Oe, disappears after -90°C. Higher values of ΔH_1 at low temperatures in liquid solution are connected with the peculiarities of the crystallization of the adsorbed and free water.

The second moments $[\Delta H]^2$ of liquid and solid polymers differ only at indoor temperatures: $[\Delta H_1]^2 = 1$ Oe^2 and $[\Delta H_{sol}]^2 = 13.5$ Oe^2. As temperature decreases, the difference between the second moments of liquid and solid polymers becomes less noticeable. Experiments were carried out to define the influence of pressure on penetrating power in coal pore spaces. Figure 3.34 shows the dependence of the penetrating power of polymer on pressure. The filling of pore spaces of coal samples of 5 cm^3 with polymer was controlled by NMR.

The differences in forms of NMR spectra and their temperature dependence may be used to determine the amounts and phase conditions of a

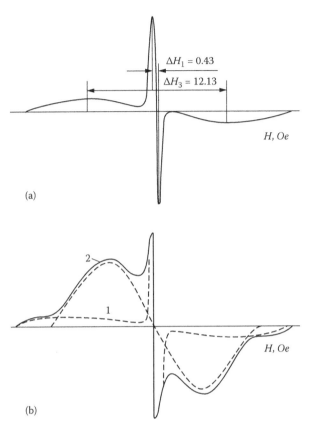

$\Delta H_1 = 0.43$

$\Delta H_3 = 12.13$

H, Oe

(a)

2

1

H, Oe

(b)

FIGURE 3.30
NMR spectra. 1. Solid polymer. 2. Coal.

polymer solution input into a coal seam. According to the intensity of the subintegral area of the narrow component of NMR spectra, the quantitative distribution of injected solution to the volume of a layer is defined.

The termination of hardening is controlled by the occurrence in a spectrum of a wide component $\Delta H_2 = 12.1$ Oe at the width of a narrow component of spectrum $\Delta H_1 = 0.4$ Oe. At a low concentration of polymer in a unit of coal volume, the termination of hardening is controlled by the increase of the value of the second moment of the NMR spectrum of the coal–polymer system.

To prove the efficiency of NMR for controlling quantity and phase condition of polymer in a seam, a water solution of 24% urea formaldehyde resin with chloride and ammonium hardener was injected into a coking coal massif at the Kalinina Mine (Artemugol) hardening was expected within 11 days. The processing of the seam by the polymer and the control of its phase condition by NMR allowed undisturbed operation of the face needed for additional actions to control gas dynamic phenomena.

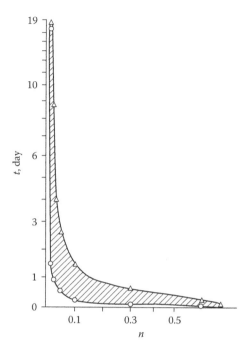

FIGURE 3.31
Dependence of gelation time t (polymerization) for 24% aqueous solution of urea formaldehyde resin with 20% ammonium chloride. Circle = start of polymerization. Triangle = end of polymerization. n = number of hardener parts per 10 parts of resin solution.

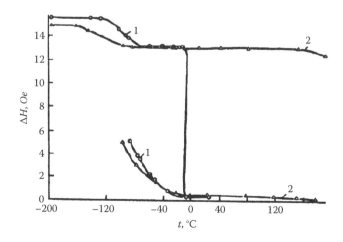

FIGURE 3.32
Dependence of line widths ΔH_1 and ΔH_2 of NMR spectra on temperature t in 24% aqueous solution of urea formaldehyde resin (1) and hardened polymer (2).

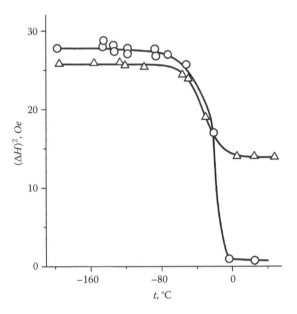

FIGURE 3.33
Dependence of the second moment of NMR $[\Delta H]^2$ spectra on temperature t in liquid (1) and solid (2) urea formaldehyde resin.

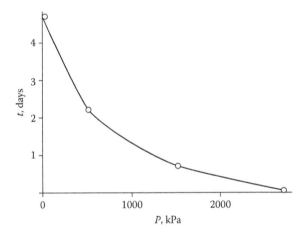

FIGURE 3.34
Influence of pressure P on speed of penetration of liquid polymer in fossil coal.

3.4.2 Binding in Rocks

The number of the complex mechanized methods for clearing faces in coal mining continues to grow. However, the coal extraction from faces has decreased slightly due to the transition to greater depths and complications arising from mining and geological conditions. As rock pressure increases, the strong top covers of coal seams lose stability. Coal extraction is accompanied by roof collapse. The coal inrush leads to down time of equipment, increased labor, decline of productivity of work and quality of coal, deterioration of conditions, and compromised safety.

In difficult mining and geological conditions, the cost of implementing and maintaining preparatory mine workings rises sharply and labor productivity decreases. Labor input represents 50% of the general expenses of sinking workings. The support and maintenance of preparatory and permanent mine workings create "bottle necks" during the tasks of sinking shafts and tunneling. About 15 to 20% of underground workers are involved in maintaining and repairing mine workings. One important way to increase the stability of top covers of coal seams and mine workings is hardening of rocks, usually via chemical or combined physical and chemical methods such as injections of chemicals and cementation.

Outside Ukraine, bonding solutions such as polyurethane, magnesia, and anhydrides (Germany), AM-9 (U.S., Canada, Great Britain), and sumisoil (Japan) are widely used. In domestic practice, magnesia solutions and solutions based on carbamide resins are mainly used. To harden unstable rocks, phenol formaldehyde, magnesia compositions, and resin TSD-9 polyurethane are used.

To develop effective methods for harden mining rocks, it was necessary to develop criteria to estimate the phase conditions of bonding solutions and their efficiencies. It is important to know the properties of bonding solutions and rocks to which they will fasten to harden effectively.

Magnesia components belong to a class of inorganic glues and cements. They are based on interactions of magnesium oxide with solutions of chloride or sulfate magnesium. Magnesia cement represents the connection of $MgCl_2 \cdot 7MgO \cdot 14H_2O$. The components of magnesia solution are nontoxic and incombustible and are obtainable at low cost. Crack widths, humidity, and dust contents of processed surfaces do not influence bonding capacities of these solutions. Their high mobility and plasticity allows them to be pumped over several hundreds of meters. The hardening process begins in 1 h; the strength of uniaxial compression $\sigma_{comp} = 10$ MPA is reached within 3 h.

The main disadvantage of the magnesia solution is fragility. To improve their plastic and adhesive properties, polymers and softeners may be added. Examples are polyvinyl acetate (PVA) emulsion and butadiene–styrene latex. These materials exerted a positive influence on the rheological properties of magnesia solutions and reduce the time of stiffening and thickening. Five solution mixtures were analyzed:

Solution I $MgCl_2$ 25 ml, PMK-83 40 g

Solution II $MgCl_2$ 25 ml, PMK-83 40 g (PVA 7 ml)

Solution III $MgCl_2$ 25 ml, PMK-83 40 g (latex 10 ml)

Solution IV   Solution II plus rocks (aleurolite with sandstone)

Solution V   Solution III plus rocks (aleurolite with sandstone)

PMK-83 is caustic magnesite powder (80% MgO, 20% CAO, SiO_2, FEO_3, SO_3 ...). All prepared solutions were analyzed on NMR equipment daily for two weeks.

Figure 3.35 illustrates the NMR spectrum of a solid magnesia solution. The spectrum consists of two lines: a wide line with width ΔH_2 and a narrow line with width ΔH_1. The hardened composition reveals two groups of hydrogen atoms. A wide line (ΔH_2) is formed by the atoms rigidly connected with the crystal composition of the substance; a narrow line (ΔH_1) is formed by mobile atoms. The wide line appeared on the second day after solution preparation and its width changed little during hardening. Only the narrow line widened. Therefore only the narrow line registered changes. Table 3.5 presents the data that lead to the following conclusions.

Solutions I and III behave approximately equally. Before the eighth day, a gradual increase of ΔH_1 appeared and the spectrum is represented by one line. On the eighth day, the spectrum split into two lines representing two groups of atoms with different mobilities. On the 11th day, the NMR spectrum again consisted of one line that increased for a time. The definitive widths of ΔH_1 for solutions I and III were 0.73 and 0.66 Oe, respectively (Table 3.5), indicating that linking of atoms was stronger in solution I than in III.

Solution II behaved a little differently. The spectrum revealed a second line on the eighth day, but it disappeared the next day. It is obvious that hardening in solution II occurred faster and ended earlier (on the 10th day). Based on final values of ΔH_1 for all solutions, protons in solutions I and II were most rigidly fixed and the least rigid protons were in solution III, proving that solution III is more plastic.

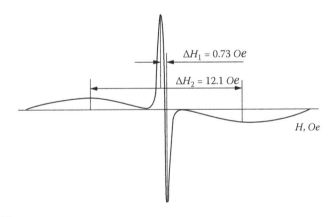

FIGURE 3.35
NMR spectrum of solid magnesia solution. ΔH_2 = width of line of rigidly connected hydrogen atoms. ΔH_1 = width of line of mobile atoms.

TABLE 3.5

Dependence of NMR ΔH (Oe) Line Width and Spectra of Magnesia
Solution on Time

Day	Solution I	Solution II	Solution III	Solution IV	Solution V
1	0.26	0.25	0.27	0.36; 0.59	0.25
2	0.39	0.29	0.42	0.36; 0.58	0.38; 0.69
3	0.54	0.31	0.45	0.43; 0.6	0.4; 0.68
4	0.57	0.35	0.49	0.5; 0.66	0.41; 0.7
5	0.62	0.36	0.5	0.36; 0.78	0.6
8	0.58; 0.82	0.51; 0.71	0.51; 0.73	0.69	0.7
9	0.56; 0.71	0.64	0.51; 0.72	0.69	0.72
10	0.54; 0.78	0.72	0.51; 0.71	0.74	0.79
11	0.71	0.72	0.65	0.74	0.79
12	0.73	0.72	0.66	0.75	0.79
13	0.73	—	0.66	—	—

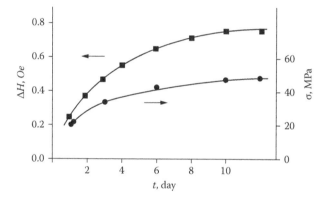

FIGURE 3.36
Dependences of ultimate strength σ and ΔH width lines of NMR spectra of magnesia solution on hardening time t.

Hardening in mixtures of magnesia solutions with rocks is somewhat different. The double line appeared on the first day for solution IV and on the second day for solution V. On the sixth and seventh days, the spectra of magnesia compositions with rocks consisted of one line that continued widening for several days. On the 11th day, the line width stabilized. ΔH_1 was 0.74 Oe for solution IV and 0.79 Oe for solution V. The linking of protons with a lattice is stronger for solution V.

Simultaneously with controlling the phase conditions of magnesia solutions under unequal compression (Chapter 1), durability testing was carried out. Figure 3.36 illustrates the dependence of line width and ultimate strength under uniaxial compression on the time of solution hardening, sample strength, and spectrum line width for 10 days. Further increases of strength and line width were not observed. Thus, the stabilization of line width confirms the maximum strength of magnesia solution.

Because solution III revealed the best durability characteristics, we focused further studies on it. The influence of temperature (5, 10, and 24°C) on magnesia composition hardening, humidity (72 and 98%) of the surroundings, and humidity of the rocks (0.6, 1.5, and 2%) was studied. At room temperature, hardening of solution III occurred faster and terminated earlier. The maximum size of spectrum line width was 0.78 Oe. At 10°C, the stabilization of spectrum line width occurred on the 12th day and was 0.66 Oe; at 5°C, the maximum value ΔH was reached on the 15th day and was 0.5 Oe. With a decrease of temperature, the mobility of protons increases and the plastic properties of solution improve but the strength decreases a little. At constant temperature and different humidity of hardened rocks, we observed that the higher the rock humidity, the smaller the widths of lines. Thus the plastic properties are better but the time of stabilization ΔH increases. At a surrounding humidity of 98%, hardening occurs more slowly for all solutions and at different temperatures (Table 3.6). Thus the plastic properties of the researched systems improved.

This research provided criteria for controlling phase conditions of magnesia solutions in seams. Over time, the strength increases in samples in a hardened magnesia solution. For pure magnesia solutions and magnesia solutions containing PVA, the division of NMR spectra into two components in five days indicates a decrease of the elasticity module.

Of these three compositions, the most plastic is the magnesia solution with latex addition. The width of a spectra line is the smallest, confirming the lower rigidity of bonds. Since pure magnesia solution is most brittle at the greatest strength, it is recommended for use with additions of polyvinyl acetate and latex. The seams hardened by these solutions can be worked

TABLE 3.6

Dependence of Line Width ΔH (Oe) of Magnesia Solutions on Time t and Surrounding Humidity W_{sur}

	ΔH (Oe)				
Day	Solution I	Solution II	Solution III	Solution IV	Solution V
1	0.2/0.16	0.2/0.14	0.19/0.15	0.17/0.15	0.16/0.13
2	0.32/0.21	0.21/0.16	0.26/0.19	0.25/0.2	0.21/0.18
3	0.36/0.25	0.27/0.21	0.33/0.21	0.31/0.25	0.25/0.21
4	0.4/0.27	0.35/0.24	0.4/0.23	0.32/0.30	0.36/0.27
5	0.42/0.29	0.37/0.26	0.45/0.26	0.42/0.32	0.4/0.30
6	0.44/0.3	0.4/0.27	0.47/0.29	0.45/0.36	0.42/0.31
10	0.47/0.3	0.45/0.3	0.55/0.32	0.53/0.42	0.47/0.36
15	0.56/0.31	0.5/0.32	0.55/0.34	0.58/0.45	0.5/0.36
20	0.6/0.31	0.55/0.33	0.55/0.34	0.6/0.45	0.55/0.36

Note: Numerator = 72% humidity. Denominator = 98% humidity.

in 10 days when they completely reach a solid condition, as confirmed by NMR spectra.

The phase conditions of urea–formaldehyde (UF) resins were studied. These resins are applied along with magnesia solutions to harden rocks. UFs contain a wide range of low cost raw materials and are economical, flame-proof, and explosion-proof. Another benefit is the water solubility of uncured UF resins that allows easy preparation of solutions from mine water and simple equipment washing after testing.

Carbamide solution consists of (1) UF resin, (2) PVA, and (3) molysite $FeCl_3$ (hardener). The proportion of components is 5 to 0.75 to 1. Careful mixing of the components yields a homogeneous yellowish, low viscosity liquid. Thickening time depends on the quality of mixing and usually occurs within 10 to 20 minutes—long enough to allow injection of the solution.

UF resin in carbamide contains the $R-NH_2$ amino group and R-CH (formaldehyde). The condensation of resins occurs during the reaction of poly-condensation under the influence of acid solutions. The subdivided and cross-linked chains of carbamide resin are formed:

$$NH \rightarrow CO \rightarrow N \rightarrow CH_2 \rightarrow NH \rightarrow CO \rightarrow N \rightarrow CH_2 \rightarrow CH_2OH \quad (3.45)$$

Chain monomers (elementary molecules) are linked by strong chemical bonds 0.1 to 0.15 nm long but between the chains there are far weaker inter-molecular bonds 0.3 to 0.4 nm long. Due to its chain structure, a polymeric molecule is capable of flexible deformation. A long, repeatedly bent molecule is capable of straightening by means of chain flexibility without cleavage.

Let us consider the process of resin hardening. To a prepared mixture of resin UF and PVA, the $FECl_3$ hardener is added and carefully mixed for 5 min. The liquid resin initially is easily soluble in water. After condensation starts, the resin becomes a solid consisting of a friable elastic gelatinous material that thickens over time. The phase condition of resin is controlled according to NMR spectra (Table 3.7).

As the UF resin hardens, the durability of the solution and increases and its plastic properties decrease. To establish the correlation of phase condition of a UF resin and its strength and deformation properties, a parallel study by NMR method focused on the effects of the strength properties of resin on the true triaxial compression unit (TTCU, see Chapter 4, Section 4.2). To learn about the behaviors of rocks and hardening resins, UF was subjected to unequal compression levels $\sigma_1 > \sigma_2 > \sigma_3$ corresponding to real conditions in a massif. The tests were conducted at minimum compression stress $\sigma_3 = 0.5$, 10, and 20 MPA, modeling the gradual sinking into a roof.

The changes of phase condition of the resin UF were noted On the first and second days after the moment of resin hardening, the lines simply consisted of a single resonant line with intensity I_1 (Figure 3.37); width was minimum, indicating the high mobility of molecules. This is explained by cross-linking of small linear molecules of the uncured resin into long linear chains at the

TABLE 3.7

Form and Width Changes of UF NMR Spectra Lines over Time

	Day 1	Day 2	Day 3	Day 4	Day 5	Day 6	Day 7
ΔH_1, Oe	0.09	0.15	0.2	0.29	0.3	0.34	0.34
ΔH_2, Oe	–	–	0.75	0.52	0.55	0.57	0.57

	Day 8	Day 9	Day 10	Day 11	Day 13	Day 20
ΔH_1, Oe	0.37	0.43	0.44	0.46	0.49	0.67
ΔH_2, Oe	0.6	0.6	–	–	–	–

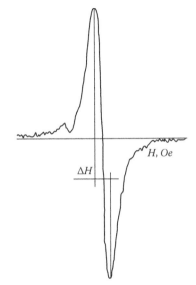

FIGURE 3.37
NMR spectrum of carbamide solution at t = 2 days.

start of hardening. The studies of TTCU demonstrated that cured resin is characterized by high plasticity (deformation up to 15%) and low, but gradually increasing strength. On the third day, a second, wider line of small intensity I_2 appeared (Figure 3.38). This indicates that the formation of cross-linking between linear chains begins in the resin and the resin represents a two-phase system in which each phase has a different mobility. The strength reaches its peak at this time and the resin is plastic enough (deformation to 8%). The relation of intensities I_2/I_1 at this moment is 0.3. By the beginning of the fourth day, a fast increase of intensity produces a wide line ΔH_2 due to the formation of a considerable number of cross-section linkings. The quantity of molecules in the mobile phase sharply decreases. The strength of the compound returns to its level before maximum but deformations decrease up to 2%, that is, embrittlement takes place.

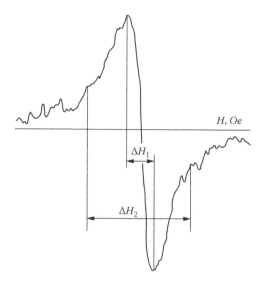

FIGURE 3.38
NMR spectrum of carbamide solution at t = 3 days. The I_2/I_1 ratio at this moment is 0.8 (Figure 3.39).

Another issue is the change of phase of a UF resin at the point of contact with rock. The phase condition of carbamide is greatly influenced by the rock and its surroundings. Sandstone rock was crushed into 1- to 2-mm particles and added to a liquid solution of UF resin and hardener. The hardened mixture was divided into 20 samples, half of which were placed into a desiccator with humidity W = 100%; the remainder were left in the open air (W = 60%). Each sample was analyzed twice in a NMR spectrometer. Samples from the surface ("a") and from the middle ("b") were taken. An hour after the beginning of hardening, a very complex and identical spectrum consisting of three components (ΔH_1 = 0.17 Oe, ΔH_2 = 0.28 Oe, ΔH_3 = 0.5 Oe) was produced for each part of the sample. The third component is the result of carbamide adsorption on the rock surface. The carbamide adsorption on the rock surface led to faster loss of atom mobility (Table 3.8). The humidity in the middle of the sample slows adsorption, as shown in the table.

Let us consider the change of the spectrum form and line width for W = 60%. In case "a" (samples from rock surface), the polymerization process was similar to the process with the pure resin. Two days later, the intensity of a wide line (ΔH_3 = 0.79 Oe) in comparison with the intensity of a narrow line (ΔH_2 = 0.24 Oe) was insignificant. This proves most molecules were still very mobile. The reduction of molecule quantity and increase of mobility (ΔH_2 = 0.29 Oe) started three days later for "a" samples as we can see from their spectrum. The width of line ΔH_2 = 0.29 Oe increased in comparison with indications for the second day ΔH_2 = 0.24 Oe. This means that the group of molecules that yield lines with width ΔH_2 undergo a gradual transition

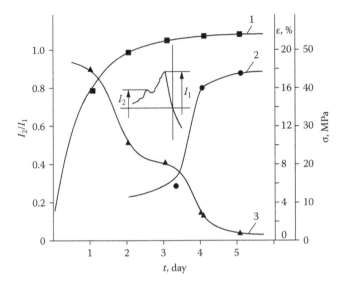

FIGURE 3.39
Dependences of ultimate compression strength σ (1), relative deformation ε (2), and relation of intensity of NMR spectra lines I_2/I_1 (3) of rock–resin UF system on hardening time t.

TABLE 3.8

Line Form and Width Changes (Oe) of NMR Spectra of
Urea–Formaldehyde Resin Caused by Time and Humidity

Day	Place of Sampling	Humidity 60%			Humidity 100%		
		ΔH_1	ΔH_2	ΔH_3	ΔH_1	ΔH_2	ΔH_3
0.05	a, b	0.17	0.28	0.5	0.17	0.28	0.5
1	a	–	0.22	0.52	0.17	0.22	0.39
1	b	0.17	0.27	0.54	–	–	–
2	a	–	0.24	0.79	–	–	–
2	b	0.18	0.27	0.54	0.18	0.26	0.63
3	a	–	0.29	0.53	–	–	–
3	b	0.15	0.26	0.53	0.22	0.32	0.54
6	a	–	0.32	0.54	–	–	–
6	b	0.16	0.32	0.54	0.2	0.3	0.54
7	a	–	0.32	0.53	0.19	0.3	0.54
7	b	–	0.32	0.54	–	–	–
8	a, b	–	0.33	0.62	0.15	0.29	0.53
9	a, b	–	0.34	0.63	0.13	0.28	0.53

Note: a = surface of sample. b = middle of sample

to a more bound state. Seven days later, the intensities of the two lines for samples from the surface are almost identical.

The hardening process for "b" samples was different. The complex three-component line with widths ΔH_1, ΔH_2, and ΔH_3 remained for six days. On the seventh day, polymerization leveled. A two-phase system existed both in the middle and on the surface of the resin. Although the NMR spectra for "a" and "b" also consisted of two lines (ΔH_1, ΔH_3), their intensity was not equal. For "b" the intensity of lines with ΔH_2 is considerably higher than the intensity for the same line for "a" samples. This means that the cross-linking of linear chains on the surface is almost finished. In the middle, the substance is mobile, i.e., only partial linking of chains occurred. After the seventh day the forms and widths of lines of all parts of the samples were identical.

Such non-uniform polymerization for "a" and "b" occurs because the top layers of the substance slow water separation from the middle during cross-linking of polymeric chains. The narrow line with width ΔH_1 proves that during polymerization, an accumulation of water for several days in the substances in the adsorbed condition slowed the further cross-linking of polymeric chains.

The spectra received from processing samples from a damp atmosphere show that water from polymerization that remains in the samples considerably slows the process of chain cross-linking, and molecules in resin are longer in the mobile state (Table 3.8).

Polymerization in the middle and on the surface of samples takes place equally. During hardening, the intensity of the resonant line with width ΔH_1 decreases and ΔH_1 increases (from 0.15 to 0.22). However, after eight days, the width of ΔH_1 again decreases (Table 3.8) because in a polymer in a damp environment, water desorption and adsorption take place simultaneously. Also the intensity of the ΔH_2 line decreases, but much more slowly, than in the case of hardening of mixture at 60% humidity. Thus, polymer hardening depends mainly on humidity.

Optimum conditions for working out the layers processed by the resin appear on the second day when the relation of intensity I_2/I_1 varies from 0.3 to 0.8. With the increase of rock and surrounding humidity, resin hardening slows (the resin is plastic but not sufficiently strong). The optimum periods may be determined based on changes of the intensity relation I_2/I_1.

3.5 Conclusions

NMR methods are used widely to study the structures and characteristics of fossil coals and rock massifs. Modern highly sensitive spectrometers allow precise qualitative and quantitative analyses of coal and pore structures, methane and water sorption, phase conditions of methane in coal during coal extractions from seams, and redistribution of methane between phases during desorption. NMR is also useful for analyzing the diffusion and filtration processes within coal pore spaces. NMR can be successfully combined with other techniques based on temperature and baric concepts and yields precise information about the changes of coal material states and fluid intrusion arising from the influences of external factors.

In practice, NMR is used to control humidity of coal seams, determine outburst dangers, and also assess the effectiveness of special solutions and water injected into coal seams to prevent outbursts. Progress in instrument engineering led to the availability of portable NMR equipment now used to analyze coal and rock characteristics directly at production sites, in laboratories, and even in mine workings. Portable equipment reduces analysis time, increases the accuracy and adequacy of results, and helps mine operators make correct conclusions about coal conditions in seam and the efficiency of anti-outburst measures. The importance of NMR for coal and rock research will continue to increase.

References

1. Synkov V.G. 1999. Design procedure of two-layer autofrettaged high pressure chambers. *Fiz. tekhn. vysok. davl.* 9, 46–50.
2. Alexeev A.D., Pestryakov B.V., Galat V.F., Bildinov K.N. 1973. Magnetic resonance in liquids under high pressure. In: *Abstr. 1st All-Union Workshop on Phys. and Tech. High Pressures*, Donetsk, Sept. 5–7, 1973, p. 148.
3. Synkov V.G. 2004. A set-up for x0 and NMR studies of methane–coal system under high pressure. *Vesti donet. gornh. inst.* 1, 95–97.
4. Alexeev A.D., Kovriga N.N., Molchanov A.N. et al. 2003. NMR-studies on kinetics of structural changes in coal substance under high pressure. *Fiz. tekhn. vysok. davl.* 13, 83–90.
5. Endru E. 1957. *Nuclear Magnetic Resonance.* Moscow: Foreign Literature Publishers.
6. Alexeev A.D., Zavrazin V.V., Melakov A.D. et al. 2002. Approximation of H^1 NMR spectra of coals. *Fiz. tekhn. vysok. davl.* 12, 71–78.
7. Franck E.U. and Wiegand G. 1995. High pressure hydrothermal combustion. In *Proc. Of Joint 15th AIRAPT and 33rd EHPRG Int. Conf.* Warsaw: World Scientific, pp. 809–814.
8. Gagarin S.G. and Skripchenko G.B. 1986. Modern conceptions of chemical structure of coals. *Khim. tverd. topl.* 3, 3–14.
9. Richards R.E. and Yorke R.W. 1960. Hydrogen resonance spectra at low temperatures of pure hydrocarbons and of selected coal samples. *J. Chem. Soc.* 6, 2489–2497.
10. Van Krevelen D.W. 1961. *Coal: Typology, Chemistry, Physics, Constitution.* Amsterdam: Elsevier.

11. Van Vleck J. H. 1948. The dipolar broadening of magnetic resonance lines in crystals. *Phys. Rev.* 74, 1168–1183.
12. Pestryakov B.V. 1982. Distribution of hydrocarbon aromatic and aliphatic groups in Donbass coals. *Khim. tverd. topl.* 4. 8.
13. Pestryakov B.V. 1986. Abstract of dissertation for a competition of candidate science degree in physics and mathematics. Krasnoyarsk: Siberian Institute for Technology.
14. Deno N.C., Curry K.W., Jones A.D. et al. 1981. Linear alkane chains in coals. *Fuel* 60, 210–212.
15. Ohtsuka Y., Nozawa T., Tomita A. et al. 1984. Application of high-field, high-resolution ^{13}C CP/MAS NMR spectroscopy to the structural analysis of Yallourn coal. *Fuel* 63, 1363–1366.
16. Neviran R.N., Davenport S.J, and Meinhold R.H. 1984. On assessment of carbon-13 solid state NMR spectroscopy for characterization of N. Zealand coals. *Chem. Div. Dep. Sci. Res. Rept.* 1, 1–20.
17. Alexeev A.D., Serebrova N.N., Ulyanova E.V. et al. 1987. High resolution ^{13}C NMR spectra of fossil coals. *Dokl. akad. nauk. USSR* 3, 3–5.
18. Alexeev A.D., Serebrova N.N., and Ulyanova E.V. 1989. Investigation of methane sorption in fossil coals by ^{1}H and C^{13} NMR high-resolution spectra. *Dokl. akad. nauk.* USSR B 9, 25–28.
19. Alexeev A.D., Zaidenvarg V.E., Sinolitskiy V.V. et al. 1992. *Radiofizika vugolnoy promyshlennosti*. Moscow: Nedra.
20. Tager A.A. 1978. *Physics and Chemistry of Polymers*. Moscow: Khimiya.
21. Frolkov G.D. et al. 1988. Structural and chemical features of organic substance of coals from sudden coal and gas outburst in hazardous areas. *Khim. tverd. topl.* 1, 9–15.
22. Gagarin S.G., Eremin I.V, and Lisurenko A.V. 1997. Structural and chemical aspects of broken fossil coals from outburst in hazardous seams. *Khim. tverd. dopl.* 3, 3–13.
23. Gibbs J.W. 1982. *Thermodynamics: Statistical Mechanics*. Moscow: Nauka.
24. Randles J. E. 1977. Physical structure at the free surface of water and aqueous electrolyte solutions. *Phys. Chem. Liquids* 7, 107–179.
25. Clifford J. 1975. Properties of water in capillaries and thin films. In *Water: A Comprehensive Treatise*, Franks F., Ed. Vol. 5. New York: Plenum, pp. 75–132.
26. Dubinin M.M. 1974. On the problem of surface and porosity of adsorbents. *Izvest. akad. nauk SSSR khim.* 5, 996–1011.
27. Croxton C. 1980. *Statistical Mechanics of the Liquid Surface*. Chichester: John Wiley & Sons.
28. Henniker J.C. 1949. The depth of the surface zone of a liquid. *Rev. Mod. Phys.* 21, 332–341.
29. Etzler F.M. and Drost-Hansen W. 1976. *Colloid and Interphase Science*, Kerker M., Ed. Vol. 8. New York: Academic Press.
30. Mank V.V. and Lebovka N.I. 1984. NMR study of structural and dynamic properties of water boundary. In *Physico-Chemical Mechanics and Lyophilicity*, Ovcharenko F.D., Ed. Kiev: Naukova Dumka, pp. 38–45.
31. Rydil E. 1971. *Development of Concepts in the Field of Catalysis*. Moscow: Mir.
32. De Boer J.H. 1962. *Dynamical Character of Adsorption*. Moscow: Foreign Literature Publishers; 1968. *Dynamic Character of Adsorption*. Oxford: Clarendon Press.

33. Scholten I.I. and Kreier S. 1973. *Structure and Properties of Adsorbents and Catalysts.* Moscow: Mir.
34. Frenkel Y. I. 1975. *Kinetic Theory of Liquids.* Leningrad: Nauka.
35. De Boer J.H. 1959. *Catalysis: Some Questions of the Theory and Technology of Organic Reactions.* Moscow: Foreign Literature Publishers.
36. Everett D.I. 1957. Some developments in the study of physical adsorption. *Proc. Chem. Soc.* 2, 38–53.
37. Gabuda S.P. and Rzhavin A.F. 1978. *Nuclear Magnetic Resonance in Crystalline Hydrates and Hydrated Proteins.* Moscow: Nauka.
38. Mank V.V. 1978. Molecular NMR spectra features in heterogeneous systems. *Ukr. khim. z.* 43, 911–918.
39. Alexeev A.D., Morozenko E.V., and Serebrova N.N. 1979. Bound water in coals with various degrees of metamorphism. *Dokl. akad. nauk. USSR,* 8, 630–634.
40. Neuman P.S., Pratt L., and Richards R.E. 1955. Proton magnetic resonance spectra of coals. *Nature* 175, 645–646.
41. Sanada Y., Honda H., and Nishioka A. 1962. Temperature dependence of nuclear magnetic resonance absorption in coal. *J Appl. Polymers* 19, 94–97.
42. Volkenstein V. 1955. *Structure and Physical Properties of Molecules.* Moscow-Leningrad: Fizmatgiz.
43. Grechishkin V.C. and Skripov F.I. 1959. Implementation of nuclear quadrupole resonance for studying the electric field gradients in some crystals. In *Paramagnetic Resonance.* Kazan, p. 160.
44. Bloembergen N., Purcell E.M., and Pound R. 1948. Relaxation effects in nuclear magnetic resonance absorption. *Phys. Rev.* 37, 679–712.
45. Uo J. and Fedin E.I. 1962. On determination of barriers of hindered rotation in solids. *Fiz. tverd. tela.* 4. 2233–2237.
46. Alexeev A.D., Krivitskaya B.V., and Pestryakov B.V. 1977. NMR study of water adsorption on a fossil coal. *Chem. Solid Fuel* 2, 94–97.
47. Bloch A.M. 1969. *Water Structures and Geologic Processes.* Moscow: Nedra.
48. D'Orasio F., Bhattacharja S., and Halperin W.P. 1990. Molecular diffusion and nuclear magnetic resonance relaxation of water in unsaturated porous silica glass. *Phys. Rev. B* 42, 9810– 9818.
49. Greg S. and Sing S. 1984. *Adsorption, Specific Surface, Porosity.* Moscow: Mir.
50. Banavar J.R. and Schwartz L.M. 1987. Magnetic resonance as a probe of permeability in porous media. *Phys. Rev. Lett.* 58, 1411–1414.
51. McCall K.R. and Guyer R.A. 1991. Fluid configurations in partially saturated porous media. *Phys. Rev. B.* 43, 808–815.
52. Alexeev A.D., Troitskii G.A., Ulyanova E.V. et al. 1999. Transversal NMR relaxation of water protons in silica gels. *Fiz. tekn. vyso. davl.* 9, 104–110.
53. Alexeev A.D., Troitskii G.A., Ulyanova E.V. et al. 1999. Studying porous structure of coals using pulse NMR technique. In *Physicotechnical Problems of Mining Production,* Alexeev A.D., Ed. Donetsk: Institute for Physics and Engineering, pp. 3–9.
54. Alexeev A.D., Ulyanova E.V., and Zavrazhin V.V. 2001. NMR study of water interaction with pore surface in silica gels and fossil coals. In *Proc. Int. Conf. on Physics of Liquid Matter,* Abstracts. Kiev: PLM, pp. 123–124.
55. Keltsev N.V. 1984. *Fundamentals of Adsorption Technique.* Moscow: Khimiya.
56. Chon S.H. and Seidel C. 1990. Relaxation times of ^{1}H NMR of H_2 in vycor glass. *Bull. Magn. Res.* 12, 107.

57. Mank V.V. and Lebovka N.I. 1988. *Spectroscopy of Nuclear Magnetic Resonance in Heterogeneous Systems*. Kiev: Naukova Dumka.
58. Giona M. and Giustiniani M. 1995. Size dependent adsorption models in microporous materials. 1. Thermodynamic consistency and theoretical analysis. *Ind. Eng. Res.* 34, 3848–3855.
59. McCall K.R., Guyer R.A., and Johnson D.L. 1993. Magnetization evolution in connected pore systems. II. Pulsed-field-gradient NMR and pore-space geometry. *Phys. Rev. B* 48, 5997–6006.
60. Gallegos D.P., Smith D.M., and Brinker C.J. 1988. An NMR technique for the analysis of pore structure: Application to mesopores and micropores. *J. Colloid Interphase* 124, 186–198.
61. Kiselev A.V. 1956. Adsorption properties of hydrocarbons. *Uspechi khim.* 25, 705–748.
62. Tarasevich Y.I. and Rak V.S. 1998. Adsorption properties of porous fossil coals. *Kolloid. z.* 60, 84–88.
63. Alexeev A D, Vasilenko T.A., and Ulyanova E.V. 1999. Measuring closed porosities in fossil coals. *Khim. tverd. topl.* 3, 39–45.
64. Alexeev A.D., Vasilenko T.A., and Ulyanova E.V. 1999. Closed porosity in fossil coals. *Fuel* 78, 635–638.
65. Vasilenko T.A., Kirillov A.K., Troitskii G.A. et al. 2008. NMR spectroscopy studies of fossil coal structure. *Fiz. tekhn. vysok. davl.* 18, 128–136.
66. Vasilenko T.A., Kirillov A.K., and Shazhko Y.V. 2007. Investigation of methane and water desorption from country rocks of coal substance. *Fiz. tekhn. problemy gorn. proiz.* 10, 39–46.
67. Carr H.J. and Purcell E.M. 1954. Effects of diffusion on free precession in nuclear magnetic resonance experiments. *Phys. Rev.* 94, 630–638.
68. Mitra P.P., Sen P.N., and Schwartz L.M. 1993. Short-time behavior of the diffusion coefficient as a geometrical probe of porous media. *Phys. Rev. B* 47, 8565–8574.
69. Sorland G.H., Djurhuus K., Wideroe H.C. et al. 2007. Absolute pore distributions from NMR. *Diffus. Fund.* 5, 4.1–4.15.
70. D'Orazio F., Bhattacharja S., Halperin W.P. et al. 1990. Molecular diffusion and nuclear magnetic resonance relaxation of water in unsaturated porous silica glass. *Phys. Rev. B* 42, 9810–9818.
71. Matthews G.P., Canonville C.F., and Moss A.K. 2006. Use of a void network model to correlate porosity, mercury porosimetry, thin section, absolute permeability, and NMR relaxation time data for sandstone rocks. *Phys. Rev. E* 73, 031307.
72. Sapoval B., Russ S., Petit D. et al. 1996. Fractal geometry impact on nuclear relaxation in irregular pores. *Magn. Reson. Imag.* 14, 863–867.
73. Vashman A.A. and Pronin I.S. 1986. *Nuclear Magnetic Relaxation Spectroscopy*. Moscow: Energoatomizdat.
74. Alexeev A D, Vasilenko T.A., and Kirillov A.K. 2008. Fractal analysis of hierarchical structure of fossil coals surface. *Fiz. tekhn. probl. razrab. polez. iskop.* 3, 14–24.
75. Talibuddin S. and Runt J.P. 1994. Reliability test of popular fractal techniques applied to small two-dimensional self-affine data sets. *J. Appl. Phys.* 76, 5070–5078.
76. Radlinski M.A., Ioannidis M.A., Hinde A.L. et al. 2004. Angstrom-to-millimeter characterization of sedimentary rock microstructure. *J. Coll. Interphase Sci.* 274, 607–612.

77. Radlinski A.P., Mastalerz M., Hinde A.L. et al. 2004. Application of SAXS and SANS in evaluation of porosity, pore size distribution and surface area of coal. *Int. J. Coal Geol.* 59, 245–271.
78. Nakashima Y. 2004. Nuclear magnetic resonance of water-rich gels of Kunigel–VI Bentonite. *J. Nucl. Sci. Technol.* 41, 981–992.
79. Cowan B. 1997. *Nuclear Magnetic Resonance and Relaxation.* New York: Cambridge University Press.
80. Vanin A.A., Piotrovskaya E.M., and Brodskaya E.N. 2003. Modeling of methane adsorption in carbon pores of various cross-sections. *Z. fiz. khim.* 77, 921–927.
81. Deryagin A.V., Churaev N.V., and Zorin Z.M. 1982. Structure and properties of the boundary layers of water. *Izvest. akad. nauk. SSSR* 8. 1689–1710.
82. Alexeev A.D., Ulyanova E.V., Starikov G.P. et al. 2004. Latent methane in fossil coals. *Fuel* 83, 1407–1411.
83. Alexeev A.D., Ulyanova E.V., and Kovriga N.N. 2001. Methane intercalation in fossil coals. *Proc. 11th Int. Symp. Intercalation Compounds.* Moscow, p. 200.
84. Alexeev A.D. et al. 2002. Desorption phenomena in fossil coals. In *Proc. Natl. Inst. Mines,* Vol. 15. Dnepropetrovsk: Nat. Mining Univ. pp. 192–197.
85. Alexeev A.D., Ulyanova E.V., and Kovriga N.N. 2001. Study of the methane phase state in coals. In Alexeev A.D., Ed. *Physicotechnical Problems of Mining Production,* Vol. 3, 3–8. Donetsk: Institute for Physics and Engineering.
86. Alexeev A.D., Vasilenko T.A., and Ulyanova E.V. 2004. Phase states of methane in fossil coals. *SSC* 130, 669–673.
87. Alexeev A.D., Shatalova G.E., Ulyanova E.V. et al. 2003. Structural features of coal–methane system. *Fiz. tekhn. vysok, davl.* 13, 100–106.
88. Alexeev A.D., Airuni A.T., Vasyuchkov Y.F. et al. June 30, 1994. *The property of coal organic substance to form metastable single-phase solid-solution-type systems with gases.* Discovery Diploma 9, Application A-016-M, Reg. 16.
89. Alexeev A.D., Starikov G.P., Vasilenko T.A. et al. 2005. Justification of methane amount and phase state determination by NMR technique. *Visti. donet. girnich. instyt.* 1, 174–177.
90. Moskalenko E.M., Alexeev A.D., Pestryakov B.V. et al. 1980. Control over the phase state of chemical solutions pumped into coal. *Fiz. tekhn. probl. razrab. polez. iskop.* 1, 117–119.

4

Behaviors of Rocks and Coals in Volumetric Fields of Compressive Stresses

In recent years, solving problems connected with analysis of the limit states of rocks has seen an essential change based on the use of physical and mechanical characteristics of rocks. The earliest indicator of limits was temporary resistance to uniaxial compression or tension. We now have the ability to simulate physical and mechanical characteristics in conditions close to the real ones. Conditions of loading are usually described by estimating the values of three main stresses.

Direct research of the mechanical processes that occur in massifs, i.e., under natural conditions, presents considerable technical difficulties. The most practical way to determine the characteristics that adequately reflect massif state is physical modeling with triaxial compression units.

4.1 Experimental Technique

Karman [1] was the first researcher who subjected solid objects (marble, sandstone) to high stress under three different positive stresses and determined the values of the stresses over the entire process of deformation. He conducted the experiment with a special unit in which he could subject a cylindrical sample to hydrostatic pressure with a simultaneous application of additional axial stress σ_1. Both stresses could be measured. The hydrostatic stress was kept constant while the axial stress was changed. In Böker's experiments, the axial stress was kept constant while the side (hydrostatic) pressure was increased gradually until plastic deformation began [2].

Bridgeman later studied solid objects under volumetric stress conditions according to Karman's scheme $\sigma_1 \neq \sigma_2 = \sigma_3$. He considerably increased the maximum pressure and found first and second order transitions in some materials [3].

At present, Karman's loading scheme is used widely to examine rocks in volumetric stress conditions both in the Ukraine and abroad. Differences in methods relate only to the way of changing the side pressure. The basic results of the research to date can be summarized as follows.

1. The strength and plasticity of rocks and minerals grow with increasing hydrostatic (side) pressure.

2. Plastic deformation of rocks is accompanied by an irreversible change of their volume (dilatancy) that causes a qualitative change of their structures. During a hydrostatic pressure increase, a specific maximum volume change based on rock features occurs. On the whole, the increase of the side pressure leads to a change (suppression) of compaction dilatancy.

3. With an increase of side pressure, the coefficient of diametrical deformation and thrust grows. The data showing changes of elasticity, deformation, and shear are controversial.

4.1.1 Literature Review

The results obtained during deformation of the rocks in accordance with Karman's scheme are significant for explaining the mechanical processes taking place in a massif. In certain situations, for example, assessing conditions in the bottom hole of a mine working where the stress is truly triaxial, it is necessary to utilize a modelling scheme with three different main stresses.

Experimental studies involving models are necessarily narrower because of the considerable technical difficulties encountered in constructing units that produce true triaxial compression. Most triaxial compression units used for testing rocks have certain weak points: (1) small sample sizes that complicate the problem of scale effect; (2) the lack of a closed chamber for samples, leading to edge effects and high friction losses; and (3) main stresses are calculated and not measured by some units.

At the Institute for Mining Telemechanics of the National Academy of Sciences of Ukraine, a device [4] was constructed to test the strengths of rock samples under true triaxial compression. A set value of lateral thrust was maintained automatically and the elements were based on the principle of dry friction. The force elements of the device were devised to provide the maximum pressure on a sample up to 30 MPa. One advantage of this device is its rigidity. Acoustic and ultrasonic emissions were used with the device to study the processes that occur inside samples.

To determine the parameters of hydrodynamic influence on coal seams [5], a series of experiments on destruction of porous objects were conducted under different process parameters using a special unit that can produce pressures up to 40 MPa. At the end of the twentieth century in the United States, laboratory research focused on the transport qualities of the rocks under conditions of triaxial stress to achieve better predictions of the efficiency of hydrocarbon reservoirs. Experimental units for volumetric loading were designed and constructed. The following activities are of special interest.

Sayers et al. [6] described a unit of triaxial loading to test cubic samples of rocks with 50 mm ribs. The device can apply compressive stress up to 120 MPa along each main direction. It was first used to measure the velocities of longitudinal (P) and polarized diametrical (S) waves and mechanical qualities of dry samples. Smart [7] later described an alternative triaxial chamber for testing cylinder samples. Several flexible pipes installed between the chamber walls and the sample casing were used to apply various minimum and medium stresses in radial directions. The differentiated radial pressure around a sample was provided by supplying liquid at different pressures to the pipes. Vertical pressure is applied along the axis to the prepared plane-parallel ends of a sample. However, the geometry limited the minimum and main mean stresses (usually the maximum difference of pressures must not exceed 15 MPa).

At the Imperial College (London), a unit of triaxial loading [8] was designed for cubic rock samples with 51-mm ribs. All three main loading components operated independently from each other at any value from zero to 115 MPa in two horizontal directions and up to 750 MPa in a vertical direction. The unit was successfully applied to create a network of breaks and microcracks oriented perpendicularly to minimum main stress and determine the permeability and velocity of ultrasonic P and polarized S vibrations, and mechanical qualities such as strength and deformability. The unit was later modified to handle injections of one- and two-phase pore liquids under the increased pressures into cubic samples with 40-mm ribs. The modified system was equipped with a device to control the liquid supply. It could develop deviatory components exceeding 200 MPa and intraporous pressures up to 145 MPa. At present, the system can measure various petrophysical characteristics such as permeability, velocities of acoustic waves, and electric conductivity in different directions under mono-phase and multi-phase intraporous filling. The coordination of compliances and high plane-parallelism of ends and supports provides homogeneity in distributing ultimate stresses with minimal end and edge effects [9]. The designation of main stresses along the sample axes are maximal (σ_1) directed along an axis z, medium (σ_2) directed along axis y, and minimal (σ_3) directed along axis x. The construction of the experimental unit [10] includes certain key elements.

- A special loading frame in which a vertical hydrocylinder produces maximal pressure (z-direction) and has a support ring of aluminum alloy with four horizontal hydrocylinders was designed to produce two orthogonal components of stress in the horizontal plane. The values of all three main components are under the control of a hydraulic servosystem with feedback capabilities. The sample deformations are measured by linear displacement transducers located in pairs on each main axis while outer stresses are conducted to the sample through the original plastic–ceramic composite supporting plates reflecting the geometry of a 40-mm cube.

- A gas accumulator (container) with a pressure control to create and maintain pressure of the intraporous medium. A sample is isolated from the atmosphere via a rubber casing placed in advance and held by the loading plates. The casing makes it possible to produce increased pressure in pores.

- Blocks for measuring capillary pressure and electric parameters including two gas and oil devices for precise measurements of changes in pore volume and liquid movement.

- A gas and liquid hydrodynamic system to measure rock permeability in the direction of axis z. The system consists of a flow meter and two pressure sensors with a numerical panel showing incoming and outgoing liquid pressures. Incoming liquid pressure is created and maintained by the gas accumulator.

- A recorder that continuously registers applied loads, intraporous pressures, and displacements.

Measurements of permeability and electric resistance were carried out in three main directions. With this purpose a cubic sample was turned with a corresponding side: its facets were marked beforehand with indices x, y or z. Each time the sample was loaded, the air was pumped out and the stresses were gradually brought to a set level. For these experiments a 5% solution of NaC1 was used as flowing liquid and the intraporous pressure was kept at the low level: at around 0.69 MPa.

After calculation of permeability and registration of electric resistance in the given direction, the stresses were slowly decreased. The sample was unloaded, turned, and loaded again. As the sample was turned, the stresses were redirected and gradually brought to a certain level. After achieving equilibrium and calculation of transport qualities in the second direction, all the procedures were repeated for the third direction.

For every selected deep and surface sample of sandstone from deposits in St. Bees and Springwell, a series of experiments was conducted. Each sample was subjected to tests at high and low stresses and under triaxial and equivalent hydrostatic stress conditions. To study low stresses, the samples were tested with an equivalent hydrostatic stress of 6.9 MPa. In triaxial stress conditions, ratios of maximum, medium, and minimum stresses were 1.0, 0.8, 0.6 and absolute values were 8.6, 6.9, and 5.2 MPa, respectively. For high loads, the hydrostatic pressure was maintained at 34.5 MPa while the ratio of components in triaxial loading was the same, corresponding to absolute values of 43.1, 34.5, and 25.9 MPa.

A 5% solution of NaC1 was used as a flowing liquid during all measurements of permeability and electric resistance. The anisotropy coefficients of permeability and electric resistance were calculated under high and low stresses while all data on electric resistance were adjusted to account for the temperature at the moment of measuring. In all cases, the hydrostatic and triaxial stress conditions were compared.

In addition to cubic samples, cylindrical core samples were measured in a Hassler chamber [10]. The samples were cut from cores in both horizontal and vertical directions and tested for permeability and specific resistance in two directions, x and z. Changes of pore volume were estimated by the amount of saline solution forced out under stress. The saline amount and the electric resistance were recorded constantly during the process of subjecting the samples to set levels of stresses. At the beginning of each experiment, a sample was subjected to a small all-round pressure of about 3 MPa. Then hydraulic lines were connected and the sample was pumped with saline solution to remove the air from the system. Stress was increased gradually in steps of ~0.69 MPa up to 6.9 MPa, then in steps of 6.9 MPa up to 34.5 MPa. At every step, volume and electric resistance were recorded. Such experiments were conducted under both hydrostatic and triaxial stress conditions.

All these systems present advantages and have proven useful in solving the problems discussed. To study rock behaviors and changes of their characteristics under all kinds of loadings (generalized according to the Lode-Nadai parameters of compression, tension, and shear), the preferred system is the TTCU (true triaxial compression unit) [11] designed at the Institute of Physics of Mining Processes of the National Academy of Sciences of Ukraine. The TTCU allows uniform stresses to be exerted on samples and predicts ultimate conditions.

A TTCU working chamber that holds a sample of natural material (rock) of prismatic form with a 60-mm rib creates stresses in three mutually perpendicular directions that are independent of each other and analogous to stress conditions in a massif. TTCU can model most uniaxial, compression, and tension stresses. By setting a Lode-Nadai parameter of the type of stress, one can create a stress state from generalized compression to generalized tension, thus modeling and determining the qualities of any part of a massif (from virgin rock to pre-working face). The results obtained with TTCU may be calibrated with results from other systems.

This ability is very important both for the estimation of data obtained from units modeling different stress states and to determine the types of rocks and stress states where the ultimate characteristics can be calculated by using the constants of materials subjected to uniaxial loading and integrating them into the corresponding cadasters.

The results of studies of rocks under conditions of true triaxial compression obtained by different researchers on different equipment often vary. For example, some authors believed that the change of the level of the intermediate main stress did not essentially affect the volumetric strength of rocks. However, TTCU experiments [12] showed that the intermediate main stress σ_2 significantly influenced the volumetric strength of the rocks under small levels of minimum main stress σ_3 and $0.5 < \sigma_3/\sigma_2 < 1$, i.e., for modeling of a pre-working face of a coal seam. Within the interval of $0.15 < \sigma_3/\sigma_2 < 0.5$ and an increase of σ_3, the effect of σ_2 on the volumetric strength of rocks decreases. At certain sufficiently high levels of spherical tensors, the main intermediate stress σ_2 does not significantly influence rock resistance to destruction (see Section 4.4).

The results of the studies of Lode-Nadai parameter μ_σ under different stress conditions were unexpected and interesting. The generally accepted idea that its value remains the same based on strength and deformation parameters was not confirmed with studies of rocks. In other words, $\mu_\sigma \neq \mu_\varepsilon$, and the connection between them is more complex and different for various types of rocks [13]. TTCU studies first revealed that during the destruction of sandstone in a volumetric field of compressive stresses in the slip plane, considerable structural reconstruction via the formation of plastic and fibrous structures up to the quartz polymorphism occurs [14]. Thus, to assess the behavior of rocks under natural conditions and to obtain data showing their physical and mechanical qualities, one studies rocks under true triaxial stresses, coordinates results with other studies, and determines new directions of research.

4.1.2 True Triaxial Compression Unit

The main component of the TTCU is the working tool (Figure 4.1 is an axonometric view). Figure 4.2 shows a cut in plane xOz. Supports (5, 11, and 16) are

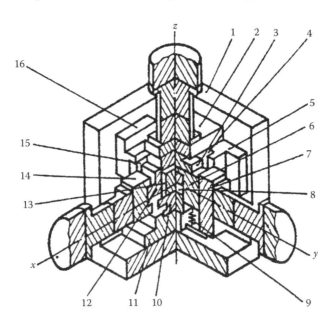

FIGURE 4.1
Axonometric view of TTCU working tool.

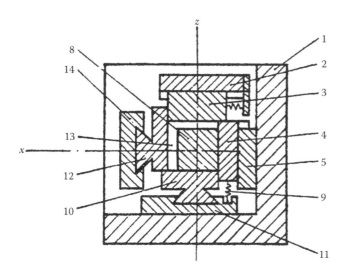

FIGURE 4.2
Cross-section of TTCU working tool in *xOz* plane.

mounted securely in a frame (1) and provide movement for the working load plates (4, 10, and 15) along the guide ways. The loading element consists of hydraulic cylinders located on three mutually perpendicular axes that link works fastened with soles (2, 7, and 14). The soles transmit stress from the loaded unit to the pressure plates (3, 6, and 12) and further onto the sample (8). They also serve as guide ways for the working pressure plates. The working load plates (4, 10, and 15) and the working pressure plates (3, 6, and 12) are equipped with springs (9).

As the rods of the hydraulic cylinders travel along three mutually perpendicular axes toward the sample (8), the working pressure plates (3, 6, and 12) stack up with each other and with the working load plates (4, 10, and 15). For example, pressure plate (12) along axis x glides on a pressure plate (6) and a load plate (10). Simultaneously, pressure plate (12) touches pressure plate (3) and load plate (15) moves them toward the guide ways of sole (2) with respect to support (16). The springs (9) shrink and by releasing force provide a constant contact of pressure plate (12) with pressure plate (3) and load plate (15). A similar process takes place during the movement of two other rods of the hydraulic cylinders (along axes y and z) and connects them with pressure plates (3 and 6) at the closed chamber (13) that contains the sample. The pressure on the sample is created by pressure plates (3, 6, and 12) on one side and by load plates (4, 10, and 15) on the other so that the whole plane of these plates is in contact with corresponding facets of the sample.

The loading unit consists of hydraulic cylinders and frames located along three mutually perpendicular axes. Each cylinder has a high pressure pump. To reduce the time of approach of rods to the sample and return to the initial position (idle running), the scheme includes a low pressure pump with high productivity. The maximum working pressure produced by each hydraulic cylinder reaches 25 MPa. The hydraulic cylinder rod diameter is 250 mm and maximum effort on the rod is $78 \cdot 10^4$ N. To decrease friction in the movable elements, the sliding friction in support-receiving or loading plates was replaced by rolling friction between two contacting surfaces. Figure 4.3 shows the design of the support-loading plate: (1) is a rolling support and (2) is a removable head with a pressure sensor.

According to calculations, the friction coefficient due to such substitution decreases 50 times. The change from rolling friction to sliding friction in the loading plates during sample deformation can accommodate any ratio among the three stresses without preliminary consideration for friction losses. In addition, the performance of the unit on the whole rises (wear of the pistons in hydraulic cylinders and gaskets decreases); as a result, the strength and deformation characteristics of samples can be determined with greater precision.

When loading and receiving elements are detached, one can use the high pressure chamber to analyze physical and mechanical features of solid objects, rocks in particular, under volumetric compression and filling with gas (methane, carbon dioxide) and model any geodynamic phenomena and design ways to prevent them. To conduct experiments on samples filled with gas, a special device [15] was constructed (Figure 4.4 and Figure 4.5). It consists of a high pressure chamber (1) that contains a cylinder case (2) with a

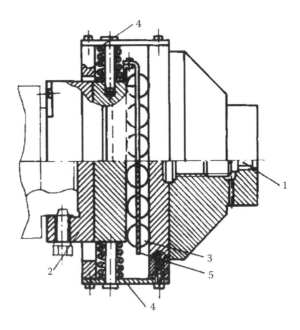

FIGURE 4.3
Design of friction assembly support.

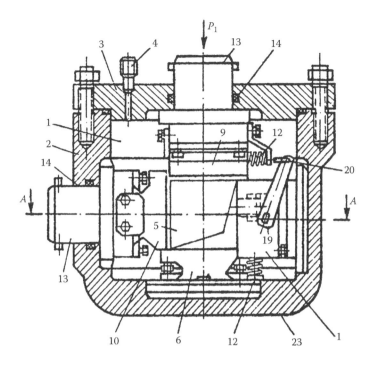

FIGURE 4.4
Device for volumetric loading of samples filled with gas.

removable cover (3). A nipple (4) supplies and removes medium from the chamber (1).

The mechanism for loading the sample (5) in the case (2) consists of three couples placed in mutually perpendicular planes of the support plates (6, 7, and 8) and corresponding pressure plates (9, 10, and 11) mounted in a stack. Springs (12) are used to tighten each pair of plates to each other. A loading device with rods (13) that pass through holes (14) in the case (2) and cover (3) and interact with pressure plates (9, 10, and 11) has a valve (15) with spring (16) placed in the support plate (7) of the chamber (17) connected with the internal volume in (1). When the spring is compressed, the valve (15) is held in the support plate (7) by a device (18) with a handle (19) from the outer side of plate (7) through the flexible link (20) with a screw (21) that serves to consolidate a hole (22) in the case (2) that connects the chamber (1) with the atmosphere. On the external surface of the case (2) are planes (23) made parallel to the corresponding support plates (6, 7, and 8) that limit the chamber (1) to transmitting (and not receiving) links in the chain of action of force loads.

The sample (5) is placed on the support plates (6, 7, and 8). The cover (3) is placed on the case (2) to pressurize the internal volume of the high pressure chamber (1). The pressure plates (9, 10, and 11) and the rods (13) are in the initial

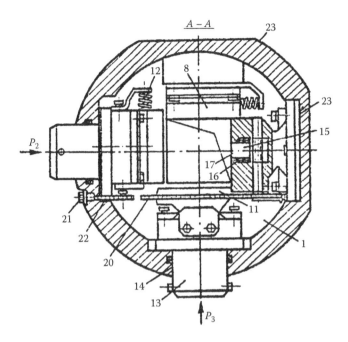

FIGURE 4.5
Cross-section A–A of unit for volumetric loading.

position (not in contact with sample). Methane gas under pressure is supplied to the chamber (1) through the nipple (4) and the sample (5) is saturated with methane. The loading devices are put into action to move the pressure plates with the help of the rods. Pressure plate (10) slides onto pressure plate (11) and support plate (6). Simultaneously pressure plate (10) comes in contact with pressure plate (9) and support plate (8), moving them in the direction of effort from rod (13). The behavior of two other pressure plates (9 and 11) during their movement is the same. Meanwhile compression of springs (12) provides permanent contact between the pressure plates (9, 10, and 11) and support plates (6, 7, and 8). All facets of the sample (5) are pressed by the pressure plates and the support plates. The sample (5) is loaded up to certain value and the loading devices are stopped. Between the support plates and the pressure plates, a conditionally airtight volume is formed where the sample compressed to a certain pressure and filled with methane is located. The tightening screw (21) is loosened and methane is removed from the chamber (1) through the outlet (22). Pressure in the chamber almost instantly become equal to atmospheric pressure.

The pressure in the sample (5) saturated with methane in the conditionally airtight volume decreases much more slowly. Using the tightening screw (21) connected to the handle (19) by a flexible link, the handle attached to the roller (18) is turned in such a way that the shaped cut in the roller is in front of the valve (15) end. The spring (15) helps the valve (15) move along the shaped cut of the roller, thus opening the chamber (17). An instant movement of the valve

(15) joins the chamber (17) with the internal volume of (1) where the pressure is equal to atmospheric pressure. Under the influence of methane in the sample (5) and the stresses formed in the sample during deformation, some of the sample is thrown into the open chamber (17). The maximum gas pressure in the high pressure chamber is 10 MPa.

The simplest physical model that conserved the physical nature and geometrical properties of elementary volumes of coal separated inside a seam served as the basis for research on TTCU. The model reproduced all basic characteristics, regularities, and effects connected with natural conditions. The main factor that defines the physical and mechanical qualities of coal in the model is the scale effect [12,16–18]. Minimal sizes $(6.0 \times 6.0 \times 6.0$ cm) [19] of cubic samples were taken from the newly exposed surface of a coal face. The time of coal exposure taking into account the interval between removal from the seam and preparation averaged 24 h. The samples were cut on a quartz-cutting tool using detachable diamond wheels with 200 mm outer diameter. The wheels were brought to the preformed blocks by hydraulic dampers. On the cubic samples, lack of parallelism did not exceed 0.05 mm, deviation of squareness of ends to the generating line was no more than 0.05 mm, and end protuberance was 0.003 mm. The precise cubic form ensured reliability of results with a minimum number of samples. To reduce the dispersion of test data, samples were chosen with deviation of volumetric weights within ±5%. The ribs were measured with Vernier calipers (point value 0.05 mm). The loading programs were set to study the influence of unloading speed on the damageability of the coal structure based on filtration parameters , the set radii of the damaged particles, and the influence of the damage mechanism on the kinetics of methane escape from coal.

Loading of the samples started with placing a sample into the working chamber. The sample was evenly loaded along three axes up to the set level and monitored with a manometer. Establishing the limit state was done by increasing value σ_1 to a level at which it was noticeable, i.e., exhibited in the sample on the lateral pressure plates (along the axes O–X and O–Y) as stretching strains. The stresses along line σ_2 appeared spontaneously or were kept at a set level. Value σ_3 corresponded to coal strength for uniaxial compression. As soon as ultimate value of σ_1 was reached, σ_3 was reduced to 0. After each 2.5 to 3.0 MPa of loading or unloading, the manometer indications were fixed. The shift of the sample facets was marked to within 0.01 mm. The sample fracture was calculated as the decrease of the highest stress and appearance of stretching strain along the line O–Y. To analyze the resulting data and determine physical and mechanical qualities of coal and degree of damageability, the following parameters were estimated [20–22].

Modulus of volume deformation:

$$K = \frac{\sigma_{mean}}{\varepsilon_{mean}}$$

where σ_{mean} and ε_{mean} are mean stresses and deformations along three axes.

Shear modulus:

$$G = \frac{1}{2}\sqrt{\frac{\sigma_1\sigma_2 + \sigma_2\cdot\sigma_3 + \sigma_1\cdot\sigma_3}{\varepsilon_1\cdot\varepsilon_2 + \varepsilon_2\cdot\varepsilon_3 + \varepsilon_1\cdot\varepsilon_3}}$$

Modulus of deformation (Young's modulus):

$$E = \frac{9K\cdot G}{3K + G}$$

Poisson ratio:

$$v = \frac{3R - 2G}{6K + 2G}$$

Relative volume decrease:

$$\frac{\Delta V}{V} = \frac{\left(L_1^i\cdot L_2^i\cdot L_3^i\right) - \left(L_1^i - L_1^1\right)\cdot\left(L_2^i - L_1^2\right)\cdot\left(L_3^i - L_1^3\right)}{L_1^i\cdot L_2^i\cdot L_3^i}, \%$$

where L_i and L'_i are initial and current sizes of the sample under deformation in direction of σ_i (I = 1, 2, 3).

Lode-Nadai parameter of type of stress:

$$\mu_\sigma = 2\frac{\left(\sigma_2 - \sigma_1\right)}{\sigma_1 - \sigma_3} - 1$$

Energy of volume change:

$$A_0 = \frac{\left(\sigma_1 + \sigma_2 + \sigma_3\right)^2}{18G}$$

Energy of change of form:

$$A_f = \frac{1 + v}{18G(1 - 2v)}\left[\left(\sigma_1 - \sigma_2\right)^2 + \left(\sigma_2 - \sigma_3\right)^2 + \left(\sigma_3 - \sigma_1\right)^2\right]$$

4.2 Strength Properties of Rocks Treated by Different Binders

One effective way to achieve preliminary (preventive) strengthening of a massif of fissured rocks is a physical–chemical method based on the forced injection of cold hardening polymeric resins that fill the fissures and form separate rock blocks into a monolith. A massif bound in this way becomes steady, thus ensuring reliable mine lining and stability of mine workings. The efficiency of strengthening solutions depends on geological, physical, and chemical factors of a mine. The most important factors influencing the process of strengthening a possibly unsteady mine roof are opening of fissures, moisture of the rocks, and temperature of the surroundings. This section presents the results of studies of the strength and deformation properties of rocks injected with magnesia and carbamide solutions and the effects of fissure opening widths, rock moisture, and temperature of the surroundings.

4.2.1 Experimental Methods and Estimates of Strengthening

The chemical and mineralogical compositions of rocks and interactions with hardening agents greatly influence the effectiveness of strengthening. Experiments were conducted on argillites of the same lithologies. We chose the most effective strengthening compositions: magnesia with latex and carbamide–formaldehyde resin with oxalic acid.

To study the strength and deformation properties of bound rocks, prismatic samples with ribs of 55 to 60 mm were cut from pieces of argillite with a stone-cutting tool. Each sample was cut into two pieces and glue layers of 1, 2, 3, 4, and 5 mm were applied. After gluing, a sample was placed into a desiccator where the humidity of the surroundings was 72 or 98%. On the basis of NMR studies of the strengthening of phase states (see Section 3.4.2), hardening factors and factors for maintaining optimal strength and deformation properties were defined. The samples bound with the magnesia solution were conditioned for 10 to 13 days; the conditioning period was 3 to 7 days for samples bound with the carbamide solution.

To estimate the influence of rock moisture, samples of argillites with natural humidity less than 0.6% were artificially moistened and the amount of humidity was determined by NMR. Samples with moisture between 0.6 and 2% were obtained. The samples were glued and placed into desiccators. Every 24 hours, the phase state of the strengthener was defined by NMR spectra.

The samples intended to be used to study the impact of the surroundings temperature on rock moisture were also placed into desiccators and then for a required time kept in refrigerators at −10, 0, 5, and 10°C and at room temperature (24°C). Temperatures were checked with a thermocouple and potentiometer. The prepared samples of strengthened rocks were deformed to destruction on the true triaxial compression unit (TTCU). Two loading schemes ($\sigma_1 > \sigma_2 > \sigma_3$) ensured the minimum levels of deformation energy: (1) $\sigma_3 = 5$ MPa, $\mu_\sigma = -0.8$ and (2) $\sigma_3 = 5$ MPa, $\mu_\sigma = 0.8$ where σ_3 is the minimum compressive stress and $\mu_\sigma = 2(\sigma_2 - \sigma_3)/(\sigma_1 - \sigma_3) - 1$ is the Lode-Nadai stressed state.

The main stresses σ_1, σ_2, σ_3 and deformations ε_1, ε_2, ε_3 were recorded. Two cases of orientation (parallel and perpendicular) of the plane of sample gluing in terms of the working direction of the prevailing compressive stress

were researched as well. The experimental results served as a basis for calculating the most typical mechanical constants characterizing strength and
deformation properties of bound rocks under true triaxial compression:
volumetric strength of the sample σ_1^*, module of volumetric compression
$K = \sigma_{mean}/\varepsilon_{mean}$, where σ_{mean} and ε_{mean} are the mean stress and deformation,
and volumetric deformation $\Delta V/V = 3\varepsilon_{mean}$.

An estimate of the orientation of the gluing planes of strengthened samples
showed that under perpendicular orientation of the gluing plane in terms of
the direction of the prevailing compressive stress σ_1, the destruction mainly
occurred on the rock. Under parallel orientation σ_1, the destruction occurred
both on the rock and at the point of contact of rock and strengthening solution. In the case of $\mu_\sigma = -0.8$, the samples were destroyed mainly along the
gluing. In further studies of the strengthened samples, only one plane of gluing relative to the direction of stress σ_1—the parallel one—(the direction of
stress σ_3 is perpendicular to the gluing plane) was studied; only one loading
scheme was realized: $\sigma_3 = 5$ MPa, $\mu_\sigma = -0.8$.

NMR spectra are complex; they consist of two lines of different intensities.
Figure 4.6 shows the change of intensity ratio I_2/I_1 of two lines depending
on time. The increase of the intensity ratio I_2/I_1 suggests a decrease of plastic
properties of the system. The figure shows that with the decrease of temperature from 24 to 5°C, a sharp increase of intensity ratio I_2/I_1, the decrease of
plasticity occurred on the third, fifth, and seventh days. Thus, the temperature decrease increases the time for preserving the plastic properties in the
carbamide resin. The increase of moisture in the strengthened rock is followed by the leveling of the line intensities in NMR spectra and depends on
moisture amount.

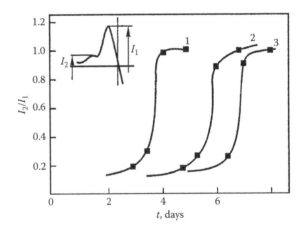

FIGURE 4.6
Dependence of intensity ratio I_2/I_1 of NMR spectra lines of carbamide–rock system on hardening time t and surrounding temperature $T°C$.

This research made it possible to find a way to calculate factors for strengthening roofs in mine workings [23], including the selection and study of the rock samples with a strengthening solution based on a carbamide–formaldehyde resin. To increase the precision of determining the time necessary for optimal strengthening, rock samples with the strengthening solution were crushed and subjected to NMR examination. The intensities of the narrow and wide lines of the spectra were defined and the degree of strengthening based on the ratio I_2/I_1 where I_1 is the intensity of the narrow line and I_2 is the intensity of the wide line of the spectra.

4.2.2 Width of Fissure Openings

The widths of fissure openings in strengthened rocks is an important factor in strengthening rocks in coal seam roofs. Laboratory studies of the thickness of the glue layer on the strength and deformation properties of the bound argillite samples allowed us to estimate the influence of the widths of fissure openings. Figure 4.7 shows the dependence of volumetric strength of samples strengthened with magnesia (curve 1) and carbamide (curve 2) solutions on the thickness of gluing (1 to 5 mm). Each point on the graph corresponds to tests of seven or eight samples with a spread not exceeding 6%.

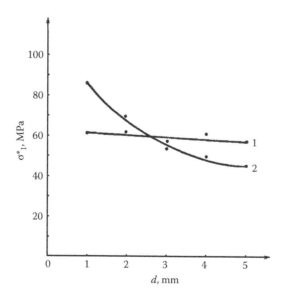

FIGURE 4.7

Dependence of volumetric strength σ_1^* of strengthened argillite samples on thickness of applied layer d. (1) Magnesia solution ($t = 10$ days). (2) Carbamide solution ($t = 3$ days). Conditions: $\sigma_3 = 5$ MPa, $\mu_\sigma = -0.8$, $W_n = 0.6\%$, $T = 24°C$.

For magnesia solution, the strength of the strengthened samples changed little despite the tendency of strength to decrease with increases of gluing thickness. As the gluing thickness increased, the strengths of the samples bound with carbamide solution reduced considerably from 86 to 45 MPa (48%). The deformation properties of the samples bound with the magnesia or carbamide solutions did not depend on the thickness of the glued layer. Thus, one can suppose that for roof rocks with large fissuring, strengthening with magnesia solutions is the most effective measure. Because the bound samples showed the highest strength properties with 1 mm glue layers, samples with this glue thickness were tested further.

4.2.3 Moisture of Rocks under Strengthening

With an increase of argillite moisture from 0.6 to 1.8%, the strength of samples bound with magnesia solutions (Figure 4.8, curve 1) reduced from 65 to 39 MPa (40%). The influence of moisture in rocks bound with a carbamide solution is not considerable. When it changes from 0.6 to 2%, the sample volumetric strength reduces from 82 to 69 Mpa (only 16%). The volumetric compression K of samples bound with a magnesia solution undergoing an increase of humidity from 0.6 to 1.8% reduced from 4.3×10^3 to 3.75×10^3 MPa (Figure 4.9, curve 1, 13%). The volumetric deformation $\Delta V/V$ (Figure 4.10, curve 1) increased from 36×10^{-3} to 45×10^{-3} (25%). The deformation properties of the samples bound with carbamide solution depend more on moisture. Thus with an increase of rock moisture from 0.6 to 2.0%, the volumetric

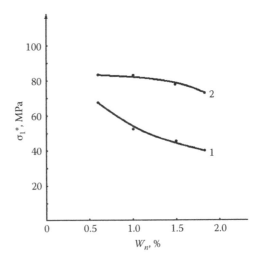

FIGURE 4.8
Dependence of volumetric strength σ_1^* of strengthened argillite samples on moisture of rock under strengthening W_n. (1) Magnesia solution ($t = 13$ days). (2) Carbamide solution ($t = 5$ days). Conditions: $\sigma_3 = 5$ MPa, $\mu_o = -0.8$, $T = 24°C$, and $d = 1$ mm.

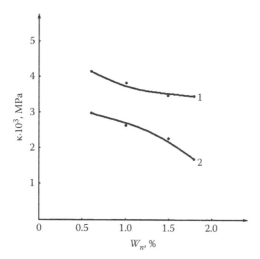

FIGURE 4.9
Dependence of volumetric compression module K of argillite samples on moisture of bound rock W_n. (1) Magnesia solution ($t = 13$ days). (2) Carbamide solution ($t = 5$ days). Conditions: $\sigma_3 = 5$ MPa, $\mu_\sigma = -0.8$, $T = 24°C$, and $d = 1$ mm.

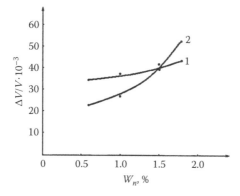

FIGURE 4.10
Dependence of volumetric deformation of argillite samples on moisture of bound rock. (1) Magnesia solution ($t = 13$ days). (2) Carbamide solution ($t = 5$ days). Conditions: $\sigma_1 = 5$ MPa, $\mu_\sigma = -0.8$, $T = 24°C$, and $d = 1$ mm.

compression of the fastened samples (Figure 4.9, curve 2) reduced from 3×10^3 to 1.4×10^3 MPa (almost half) while the volumetric deformation (Figure 4.10, curve 2) simultaneously increased from 24×10^{-3} to 65×10^{-3} (2.7 times).

Based on research, it is possible to conclude that the moisture in bound rocks influences more considerably the strength properties of samples bound with magnesia solution and the deformation properties of samples bound with carbamide solution. Thus, one can suppose that based on the increased moisture in bound samples, the strengthening by carbamide solution is more effective.

4.2.4 Temperature of Surroundings

With the reduction of the surrounding temperature from 24 to –10°C, the volumetric strength of the samples bound with magnesia solution varied little (Figure 4.11, curve 1). With carbamide solution, the volumetric strength of the samples decreased from 82 to 50 MPa (Figure 4.12, curve 2, 39%). With the reduction of temperature, the deformation properties of the samples bound with magnesia or carbamide solutions, plasticity increased. For example, with a change of the surrounding temperature from 24 to –10°C, the volumetric compression K of the samples glued with magnesia solution decreased from 4.4×10^3 to 2.6×10^3 MPa (Figure 4.12, curve 1, 1.7 times). The volumetric deformation $\Delta V / V$ increased from 31×10^{-3} to 46×10^{-3} (Figure 4.13, curve 1, 48%). With a decrease of temperature, the volumetric compression K of samples bound with carbamide solution decreased from 3×10^3 to 1.6×10^3 MPa (Figure 4.12, curve 2, 1.87 times less) while the volumetric deformation increased from 21×10^{-3} to 45×10^{-3} (Figure 4.13, curve 2, more than double). With an increase of the surrounding moisture from 72 to 98%, the influence of temperature on the deformation properties of bound samples grows considerably, especially in samples strengthened with carbamide solution. Thus, a reduction of the surrounding temperature decreases the plastic properties of samples strengthened with magnesia and carbamide solutions and considerably influences the strength properties of samples strengthened with carbamide. Under conditions of lower temperatures of surroundings (for example, permafrost conditions), strengthening with magnesia solution is the more effective method.

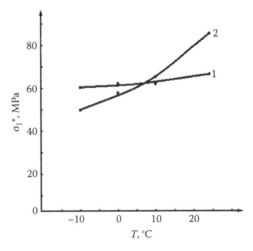

FIGURE 4.11
Dependence of volumetric strength σ_1^* of strengthened argillite samples on temperature of surroundings. (1) Magnesia solution ($t = 13$ days). (2) Carbamide solution ($t = 5$ days). Conditions: $\sigma_3 = 5$ MPa, $\mu_\sigma = -0.8$, $T = 24$°C, and $d = 1$ mm.

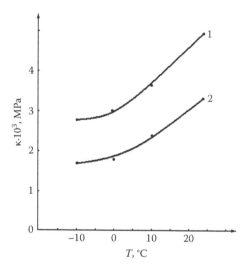

FIGURE 4.12

Dependence of volumetric compression of strengthened samples of argillite on temperature of surroundings. (1) Magnesia solution ($t = 13$ days). (2) Carbamide solution ($t = 5$ days). Conditions: $\sigma_3 = 5$ MPa, $\mu_\sigma = -0.8$, $W_n = 0.6\%$, and $d = 1$ mm.

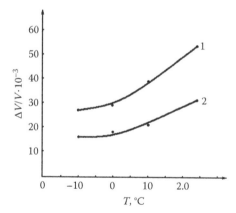

FIGURE 4.13

Dependence of volumetric deformation of strengthened samples of argillite on temperature of surroundings. (1) Magnesia solution ($t = 13$ days). (2) Carbamide solution ($t = 5$ days). Conditions: $\sigma_3 = 5$ MPa, $\mu_\sigma = -0.8$, $W_n = 0.6\%$, and $d = 1$ mm.

4.2.5 Conclusions

Techniques for determining the degree of strengthening minimally steady roofs in mine workings made it possible to define the time when bound rocks exhibit optimal properties. The time between hardening and achieving optimal properties is 3 to 7 days for the carbamide solution and 10 to 13 days for the magnesia solution.

The increase of moisture of rocks by strengthening and decreasing the temperature of the surroundings decelerates the strengthening processes of the solutions studied.

Increasing the thickness of the applied glue layer (changing the width of fissure opening) did not cause considerable variation of the strength and deformation properties of the rock samples strengthened with magnesia solution. In the samples strengthened with carbamide solution, the decrease of sample strength was evident.

The increase of sample moisture under strengthening led to a considerable decrease in the strength of the rock samples strengthened with magnesia solution and an insignificant decrease in the strength of the samples glued with carbamide. With the increase of the rock moisture, the plastic properties of the samples improved with both carbamide and magnesia solutions.

The decrease of the temperature of the surroundings led to a decrease in strength of the rock samples strengthened with carbamide solution and did not influence the strength of the samples glued with magnesia. It also improved the plastic properties of the samples strengthened with both solutions.

4.3 Influence of Loading Method and Loading History on Volumetric Strength

4.3.1 Loading Method

From the view of outburst danger, it is useful to study the behaviors of rocks destroyed under stress conditions by modeling the pre-coal face zone (σ_3 is small). It is important to estimate the influence of the approach to such stress condition on the strength and energy capacity of the rock destruction. The difference in the nature of the rock destruction from loading and unloading is the main interest. To estimate the difference, we considered (1) a gradual increase of the main stress σ_1 under fixed σ_2 and σ_3 to destruction and (2) gradual unloading of a preliminarily loaded sample along the axis of the least main stress σ_3 to destruction.

The destruction of sandstone under minimal compressive stress $\sigma_3 = 0$ and 5 MPa was analyzed. Results indicated that the energy capacity of destruction and its nature depend largely on the approach to the stress condition even though sandstone strength does not depend on the method of loading. Figure 4.14 and Figure 4.15 show dependences of $\sigma_i - \varepsilon_i$ and $\sigma_{mean} - \varepsilon_{mean}$ on destruction via loading and unloading. The mean stresses and stress deviators are almost equal in both cases. Figure 4.14 shows dependence $\sigma_i - \varepsilon_i$ in the direction of minimum compressive (effective stretching) stress. In both cases, the effective stretching stresses are equal but the stretching deformations in

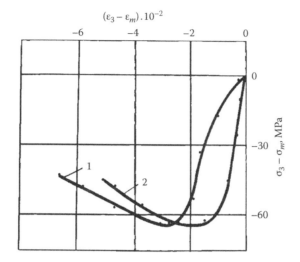

FIGURE 4.14
Dependence of $\sigma_i - \varepsilon_i$ on direction of minimum compressive stress during loading (1) and unloading (2).

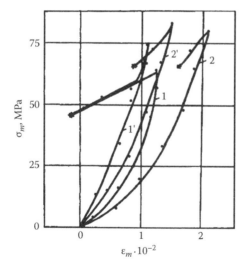

FIGURE 4.15
Dependence of $\sigma_{mean} - \varepsilon_{mean}$ on destruction of sandstone. 1 and 2. At loading $\sigma_3 = 0$ and 5 MPa, respectively. 1' and 2'. At unloading.

the direction of minimum compressive stress σ_3 during unloading are much less than at loading.

During loading and unloading, the deformation characteristics differ considerably. The compression during unloading is 1.1 to 1.2 times more than during loading (0.66×10^4 MPa and 0.8×10^4 MPa for $\sigma_3 = 0$; 0.6×10^4 MPa and

0.48×10^4 MPa for $\sigma_3 = 5$ MPa). The shear module during unloading is more as well (0.29×10^4 MPa and 0.19×10^4 MPa for $\sigma_3 = 0$; 0.16×10^4 and 0.15×10^4 MPa for $\sigma_3 = 5$ MPa). During unloading, the Poisson ratio decreases (0.31 and 0.37 for $\sigma_3 = 0$; 0.38 and 0.39 for $\sigma_3 = 5$ MPa). These data indicate that as σ_3 increases, the differences of the rigidity module and Poisson ratio are smoothed.

The typical difference of destruction during loading and unloading is the unequal deformation state formed before destruction. At loading, the deformation state does not correspond to the stress condition (deformation advances it) and it appears that $\mu_\varepsilon = 0$, i.e., real displacement is noticed. Meanwhile the planes of destruction are directed at an angle of about 45 degrees to the biggest compressive stress. During unloading, the deformation state corresponds well to stress condition $\mu_\sigma = \mu_\varepsilon$ and is very close to generalized compression. As shown above, fissures of diametrical and out-of-plane shifts are formed. The planes of destruction are directed at various angles to the maximum compressive stress; some are almost parallel to the unloading plane.

The energy capacity of destruction (see Section 4.6.2) by way of unloading is lower only when $\sigma_3 = 0$, but even when $\sigma_3 = 5$ Mpa, the density of deformation power during unloading is higher than during loading (6.6 and 6.14 MJ/m³) and coming off the free surface is not explicitly manifested. However, the quantities of dilatancy (Figure 4.16) and newly formed surface (Table 4.1) during unloading in this case are higher.

The principal increase of surface during unloading is provided by large fractions. The degree of grinding at unloading is higher than at loading

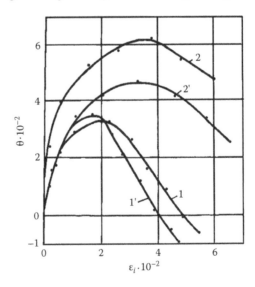

FIGURE 4.16
Dependence of volumetric deformation θ on deformation deviator ε_i. 1 and 2. At loading $\sigma_3 = 0$ and 5 MPa, respectively. 1′ and 2′. At unloading.

TABLE 4.1

Sizes of Newly Formed Surfaces

Fraction (mm)	Surface at Unloading (cm²)	Surface at Loading (cm²)
10	940	200
7	41	17
5.5	18	10
4.5	15	6
3.5	13	8
3.25	3	3
3	2	2
2.5	6	4
2	10	7
1	40	18
0.5	31	15
0.25	127	49
0.2	18	10
<0.2	599	264

(0.55×105 and 0.34×105 m^{-1}, respectively), that is, the increase of energy capacity at unloading is caused by an increase of grinding. The increase of grinding can be explained by the redistribution of stresses in a sample with the reduction of the least compressive stress σ_3. By reducing σ_3, we reduce the yield strength, leading to the crushing yield as the decrease of stresses is not significant. Breaks and redistribution of stresses take place, and new breaks appear in totally different directions [24]. At unloading, fissures appear in one primary direction.

Thus, the main difference in sample behavior during unloading and loading lies in the amount of accumulated power and the deformation state in the sample. At unloading, due to the spread of fissures in diametrical and longitudinal shear, material grinding increases considerably.

4.3.2 Loading History

Rocks in mining areas are already under natural stress from evolution of massifs during geological times. This stress is based on the interaction of independent gravitational and tectonic force fields. Mining changes the natural stresses by deforming and destroying rocks. Zones of increased and decreased stress conditions (in comparison with natural conditions) form around working areas. Hence, the rocks under natural stress conditions are further subjected to the influences on the force fields resulting from mining, that is, the unloading stress works on the so-called background of natural stresses. During strength testing, samples are subject to the influence of the destruction stresses starting from an initially unloaded state. The acceptability of the results of such tests for solving particular mining and technical tasks appears questionable.

The issue is important as it directly influences the field value of stresses and sizes of destroyed rock zones around mine workings. As noted by Kartashov et al. [25], a researcher who does not know the history of sample loading may assume he or she is dealing with totally different rocks. In other studies [26], the authors referred to results obtained by I.V. Rodin during optical modeling of stress-deformed conditions around rectangular workings and works of Karman and Böker who point out that the real characteristics of a massif under the compressive load can be determined only when previously loaded samples are tested.

Studies to determine the ultimate strengths of rocks depending on their loading history was carried out on the TTCU system [11]. Cubic sandstone samples ($55 \times 55 \times 55$ mm) were prepared by a stone-cutting tool, then divided into two groups of five samples each. The experiments to find the ultimate strengths of samples in the first group were carried out under true triaxial compression ($\sigma_1 \neq \sigma_2 \neq \sigma_3$) from the initial unloaded state to the moment of destruction. A diagram of stress deformation was registered along each axis until the moment of sample destruction. The samples from the second series were compressed (loaded) up to $\sigma_1 = \sigma_1 = \sigma_3 = \sigma_m$, which was considered equal to the spherical tensor $\left(\sigma_m = (\sigma_1^* + \sigma_2^* + \sigma_3^*)/3\right)$ obtained during tests of samples from the first series at the moment of destruction. After simultaneous loading along the axes σ_2 and σ_3 (intermediate and minimum stress, respectively), unloading of the samples was carried out until their values in the first series of experiments ($\sigma_2 \rightarrow \sigma_2^*$ and $\sigma_3 \rightarrow \sigma_3^*$) were reached. Along axis σ_1 (the greatest compression stress), the stress was applied until the moment of destruction. During the test of the second series. the diagram of stress deformation was registered at compression of the samples, at further unloading, and loading until the moment of destruction.

Table 4.2, Figure 4.17, and Figure 4.18 present the testing results. The figures show the dependencies of the stress deviator ($\sigma_1 - \sigma_m$) and deformation

TABLE 4.2

Results of Tests of Rock Samples

Test Type	Preliminary Compression $\sigma_1 - \sigma_m$ (MPa)	Ultimate Destruction Stress (MPa)			Ultimate Relative Deformation				
		σ_1	σ_2	σ_2	ε_m	ε_1	ε_2	ε_3	ε_m
Without preliminary compression	–	201.0	8.9	6.2	76.85	0.090	0.004	−0.087	0.0023
After preliminary compression	84.5	215.6	19.5	6.1	79.6	0.0865	0.014	−0.051	0.0165

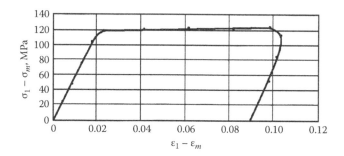

FIGURE 4.17
Dependence of stress deviator on deformation deviator for samples without pre-compression.

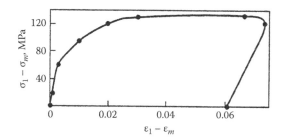

FIGURE 4.18
Dependence of stress deviator on deformation deviator for pre-compressed samples.

deviator $(\varepsilon_1 - \varepsilon_m)$ for samples of the first and second series. The results suggest that the history of loading exerted an insignificant influence on the value of destroying stresses. In the studies, a slight increase of maximum stresses (7 to 10%) was registered for sandstone samples, preliminarily those compressed at the set values of intermediate and minimum stresses.

Simultaneously the decrease of dilatancy effect of the same samples compared with the earlier loaded ones was noted. Relative volumetric deformation for the samples of the first series at the moment of destruction was $\Theta^p = 0.003 \div 0.01$; after unloading $\Theta^p = -0.015 \div -0.032$, and for the samples of the second series these values were $\Theta^p = 0.003 \div 0.01$ and $\Theta^p = 0.003 \div 0.01$, respectively (compression deformations were considered positive). Intensive fissure formation preceding the moment of sample destruction started at the level of mean stress for the samples of the first series (70 MPa) and for the samples of the second series this value was 84 MPa.

However, under compression, the level of spherical tensor influences the material destruction. Destruction starts, as a rule, at a relatively high main stress difference that causes the change of the sample form. The maximum tangential and octahedral stresses that take into account this difference were at the moment of destruction for the samples of the first and second series $\tau^1_{max} = 97.4$ MPa and $\sigma^1_{oc} = 89.0$ MPa; $\tau^2_{max} = 104.0$ MPa and $\sigma^2_{oc} = 95.7$ MPa, respectively. During loading of samples from the initial no-load state, the necessary difference between the main stresses is achieved gradually and at lower values of maximum compression, as the level of lateral thrust prevents to a lesser degree the change of the sample form. After pre-compression of the sample to a sufficiently high stress level, although at the increase of the maximum (vertical) stress the unloading of its lateral face took place, the level of lateral pressure was high enough to prevent the change of form that occurred in the first case. That is why the ultimate deformation of the sample decreases and destruction can be more rapid [13].

In addition, as noted above, at this level of sample compression the ultimate destructive stresses differed insignificantly from these values in experiments on samples without pre-loading. The conclusion is that no significant change of strength takes place and the results of the tests on the samples without pre-compression are acceptable for technical mining calculations. Their slight increase can be attributed to strength margin.

4.3.3 Conclusions

A diagram of stress deformation helps estimate energy capacity and the characteristics of destruction, retained strength, and bearing capacity of rock—valuable information for solving several mining problems.

During rock destruction in a true triaxial field of compressing stresses, the decreases of module maximum and dilatancy maxima coincide. Brittle failure with dynamic effect and significant tension drop is visible during the spreading of fissures of in-plane and out-of-plane shear and are accompanied by the most scrubbing and grinding of the material. The spreading of fissures of normal fracture in rocks leads to minimum scrubbing and grinding of the material, the least tension drop, and the least value of decrease.

The main difference in rock sample behavior during unloading and surcharging lies in energy accumulation in the form of deformation state in a sample. During unloading, the rate of material grinding increases significantly due to formation of in-plane and out-of-plane shear fissures. The ultimate destruction stress of pre-compressed samples of rocks and samples without pre-compression barely differs. The results of tests on rock samples without pre-compression are acceptable for technical mining calculations; the influence of sample loading history can be disregarded.

4.4 Coal Destruction

This section analyzes the conditions and peculiarities of coal destruction connected with the formation of outburst-dangerous zones and sudden coal and gas outbursts.

4.4.1 Mechanism of Coal Destruction

The main condition for formation of coal and gas outburst is the unloading of part of a gas-saturated coal massif from rock stress. The unloading applies tension stress along the line of minimum stress to produce orthogonal maximum stress. The optimal condition contributing to the development of the unloading wave is the ratio of main stresses in the coal massif that form generalized tension ($\sigma_1 = \sigma_2$, $\sigma_3 = 0$). Otherwise additional tension stresses will create conditions $\sigma_1 > \sigma_2$, $\sigma_3 \neq 0$ and form a geo-mechanical state of generalized shear.

The parameters of the change of mechanical state of the physical coal seam system in a pre-face zone are three main stresses σ_1, σ_2, σ_3 (σ_1 is the maximum compression stress, active along the normal line direction toward the seam, σ_2 is the intermediate main stress active along the face line, and σ_3 is the minimum compression stress active along the normal line toward the seam plane) and three main deformations ε_1, ε_2, ε_3, active along the same directions. They are complemented by elastic constants (K, E, and G), Lode-Nadai parameters μ_σ and μ_ε, characterizing the stressed and deformed states, and energetic parameters (A_v represents the work of volume change forces and A_f represents the work of form-changing forces) [27].

On the basis of experimental data and theoretical research [28,29] on loading and deformation of a coal seam in the pre-face zone, three typical sections of different volumetric loading and different mechanical conditions exist (Figure 4.19).

The first section of volumetric loading of a seam in the depth of a massif outside the zone of a mine working influence is characterized by the ratio connecting the three main stresses ($\sigma_1 = p_1 > 0$, $\sigma_2 = \sigma_3 = \lambda p_1$). p_1 is the value of the maximum compression stress on a seam in the zone of a "virgin" massif and λ is the coefficient of the lateral pressure. The hypothesis of the geostatic compressed state realized in the zone of the "virgin" massif is used. Based on the relations between the main stresses, the Lode-Nadai parameter is –1,

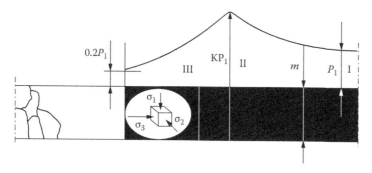

FIGURE 4.19
Distribution of stresses in zone of limit.

which suggests the type of volumetric stressed condition of the generalized compression for this loading zone.

The second section of the volumetric loading of a coal seam in the depth of the massif where partial destruction starts and the compression stresses σ_1, σ_2 and σ_3 are at their maximum is characterized by the maximum compression stress σ_1 value KP_1, where K is a stress concentrator; the minimum compression stress σ_3 changes within λP_1 and $K\lambda P_1$, and intermediate main stress σ_2 will reach half the sum of σ_1 and σ_3. The value μ_σ differs significantly from 1 and approaches 0. This confirms that in the second section of coal seam loading, the type of stress is close to generalized shear.

It is natural that geo-mechanical processes take place in the first and second sections of loading and also occur continually in each element of the coal seam between them. This part of a coal seam can be called the zone of mine working influence at active loading. Loading and deformation in a seam take place in such a way that the system load constantly increases (components of compression stress condition increase) and the type of volumetric stress condition changes from generalized compression ($\mu_\sigma = -1$) to generalized shear ($\mu_\sigma = 0$). If we consider the geostatic stress condition active in the "virgin" massif in the whole zone of the mine working influence at active loading, the coal seam will be loaded by volumetric stress of the generalized shear. Analysis of work loading from generalized compression to generalized shear presents a common case (from the position of change of type of volumetric stress condition).

The third section of volumetric loading of the coal seam is part of the face where minimum compression stress tends to 0 and the relation between two other main stresses depends on the conditions of coal seam destruction in the section between the second and third loading zones. For coal seams, the maximum compression stress σ_1 on a seam edge in some cases can decrease to the level of the intermediate main stress σ_2, confirming the presence of generalized tension.

The zone of the coal seam between the second and the third loading sections is called the zone of the limit state. The mechanical state and strength of the seam in this zone are determined by the conditions of the transition of pre-face zone of the massif from generalized shear to generalized tension. Thus, depending on the set task—examining the mechanism of deformation and destruction of coal and simultaneous actions of mechanical stresses and gases—the defining factors are

- The mechanisms of stress component change in the zone of change of mechanic stresses (support pressure)
- The conditions of formation of coal limit state in this zone
- The character of coal deformation and destruction at various ratios of stress components under volumetric compression
- The influence of methane in a porous volume of coal on its mechanic characteristics and mechanism of destruction

In this connection, programs of triaxial loading were built by two schemes:

- The analysis of behavior and mechanism of destruction of coal in the massif, remote from the working face, that is, at various values of the set lateral thrust (minimum main stress $\sigma_3 = \sigma_{comp}$, where σ_{comp} is coal strength on uniaxial compression) with subsequent decrease $\sigma_3 = 0$

- The study of the mechanism of coal deformation in the pre-face part of a working where due to the face advance and destruction of the seam edge zone the lateral thrust (σ_3) decreases to zero.

The first program was carried out in the following way. The values of stress σ_3 were set from 1 to 20 MPa with intervals of 2.5 MPa, and from 10 to 20 MPa with intervals of 5 MPa. In the beginning, the compression of a sample was conducted along three axes simultaneously (spherical tensor of stresses and deformations) until the level of the set least stress σ_3 was achieved, i.e., according to $\sigma_1 = \sigma_2 = \sigma_3$. Then pressure σ_1 was raised continually along axis O–Z, and the value σ_3 was kept on the same (set) level. The value of the intermediate main stress σ_2 (along the axis O–Y) was formed spontaneously by sample deformation. The formation of the limit state was registered as the termination of pressure increase along axis O–Z (σ_1).

Coal from the outburst-dangerous seam h'_6 Smolyaninovsky at level 1312 m of the Skochinskogo Mine was taken. Starikov provided a detailed description of the experiments [30]. Definitions of methane-saturation level of coal samples, elastic and deformation properties and destruction mechanism were carried out. Figure 4.20 and Figure 4.21 graph the results of tests of gas-saturated ($Q_g = 12$ to 14 cm^3/g) and degassed samples ($Q_g = 0.1$ to 0.3 cm^3/g), respectively, in coordinates $\sigma_1 = f(\varepsilon_1)$. All graphs showing various values of least main stresses σ_3 are characterized by a common nonlinear dependence between stresses and deformations—an increase of coal strength with a rise

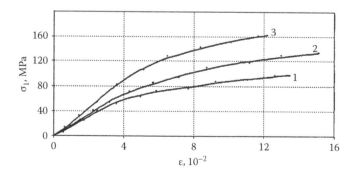

FIGURE 4.20
Dependence $\sigma_1 = f(\varepsilon_1)$ for gas-saturated coal at various values of minimum compression stress σ_3. MPa: 1 = 5.0; 2 = 10; and 3 = 20.0.

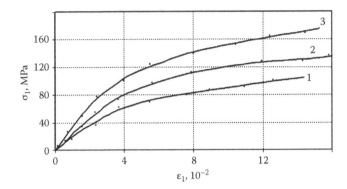

FIGURE 4.21
Dependence $\sigma_1 = f(\varepsilon_1)$ for degassed coal at various values of minimum compression stress σ_3. MPa: $1 = 5.0$; $2 = 10$; and $3 = 20.0$.

of lateral pressure. All graphs have three typical sectors showing various intensities of stress increases and deformation character.

In the first sector deformation module, shear module, and Poisson ratio are stable, indicating elastic deformation of coal samples. The second (nonlinear) sector is characterized by a more intense increase of compression deformation along O–Z axis, a decrease of shear module by 25 to 30%, and an increase of deformation module and Poisson ratio by 15 to 20%. Along the O–X axis, tensile deformation is registered, indicating the beginning of destruction and loosening of the sample. In the third sector, the destruction of a sample is registered—an increase of volumetric deformation of 1.5 to 2 times, a decrease of shear module of 2.5 to 3 times, and a Poisson ratio close to 0.5.

The influence of gas saturation on deformation, physical and mechanical properties (volumetric deformation, shear, Poisson coefficient), and stress deformed state of a sample can be explained by analysis of the results at $\sigma_3 = 5.0, 10.0$, and 20.0 MPa. Table 4.3 presents data of measured and calculated properties, reflecting coal state in the area of pre-destruction (upper line) and in the area of destruction (lower line).

Based on the results, ultimate coal strength depends insignificantly on the degree of gas saturation and cannot work as a criterion for estimating methane influence on the properties of coal. The destruction of gas-saturated coals under conditions of volumetric loading (transition to $\sigma_1 = f(\varepsilon_1)$) begins at stresses 11 to 23 MPa less than degassed ones.

The most important moment of deformation in the area of pre-destruction is the approximate equality of energy of formation indicating the destruction rate of the structure and the volume change energy. During coal change into the limit state, the shear and elasticity modules of both gas-saturated, and degassed samples decrease 2.5 to 3.2 times, and the shear module (G) and volume compression (K) levels of gas-saturated samples are 8 to 24 % higher than levels in degassed samples. This effect is amplified by the growth of σ_3. At the

TABLE 4.3

Measured and Calculated Properties of Deformed Coals

σ_3 (MPa)	Methane Presence	$K_m \times 10^2$ (MPa)	$G_m \times 10^2$ (MPa)	v_m	σ_{1m} (MPa)	σ_{2m} (MPa)	$E_m \times 10^2$ (MPa)	$A_{v,m}$ (J/m³)	$A_{f,m}$ (J/m³)
5.0	GS	6.5	7.4	0.42	54.3	6.4	11.8	3.6	4.7
		13.7	2.5	0.49	98.2	14.1	2.7	5.1	56.0
	D	10.0	8.3	0.44	69.8	5.8	11.5	3.6	3.5
		13.0	2.3	0.48	103.8	10.5	3.9	4.5	37.0
10.0	GS	7.5	7.9	0.41	59.7	12.2	11.2	6.0	4.9
		19.3	2.7	0.49	133.7	29.6	3.2	8.7	90.0
	D	9.1	8.9	0.43	83.4	10.9	11.2	6.0	7.7
		16.7	2.5	0.49	136.2	28.6	4.7	6.9	67.3
20.0	GS	10.3	8.8	0.44	110.4	29.1	12.3	11.9	11.1
		24.1	4.1	0.49	162.1	40.5	4.1	13.0	74.0
	D	11.3	9.8	0.44	121.4	23.1	12.6	18.4	10.5
		21.4	3.3	0.49	175.0	36.4	6.6	11.2	58.9

Note: Numerator presents data before destruction. Denominator indicates data during destruction. GS = gas-saturated. D = degassed.

TABLE 4.4

Determination of Ultimate Deformations and Calculations of Stress and Deformation

σ_3 (MPa)	Methane	$\varepsilon_1 \times 10^{-2}$	$\varepsilon_2 \times 10^{-2}$	$\varepsilon_3 \times 10^{-2}$	μ_σ	μ_ε
5.0	GS	13.4	−0.96	−9.7	−0.8	−02
	D	14.3	−1.20	−9.9	−0.87	−0.1
10.0	GS	15.1	0	−121	−0.7	0.11
	D	16.4	−1.60	−113	−0.7	0
20.0	GS	12.1	−0.58	−8.4	−0.7	0.17
	D	15.2	−0.60	−12.0	−0.75	0

Note: GS = gas-saturated. D = degassed.

same time, the elasticity of gas-saturated samples is 45 to 60% lower. The most essential changes occur with the form energy of samples that increases 5 to 10 times in comparison with the volume energy.

In addition, the energy of form change in gas-saturated samples is 25 to 51% higher than in degassed ones. The essential growth of formation energy in gas-saturated coal samples relates to the high damage rate of closed pores and methane transition into the system of fissures. Table 4.4 presents stress and deformation data for gas-saturated and degassed coals in the limit state.

Regardless of the degree of gas saturation of coal in the limit state, the kind of stress condition (μ_σ) does not correspond to the kind of deformation condition. Generalized stretching is minimal in samples with low levels of gas saturation (below 6 m³/t) at decrease of σ_3 to 0; the destruction mechanism

FIGURE 4.22
Mechanisms of coal sample destruction at $\sigma_1 \neq \sigma_2 \neq \sigma_3$, $\sigma_3 \to 0$. (a) Shear destruction: $Q_r = 3.5 -$ 5.4 m^3/t, $\mu_\sigma = 0$, $\mu_\varepsilon = 0$. (b) Destruction by shear and separation: $Q_r = 14 - 17$ m^3/t, $\mu_\sigma = 0.45$, $\mu_\varepsilon = 0$. (c) Separation destruction: $Q_r = 22$ m^3/t, $\mu_\sigma = 0.6$, $\mu_\varepsilon = 0.12$.

results from shear deformation (Figure 4.22). It is important to note the influence of the adsorbed moisture on the mechanism of deformation under condition of unloading σ_3. The critical content of physically bound moisture [31–33] of coal in the limit state changes to generalized stretching but at that σ_1, the σ_2 and shear module are 2.5 to 3 times less than in gas-saturated samples and destruction occurs only by means of the shear. These studies of the influence of methane on mechanisms of coal deformation in limit state due to unloading of minimum compressive stress are general (not based on particular depths). They are useful for analyzing outburst-dangerous zones near geological failures.

The second program simulated conditions in a pre-face zone of coal seam with minimum coefficient of stress concentration. A cubic sample placed in a working chamber was loaded at regular intervals to a level corresponding to the depth of the seam attitude: $\sigma_1 = \sigma_2 = \sigma_3 = \gamma H$ ($H = 800$ to 3000 m). The largest main stress was raised to $\sigma_1 = \sigma_3/\lambda$; λ is the factor of the lateral thrust of coal in the elastic loading area ($\lambda = \nu/(1-\nu)$ where ν is the Poisson ratio; σ_2 forms spontaneously. After the simulation, σ_3 was reduced to 0 at a rate of 0.1 to 0.5 MPa/sec with the registration of pressures and displacements along the three axes. Figure 4.23 shows results in the form of dependences $\sigma_1 = f(\varepsilon_1)$. The maximum spread of experimental points did not exceed 6%.

Curve 1 corresponds to the simulated depth $H = 800$ m and consists of three sections. The stress and deformation conditions of coal at these depths remain under generalized compression even when σ_3 decreases to zero. Minor alterations are noticed at a depth of 1200 m. At pressures corresponding to a depth of 3000 m in zones of raised rock pressure, the dependence $\sigma_1 = f(\varepsilon_1)$ includes three sectors with various intensities of pressure growth and deformations at the expense of an increase in coefficient of stress concentration.

We analyzed the influence of unloading by means of a decrease of σ_3 to 0 on deformation properties on coal samples selected from beds h_6' Smoljaninovsky

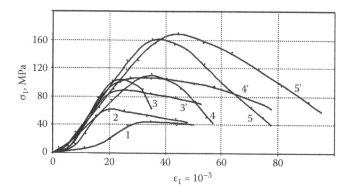

FIGURE 4.23
Dependence $\sigma_1 = f(\varepsilon_1)$ at simulated depths. (1) 800 m. (2) 1200 m. (3) 1600 m (gas-saturated). (3′) 1600 m. (4) 2000 m (gas-saturated). (4′) 2000 m. (5) 3000 m (gas-saturated). (5′) 3000 m.

and h_{10} Proskovievsky of the A.A. Skochinskogo and Glubokaya Mines. The only difference from the previous experiment was the degree of gas saturation. Coal with maximum methane capacity ($Q = 21$ to 28 m³/t) was saturated under a pressure of 10.0 MPa for more than 60 to 80 days. The results [30] show that at a decrease of σ_3 to 0, σ_1 decreases to σ_2, and the shear module decreases 5.5 times on average due to considerable stretching deformation. In general. as a result of unloading, the stress and deformation states change and generalized stretching both on pressure and on deformations is formed. Figure 4.22 shows results. The first section shows elastic deformation of coal samples with a stable increase of elastic coefficient for both gas-saturated and degassed samples. The second section is located parallel to the deformation axis (σ_1 is almost constant); essential changes in properties are not registered. The third section on the descending part of the curve $\sigma_1 = f(\varepsilon_1)$ shows a decrease in the elastic characteristics and the transition from generalized compression to generalized stretching of μ_σ and a generalized shear on μ_ε. In the final condition ($\sigma_3 = 0$), the shear module in gas-saturated coals is 10 to 12% more than in degassed ones. The Lode-Nadai parameter for gas-saturated coal indicates a greater degree of generalized stretching.

It is necessary to mention that during modeling of the face zone based on the depth of a coal seam, unloading along the line of minimum stress leads to changes of the stress and deformed conditions similar to the conditions of modeling $\sigma_1 > \sigma_2 > \sigma_3 = \sigma_{comp}$ (see Table 4.3). However, the destruction rate on change A_f/A_0 and G is much less. The results of analysis of change of sample volume at deformation in the form of dependence $\sigma_3 = f(\Delta V/V)$ in Figure 4.24 give more detailed information.

The graphs show the degree of compression and weakening of gas-saturated and degassed coal at various levels of lateral pressure on axis O–X (σ_3). Curves 1 through 6 for gas-saturated and degassed coals at initial uniform compression of 40, 50, and 75 MPa have four typical segments defining the initial

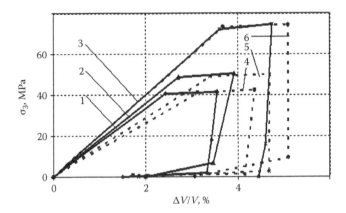

FIGURE 4.24
Influence of unloading σ_3 on volumetric deformation . 1–3: Gas-saturated samples. 4–6: Non-gas-saturated samples ($H = 1600$ to 3000).

and final conditions of the deformed samples. The first and second segments reveal the maximum compressibility depending on σ_3. Here maximum values of volume, shear deformation, and volume change energy are formed. The sample is deformed without structural failures, indicating a low level of form-changing energy 3.5 to 4 times less than the volume change energy.

The third segment, which is almost parallel to the vertical axis, is formed at the decrease of σ_3 and a constant value of σ_1. It characterizes the structural changes in coal at constant volume. The compressive deformations are actively formed in the direction σ_1 and stretching deformation in the direction of an unloaded side along axis O–X while volume deformation decreases by 20 to 25% and shear by 17 to 35%. The sample in the condition reflected by the termination of this segment where volume $\Delta V/V$ starts to decrease shows residual strength at the minimum value of lateral pressure σ_3, with an increase of energy of form 6 to 16% greater than the energy of volume change.

The last (fourth) segment is characterized by the decrease of σ_3 to 0. At the end of the segment, the shear module decreases 5.5 to 8 times, the form changing energy increases in gas-saturated samples by an average 8.1 times (6.3 times in non-gas-saturated samples), and the energy of volume change decreases 4.5 times on average. One basic finding was that in gas-saturated samples the relative change of volume was 25 to 30% less than in degassed ones. It is explained by the fact that volumetric deformation and shear in gas-saturated samples are 14 to 25% higher. The destruction of gas-saturated samples occurs at a smaller degree of strengthening in the pre-destruction stage and confirms that the destruction of gas-saturated samples by shear is less probable than destruction in degassed ones.

The most important conclusion about coal behavior at unloading is that the destruction of gas-saturated coal samples at $\sigma_1 = \sigma_2 = \sigma_3 = 40$, 50, and 75 MPa occurs at the moment when the energy of form-changing is 6 to 16% more

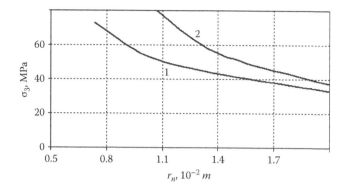

FIGURE 4.25
Change of reduced radii of destroyed coal particles at unloading of minimum compressing pressure. (1) Gas-saturated coals. (2) Non-gas-saturated coals.

than the energy of volume change. In gas-saturated samples, a large amount of deformation remains even when σ_3 decreases to 0 under generalized stretching. The degree of destruction (grinding) of gas-saturated samples (Figure 4.25) grows with the increase of σ_3 and is higher than in degassed samples. The radius of destruction r_n of particles at $\sigma_3 = 40$ MPa was $1.58 \cdot 10^{-2}$ m for gas-saturated samples and $1.91 \cdot 10^{-2}$ m for degassed ones. At $\sigma_3 = 50$ MPa, $r_n = 1.12 \cdot 10^{-2}$ m and $1.56 \cdot 10^{-2}$ m; and at $\sigma_3 = 75$ MPa, $r_n = 0.69 \cdot 10^{-2}$ m and $1.07 \cdot 10^{-2}$ m for gas-saturated and degassed samples, respectively.

Degassed samples generally contain fissures located at an angle of 45 to 50 degrees from the operating breaking point σ_1 (Figure 4.22a). In gas-saturated samples, the fissures are located 65 to 70 degrees from the line of action of the breaking point σ_1 and are typical of co-destruction by shear and separation (Figure 4.22b).

Because gas-saturated samples reveal low probability for the development of sliding fissures, their destruction is shown in the area adjoining the unloaded side of the sample on axis O–X (axis of stress σ_3 influence). This destruction is shown as separation of a part of a sample caused by changes of characteristics and volume. Generally, an increase of gas saturation of a coal massif at unloading causes changes of stress and deformation conditions indicated by generalized stretching with shear. Since destruction by shear is not the prevailing factor, the probability of destruction of part of a coal massif by separation increases dramatically (Figure 4.22c). Structural failures and destruction, as a rule, intensify methane desorption, raising its pressure on the walls of the developing fissures and promoting coal grinding at destruction.

4.4.2 Filtration Properties

Based on levels and proportions of stress values along three main axes, the filtration properties of coal can be determined by the presence and behavior of the fissures that form systems of filtration channels. The analysis of various methods for determining coal permeability in natural deposits classified them into three groups. The first group includes analytical, graph-analytical, and other methods [34] utilizing filtration theory in the context of coal seams. The results allow us to estimate the filtration parameters of a simulated section of a seam. The methods require the use of complex mathematical techniques. Many systems of equations do not yield accurate analytical solutions and must be adjusted by using dependencies that can be determined experimentally. One of the main disadvantages of analytical methods is that the solutions describe two-dimensional models of a coal massif and cannot estimate the influence of stress fields on fluid filtration in fissured and porous structures of a coal seam.

The second group of methods analyzes rock permeability [35,36]. Using cylindrical samples 5 to 15 cm in diameter and 5 to 20 cm in length, the quantity of fluid that permeates samples at a specific pressure drop over a certain time can be determined. The testing is performed in a working chamber that allows a sample to be subjected to lateral pressure. These methods make it possible to estimate the influence of absolute values of tension on fluid permeability.

The third group utilizes results of mine testing [37]. The methods are based on the restoration of gas pressure after a short-decrease or the dependence of the pressure in a well on time of steady filtration. The advantage of these methods is that data for the calculations are obtained under mining conditions and thus produce reliable results.

Research of rock permeability also presents considerable difficulties such as a lack of reliable equipment, questionable accuracy of results, and other factors present in extraction zones. The most reliable method of coal permeability research that simulates conditions in a massif under rock pressure is testing samples on a unit that produces triaxial stresses [36]. This technique can model any combination of main stresses that damage coal structures based on the amount and speed of reduction of intermediate stresses.

To research the filtration properties of rock, cubic samples with 6.0 cm edges were used. Each sample was covered with glue BF-2 so that 7.0 cm² of each edge was free of glue and then transferred to the working chamber of the unit. Air was blown through three movable (pushing) plates via hoses and nipples and to recording gages through stationary (receiving) plates. To ensure that the pushing and receiving plates allow afflux and reception of fluid through samples, holes of 0.1 cm diameter were located in concentric circles on the plates. During loading and unloading, compressed air was blown through a sample at certain levels of pressure in three mutually perpendicular directions and the amount measured. Hermetic sealing between the plates and coal sample is achieved by special gaskets of vacuum rubber 0.1 to 0.3 mm thick. Other materials used included an air bomb, reduction gear box RDD-5, wet gas meter GSB-400, stopwatch, standard pressure gage

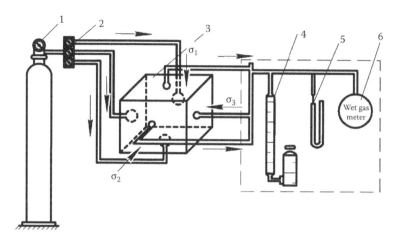

FIGURE 4.26
Apparatus for studying filtration properties of coal. 1. Air bomb. 2. Reduction gear box. 3. Sample. 4. Drop meter. 5. V-shaped water meter. 6. Wet gas meter.

of 0.15 class, aneroid barometer, thermometer, drop meter, V-shaped water meter, and sets of two- and three-way stop cocks.

Figure 4.26 depicts the unit for determining the gas permeability of coal. The compressed gas from the bomb (1) goes through the reduction gear box (2). The tube into the pushing plate that contacts the edge goes through the sample (3). Gas at the outlet is measured by a drop-meter (4) or wet gas meter (6), depending on air consumption. The gas permeability is determined at a pressure of 1.0 MPa. The coefficient of gas permeability [35] is calculated as

$$K = \frac{Q \cdot \mu \cdot L \cdot Z \cdot 10^3}{F \cdot P_1} \mu g \qquad (4.1)$$

where Q is the rate of air flow through the sample $Q = V/\tau$ per second, sm³/s (V is the volume of air measured by the wet gas meter, sm³ and τ is the time of gas emission, sec); μ is dynamic air viscosity, kg · s/m²; L is the length of the sample, cm; Z is the coefficient of the recalculation of the volume rate of air flow under experiment conditions:

$$Z = \frac{2P_b}{P_1 + P_2 + 2P_b},$$

where P_b is barometric pressure, MPa; P_1, P_2 is air pressure before and after the sample, respectively, MPa; and F is the area of the cross section of the sample, cm².

The testing was carried out with two or three pressure drops, depending on the permeability of the sample. If the calculated value of the gas permeability coefficient was not more than ±5% different, the arithmetic mean value (target value of gas permeability coefficient) was calculated. Large differences of gas permeability coefficient suggested that the sealing was faulty or the gas flow did not follow the law of straight-line relation and the experiments were repeated at lower pressures (0.3 to 0.4 MPa).

The sample was inserted into the working chamber of the TTCU. As it was loaded or unloaded, compressed air was blown in at set pressure intervals along three axes of loading and its consumption was recorded. After completing the test, the permeability coefficient was calculated by the volume of the inlet air and the dependence $K_i = f(\sigma_i)$ was calculated taking into account the value and speed of unloading σ_3. Some values of the calculated coefficient of permeability are given in the following section.

4.4.3 Changes of Fissured and Porous Structures

The results cited above show that the unloading speed of a simulated coal seam [38] significantly influences the depth of spreading of unloading wave. This is an important factor for coal–gas systems with low values of viscosity because this parameter considers changes of elasticity and strength characteristics. It also reveals the intensification of the processes of opening of the closed pores, their joining with fissures, and the change of the mechanism of methane desorption from diffusion to filtration.

Some research [39–43] focused on the influence of loading speed on rock and coal strength and deformation. However, in the context of coal seams, the main issue is determining the changes of fissured and porous structures of coals based on the unloading speed of one of the transitory components of the stress tensor [44]. Knowledge of the changes is important because coal is a natural methane collector and the kinetics of methane desorption into stopes and development workings will depend on whether loading changes or does not change coal structures. For that reason, we attempted to determine the influence of the speed of coal massif unloading on changes of fissured and porous structures of coal under conditions simulating the state of the face zone at the current depth of excavation. The studies were conducted on the TTCU and we controlled the degree of destruction relative to the changes of filtration properties determined according to Subsection 4.4.2.

Filtration was measured as the sample was subjected to stress at 3 MPa. After the limit state was reached ($\sigma_1 = 50$ to 60 MPa, $\sigma_2 = 25$ to 30 MPa, $\sigma_3 = 10$ to 12 MPa), unloading occurred along the line of the smallest compression stress $\sigma_3 = 0$ at speeds of 1 MPa/sec and 10 MPa/sec and the filtration parameters were determined in three orthogonally related directions. Based on stress results, the coefficient dependences of gas permeability on stresses $K_1 = f(\sigma_1)$, $K_2 = f(\sigma_2)$, and $K_3 = f(\sigma_3)$ as shown in Figure 4.27 were determined. Analysis of the results shows that during the loading of samples according to the program simulating the limit state of the face zone, $K_1 K_2 K_3$ is reduced to 10^{-6} to 10^{-7} D compared with 1.2 to 1.8×10^{-4} D in the non-destroyed state [44,45].

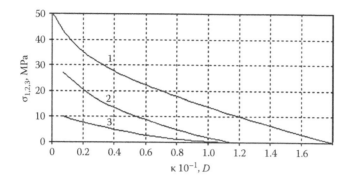

FIGURE 4.27
Changes of gas permeability of coal samples at loading according to $\sigma_1 > \sigma_2 > \sigma_3$. 1. $K_1 = f(\sigma_1)$. 2. $K_2 = f(\sigma_2)$. 3. $K_3 = f(\sigma_3)$.

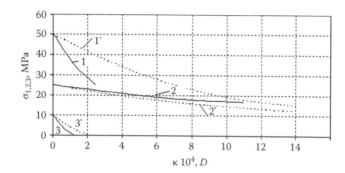

FIGURE 4.28
Changes of gas permeability when $\sigma_3 \to 0$. 1. $K_1 = f(\sigma_1)$. 2. $K_2 = f(\sigma_2)$. 3. $K_3 = f(\sigma_3)$ at unloading speed of 1 MPa/sec. 1′. $K_1 = f(\sigma_1)$. 2′. $K_2 = f(\sigma_2)$. 3′. $K_3 = f(\sigma_3)$ at unloading speed of 10 MPa/sec.

The changes of permeability by reducing the minimum compression stress σ_3 to 0 are of special interest. As shown in Figure 4.28, at an unloading speed of 1.0 MPa/sec, anisotropy of gas permeability coefficients occurs. The largest increase is in direction σ_2, while K_2 at unloading is 10 times the gas permeability of the unstressed sample. At the same time, K_1 and K_3 are at the same levels shown by unstressed coal.

These results allow us to consider the formation of planes of destruction in the σ_3 direction that are at an $\arctan(\rho/2)$ angle toward stress σ_1. At an unloading speed of 10 MPa/sec, $K_1 = K_2$ reaches 12 to $14 \times 10^{-4} D$ and becomes one order higher than the gas permeability coefficient of the unstressed coal; K_3 stays almost the same. This indicates the formation of destruction planes parallel to the unloaded side. The formation of the systems of secondary fissures that sharply change coal permeability primarily relates to the opening of some closed pores which, for the coal of seam h'_6 Smolyaninovskiy was 0.25 m³/m³ on average [46].

4.5 Limit State

A solid reaches its limit state when a small increase of any stress (mechanical, thermal, electromagnetic, etc.) leads to the loss of stability (destruction). Scientists have studied the deformations of solid bodies during volumetric loading which leads to limit state and destruction for a long time. A review of methods and criteria obtained from studying rocks and coals by the above methods can be found in a monograph [12]. We will present a brief review of studies we used in analyzing the factors surrounding deformation of gas-saturated coals and destruction in the volumetric stressed state.

Galileo started studying destruction mechanics and Coulomb, Saint-Venant, and Mohr developed the concept further. Destruction mechanics involves analysis of deformational properties of solid bodies and developing different hypotheses of strength to create criteria of deformation. These early scientists created the underlying principle: destruction of a solid body requires ultimate stress, reaching a deformation state, or a combination of both factors. Well known theories of large linear deformations and normal and tangential stresses [47,48] are not applicable to rocks that exhibit great differences of ultimate strength during extension and compression.

Among the classical theories applied to the rocks, those related to energy are of interest. Based on the theory of ultimate full-energy deformation [49,50], a dangerous condition of a material can be observed when specific potential energy reaches a certain limit. If we consider reaching the viscosity limit by the energy of form change as a critical state of a material, the above theory becomes a theory of supreme energy of form changing (fourth classical theory of strength). The condition of strength $(1/\sqrt{2})\sqrt{(\sigma_1 - \sigma_2)^2 + (\sigma_2 - \sigma_3)^2 + (\sigma_3 - \sigma_1)^2} \leq \sigma_T$ satisfactorily shows the start of plastic deformations. That criterion can be expressed as

$$\tau_{\text{shear}} \leq \frac{\sqrt{2}}{3}\sigma_y$$

where τ_{shear} is tangential stress on the octahedral plane and σ_y is the yield strength of the material. Coulomb and Navier suggested that the S_0 shearing strength increases at a value proportional to normal stress in the shear plane:

$$|\tau| = S_0 - \mu S$$

where μ is a constant. Anderson [51] used this variant to explain geological failures.

The most popular concept for rocks was the criterion of viscosity of Coulomb and Mohr, which creates on the shear plane a functional dependence between tangential τ and normal σ stresses and estimates ultimate strength during ordinary loading and under volumetric stress. Researchers

[52] found out that adhesion and the angle of internal friction depend on the stresses but the maximum angle of internal friction and adhesion coefficient are formed at different levels of axial deformations, i.e., they cannot be used for exact determinations of shearing strength. The Coulomb–Mohr theory does not consider in the volumetric stress state the influence of the intermediate state σ_2, whose level is decisive during unloading of coals and rocks. It considerably decreases the perspective and narrows the area of use. This disadvantage was eliminated in a later work [53]. To estimate the influences of all three main stresses, we accepted the condition that a proportion between normal σ_K and tangential τ_K stresses on the shear plane exists at the moment of destruction. These stresses can be presented by well known formulas:

$$\sigma_K = \sigma_1 l^2 + \sigma_2 m^2 + \sigma_3 n^2$$
$$\tau_K = \sqrt{\sigma_1^2 l^2 + \sigma_2^2 m^2 + \sigma_3^2 n^2 - \sigma_K^2}$$

(4.2)

where l, m and n are direction cosines of angles formed normally toward the shear plane with vector directions σ_1, σ_2, and σ_3, respectively. When $m = 0$, the stress σ_2 falls out of the stress condition, in agreement with the theory of Mohr. Geometrically the shear plane goes through the direction of the main stress σ_2. In general at $m \neq 0$, stress σ_2 is within the strength condition and the plane where the shear is realized does not coincide with direction σ_2. Let us demonstrate this dependence in the form of the initial approximation $\tau_K(\sigma_K)$:

$$\tau_K = K + \sigma_K tg\rho$$

(4.3)

where K is coefficient of adhesion and ρ is angle of internal friction of the material. The stresses σ_K and τ_K in Equation (4.3) are supposed to be dependent on σ_2, i.e., cosine m in Equation (4.2) is accepted as not equal to zero. Inserting Equation (4.2) into (4.3) we receive a quadratic equation in reference to σ_1:

$$a\sigma_1^2 - 2b\sigma_1 + c = 0$$

$$a = l^2\left(\cos^2 \rho - l^2\right)$$
$$b = l^2\left(\sigma_2 m^2 + \sigma_3 n^2 + K\sin\rho\cos\rho\right)$$
$$c = \sigma_2^2 m^2\left(\cos^2 \rho - m^2\right) + \sigma_3^2 n^2\left(\cos^2 \rho - n^2\right) - 2\sigma_2\sigma_3 m^2 n^2$$
$$\quad - K\sin 2\rho\left(\sigma_2 m^2 + \sigma_3 n^2\right) - K^2 \cos^2 \rho$$

(4.4)

From Equation (4.4) it follows that

$$\sigma_1 = \frac{1}{a}\left(b \pm \sqrt{b^2 - ac}\right)$$

(4.5)

In Equations (4.4) and (4.5), σ_1 is considered as any of three (not necessarily the largest) stress. As

$$l^2 + m^2 + n^2 = 1 \tag{4.6}$$

for determination of σ_1, it is necessary to know the value of two direction cosines in addition to K, ρ, σ_2, and σ_3. The l represents the constant of the given material and does not depend on stresses σ_2 and σ_3:

$$l = \cos\left(45° + \frac{\rho}{2}\right) \tag{4.7}$$

Here σ_1 is a greater stress. Equation (4.7) is based on the fact that under uniaxial and two-axial loading, the shear plane deviates from the direction of greater main stress at $(45° - \rho/2)$. The results [54] of examining the destruction of rocks according to the scheme $\sigma_1 > \sigma_2 = \sigma_3$ also proves the acceptability of Equation (4.7). If we insert (4.6) and (4.7) into (4.4), after transformation we will receive:

$$a = \frac{1}{4}(1 + 2\sin\rho)(1 - \sin\rho)^2;$$

$$b = \frac{1}{4}(1 - \sin\rho)[K\sin 2\rho + \sigma_2(1 + \sin\rho) - 2v(\sigma_2 - \sigma_3)];$$

$$c = \frac{1}{4}\sigma_2^2(1 + \sin\rho - 2v)(1 - \sin\rho - 2\sin^2\rho + 2v) + \sigma_3^2 v(\cos^2\rho - v)$$

$$- \sigma_2\sigma_3 v(1 + \sin\rho - 2v) - K\sin\rho\cos\rho[\sigma_2(1 + \sin\rho) - 2v(\sigma_2 - \sigma_3)] - K^2\cos^2\rho \tag{4.8}$$

where $v = n^2$. Let us show that

$$\frac{1}{4}(1 + \sin\rho) \le v \le \frac{1}{2}(1 + \sin\rho) \tag{4.9}$$

We will insert in accordance with the v values a non-dimensional index f:

$$f = \frac{\sigma_3}{\sigma_2} \tag{4.10}$$

assuming that the orientation of the shear plane in the space is determined by both the properties of the material according to Equation (4.7) and the proportion of the main stresses. As the area of compression stresses is considered, we can see from Equation (4.10) that

$$0 \leq f \leq 1 \tag{4.11}$$

The extreme values f correspond to values v. At $f = 1$ ($\sigma_2 = \sigma_3$) one can conclude on symmetry grounds that the shear plane declines to equal angles with directions σ_2 and σ_3. As

$$\mu + v = 1 - \cos^2\left(45° + \frac{\rho}{2}\right) = \frac{1}{2}(1 + \sin\rho)$$

we have at $f = 1$

$$\mu = v = \frac{1}{4}(1 + \sin\rho) = v_1 \tag{4.12}$$

where $\mu = m^2$. When $f = 0$ ($\sigma_3 = 0$). the shear plane obviously deviates from the medium position (corresponding to $\sigma_2 = \sigma_3$). Within the limit, this plane coincides with σ_2 and σ_3. However, as we know from the experimental data, the destruction does not occur in the plane of σ_3 action, but it can coincide with σ_2. That is why when $f = 0$:

$$\mu = 0; \quad v = \frac{1}{2}(1 + \sin\rho) = v_0 = 2v_1 \tag{4.13}$$

We obtained the necessary change limits of v given in Equation (4.9). Based on Mohr's theory of strength:

$$\left.\begin{array}{l} \mu = 0 \\ v = \dfrac{1}{2}(1 + \sin\rho) \end{array}\right\} \tag{4.14}$$

Let us define the dependence $v(f)$ in the form of

$$v = \left(2 - f^{\frac{1}{\alpha}}\right)v_1 \tag{4.15}$$

where α is a non-dimensional parameter. Figure 4.29 shows the limits of α change. Horizontal lines 1 and 8 correspond to the conditions of Equation (4.14), curves 2 and 4 are built according to Equation (4.15) and therefore are related to parameter v. Curves 5 and 7 are related to parameter μ. The case

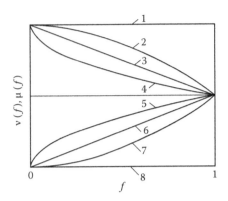

FIGURE 4.29
Graph of dependences $v(f)$ and $\mu(f)$.

$0 < \alpha < 1$ corresponds to curves 2 and 7; $\mu = 1$ to curves 3 and 6; and $1 < \alpha < \infty$ to curves 4 and 5. The figure shows that at $\alpha = 0$, curves 2 and 7 are merged with 1 and 8 and Equation (4.15) coincides with (4.14). Hence we can assume that the most probable values of α are ~0 and do not exceed 1 (as the behavior of curves $v(f)$ and $\mu(f)$ has a sharp distinction at $\alpha < 1$ or $\alpha > 1$):

$$0 \leq \alpha < 1 \tag{4.16}$$

Inserting Equation (4.15) into (4.8) we obtain:

$$a = \frac{1}{4}(1 + 2\sin\rho)(1 - \sin\rho)^2;$$

$$b = \frac{1}{8}(1 - \sin\rho)\left[\sigma_3(1 + \sin\rho)\left(2 - f^{\frac{1}{\alpha}} + f^{\frac{1-\alpha}{\alpha}}\right) + 2K\sin 2\rho\right];$$

$$c = \frac{1}{16}\sigma_3^2(1 + \sin\rho)^2\left[\left(2 - f^{\frac{1}{\alpha}}\right)\left(2 - 4\sin\rho + f^{\frac{1}{\alpha}} - 2f^{\frac{1-\alpha}{\alpha}}\right) \tag{4.17}\right.$$

$$\left. + \left(4 - 4\sin\rho - f^{\frac{1}{\alpha}}\right)f^{\frac{1-2\alpha}{\alpha}}\right] -$$

$$-\frac{1}{4}K\sigma_3(1 + \sin\rho)\sin 2\rho\left(2 - f^{\frac{1}{\alpha}} + f^{\frac{1-\alpha}{\alpha}}\right) - K^2\cos^2\rho$$

Finally stress σ_1 is calculated by Equation (4.5); before the root, a plus sign is used (as σ_1 is a higher stress). The coefficient α can be found, for example, by back-recalculation from Equations (4.5) and (4.17) on the known destruction stresses. Tentatively assuming α is the same for different rocks and equals the average upon the interval of Equation (4.16), i.e., taking $\alpha = 0.5$ from (4.17) we obtain:

$$b = \frac{1}{8}(1 - \sin\rho)\left[\sigma_3(1 + \sin\rho)(2 - f + f^2) + 2K\sin 2\rho\right];$$

$$c = \frac{1}{16}\sigma_3^2(1 + \sin\rho)^2\left[4 - 4\sin\rho - f^2 + (2 - f^2)(2 - 2f - 4\sin\rho + f^2)\right] \qquad (4.18)$$

$$-\frac{1}{4}K\sigma_3(1 + \sin\rho)\sin 2\rho(2 + f - f^2) - K^2\cos^2\rho$$

When $\sigma_2 = \sigma_3 = \sigma_0$, i.e. $f = 1$, Equations (4.17) and (4.18) are simplified:

$$b = \frac{1}{4}(1 - \sin\rho)\left[\sigma_0(1 + \sin\rho) + 2K\sin 2\rho\right];$$

$$\qquad (4.19)$$

$$c = \frac{1}{4}\sigma_0^2(1 + \sin\rho)^2(1 - 2\sin\rho) - \frac{1}{2}K\sigma_0(1 + \sin\rho)\sin 2\rho - K^2\cos^2\rho$$

If we insert Equation (4.19) into (4.5), we obtain the Coulomb–Mohr formula:

$$\sigma_1 = \sigma_0\frac{1 + \sin\rho}{1 - \sin\rho} + 2K\frac{\cos\rho}{1 - \sin\rho}.$$

We can show that at $\sigma_2 = \sigma_3$ and also at $\sigma_1 = \sigma_2$, i.e., at the equality of any two main stresses. The third is defined by the Coulomb-Mohr formula. Generally, where $\sigma_2 \neq \sigma_3$, Equations (4.5) and (4.17) should be used. The result can serve as a basis for further improvement of the mechanical theory of strength for calculating the influence of intermediate stress σ_2 on the ultimate strength of rocks under true triaxial loading. TTCU studies [12] showed that σ_2 does not affect the strength of argillite when $\sigma_0 = (\sigma_1 + \sigma_2 + \sigma_3)/3$ is more than 163 MPa or coals at $\sigma_0 > 60$ MPa. According to Doshchinskii [55], the limit state of the material is determined by absolute values of the deformation components and is expressed as the mean square value of deformation components ε_1, ε_2, and ε_3:

$$\varepsilon_m^{squ} = \sqrt{\frac{1}{3}\cdot(\varepsilon_1^2 + \varepsilon_2^2 + \varepsilon_3^2)}$$

ε_m^{squ} is considered linearly dependent on volumetric deformation for brittle materials. Studies of triaxial compression show that the materials including rocks that exhibit the same ultimate strength have different values of deformation.

These theories describe the limit states of solid bodies in a field of pure separation *or* cut. An attempt to describe material behavior at a cut *and* separation by synthesis of the known theories did not give the expected results. These strength theories do not take into account the defects arising from local perturbation of stress fields and the resulting fluctuation of mechanical properties in the most stressed or most defective areas. The model of continuous medium is inadequate because the areas of concentration of stress fields are not distributed accidentally; this led to the creation of the theory of statistical treatment of strength.

A.P. Aleksandrov and S.N. Zhurkov [56] were the first physicists to assume the static nature of the strength of solids. The concept was later developed by Weibull [57], Kontorova and Frenkel [58], Volkov [59], Bolotin [60], and others. The analysis of formulas of these authors shows that the strength of materials having large numbers of defects is primarily determined by maximum principal stress σ_1 and depends very little on other stresses. In reality, under volumetric compression, the strength of a material is highly dependent on lateral stress σ_2 and σ_3. All static theories consider that the strength of a material is determined by its local strength. However, in real solids, this condition is not met. In real solids, especially in rocks, due to microplastic deformations, local stresses are redistributed and do not cause destruction. Therefore, the static theories of strength were not developed to describe the limit states of coal and rock. Analysis of the literature suggests two basic conclusions:

1. In areas of a massif exposed to mining or geological disturbances, coal exists under true triaxial compression with different ratios of stress and deformation.

2. Existing hypotheses and strength criteria do not allow reliable and unambiguous determinations of limit states and destruction mechanisms of such defective solids as rocks and gas-saturated coals.

The closest theory to the description of the limit state of rock and coal is a modification of the concepts of Coulomb and Mohr described earlier [12]. The theory best justified on physical principles (loss of stability of a solid on the basis of the fissure spreading) was formulated by Griffiths. He devised a basic (energetic) principle of avalanche-like fissure spreading. At the edges of fissures, elastic energy required for the formation of new surfaces accumulates and is then released on the edges due to relaxation of stresses and deformations as the fissure opens. Elastic energy accumulated at the mouth of the fissure is converted to surface energy. As a small fissure grows, it consumes more energy than it produces, and at some critical point the process reverses. Griffiths defined fissures growth by the following energy criterion:

$$\frac{\partial}{\partial l}\left[U(l) - W(l, P_*)\right] = 0 \tag{4.20}$$

where $U(l)$ is the surface energy of a fissure, $W(l, P_*)$ is the energy of elastic deformation during the growth of fissure at length $2l$ at the moment of the external load impact on the body, and P, P_* is the ultimate value of load P. Griffith's approach was further developed by Irwin and Orowan [61,62]. Great contributions to the development of the theory of fissures were made by Russian and Ukrainian scientists such as S.A. Khristianovich, G.I. Barenblatt, M.J. Leonov, D.D. Ivlev, V.V. Panasyuk, G.P. Cherepanov, V.Z. Parton, E.M. Morozov, L.V. Ershov, A.E. Andreikiv, L.T. Berezhnytsky, A.A. Kaminsky, V.P. Naumenko, and others.

Practical application of the theory of fissures required solutions to a number of problems for determining critical loads in uniaxial, plane, and axisymmetric three-dimensional stress states. Effective surface energy was experimentally measured to use the theory of fissures to calculate rock stability (160 J/m² for sandstone, up to 65 J/m² for argillite, and up to 14 J/m² for coal [13].

However, the theory of fissures is rarely used to calculate the strengths of rocks and coals in true triaxial stress fields ($\sigma_1 \neq \sigma_2 \neq \sigma_3$) [53]. A detailed bibliography and results of using the fissure theory to describe the limit state of rocks can be found in [12].

4.6 Post-Limit State

All rocks and coals after destruction by three dimensional stresses exhibit residual strength, i.e., they continue to bear loads that are considerably smaller. Mining engineers use this concept to calculate the bearing capacities of roof supports and coal pillars.

The first step is obtaining complete stress–strain diagrams that show part of the curve beyond the maximum load to evaluate the energy and character of the material destruction under stress. Without considering post-limit deformation, we assume that all the energy of the ultimate deformation of the material destroys it and destruction is always brittle. However, in many cases of destruction, the load does not relax to zero; in some cases, no load-off occurs. Since plastic deformation is destruction without breaking bonds, a beyond-the-elastic limit (or compressibility) term should be added to the concept of post-limit deformation. Conventional ultimate strength should be considered the starting point of dilatancy on the curve of mean stress volumetric deformation (Figure 4.30). In this case the bearing capacity of a material beyond ultimate (residual) strength may be either smaller (softening material) or greater than ultimate strength (strain-hardening material). In creep (increasing deformation under constant pressure), residual strength is equal to ultimate strength. The coefficient of proportionality between stress and deformation (rigidity) in the post-limit area for

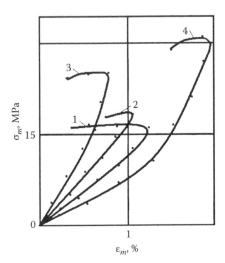

FIGURE 4.30
Dependence $\sigma_{mean} - \varepsilon_{mean}$ for samples of coal at $\sigma_3 = 1.3$ MPa (1); $\sigma_3 = 1.8$ MPa (2); $\sigma_3 = 2.5$ MPa (3); and $\sigma_3 = 5$ MPa (4).

softening materials is the decrease module; for strain-hardening materials, it is the hardening module.

4.6.1 Essence of Problem

In mining, the study of rock behavior beyond the ultimate strength is important for solving the problems of stability of a rock massif, coal, rock and gas outbursts, mining strikes, and earthquakes. Fissure spreading in rocks during destruction can be stable and occur at the expense of elastic energy accumulated in the material during its deformation. Spreading may also be unstable and require additional energy from the loading device for destruction. For stable spreading, controlled testing of the behavior of rock beyond its ultimate strength can be performed only by special testing units. Even with a loading device of infinite rigidity, controlled testing with a conventional unit is impossible (Figure 4.31). For unstable spreading, the criterion of a controlled test is the expression $E_{out.} > E_{un}$ or $|dK_{un}| > |dK_0|$, where E_{out} is the post-limit deformation energy of the sample; E_{un} is the energy accumulated in the unit, $E_{un} = p^2/2K_{un}$; K_{un} is stiffness of the unit, $K_{un} = P/\Delta l_{un}$; K_0 is stiffness of the sample beyond the ultimate strength; Δl_{un} is elongation of the unit; and P is applied load.

If the post-limit deformation energy of the sample is less than the energy accumulated in the unit, dynamic destruction of the material occurs. In many cases, explosive destruction does not reveal the properties of rocks. To obtain accurate data about rock properties, the rigidity of testing units may be increased in various ways [63,64].

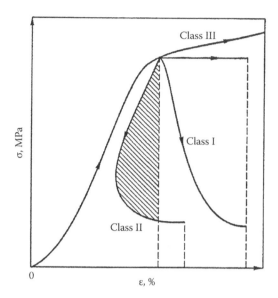

FIGURE 4.31
Stress and deformation under different conditions of destruction.

Regarding data obtained by rigid testing units, it should be noted that experiments on the post-limit states of rocks were carried out mostly under uniaxial compression and synthesis where $\sigma_2 = \sigma_3$. Under uniaxial compression, the residual strength of rock is not more than 5% of the maximum strength and the module of decrease significantly exceeds the module of elasticity. The absolute increase of volume at the total loss of load-bearing capacity is 2 to 15%.

The nature of post-limit deformation usually depends on the levels of three-dimensional stresses. With an increase of lateral pressure $\sigma_2 = \sigma_3$, a decrease in the module of decrease and increases of the post-limit deformation value and residual strength of plastic rocks were noted. For brittle rocks, the module of decrease with increasing lateral pressure remains almost constant and post-limit deformation does not depend on the level of three-dimensional stress state. Lateral pressure exerts greater effects with more plastic rocks. With an increase of lateral pressure, a significant reduction in the effect of dilatancy appears. At a certain value of lateral pressure, the maximum effect of dilatancy is observed [63].

Taking into account the parameters of post-limit deformation, a number of fragility coefficients were proposed, most of which have no physical value and do not affect the residual strength of rocks [64]. The tendency of a rock to dynamic destruction may be based on the differences between energy stored in the unit and post-limit deformation energy along with the module of decrease. We drew a number of conclusions after analyzing several studies of post-limit states and evaluating their main conclusions.

Most works classify rocks as brittle or plastic, and the results of analyzing both types differ sharply. Based on the physics of solids, we know that the rock brittleness (viscosity) is determined by thermodynamic parameters (pressure and temperature). At certain values of pressure and temperature, a material makes a transition from viscous to brittle or brittle to plastic. No studies on post-limit state yielded data that prove rocks that are brittle under uniaxial compression become viscous with an increase of lateral pressure. The conclusions that (1) as lateral pressure in plastic rocks increases, the parameters of post-limit deformation (residual strength, module of decrease) change gradually and (2) stress deformation diagrams for brittle rocks are invariant under uniaxial and volumetric compression are somewhat doubtful. Several studies of stress states in rock massifs have shown that even in virgin massif, and particularly in pre-face zones of workings, compression should be considered true triaxial, i.e., $\sigma_1 \neq \sigma_2 \neq \sigma_3$. No studies of post-limit deformation of rocks under true triaxial compression are known.

Evaluation of brittleness and tendency of rocks to dynamic destruction without considering residual strength, in our opinion, does not reflect the true expenditure of energy for destruction and thus does not reflect the essence of the phenomenon. For these reasons, we consider studies of post-limit state under conditions of true triaxial compression very important.

4.6.2 Post-Limit Deformation under True Triaxial Compression

The true triaxial compression unit (TTCU) allows fast decreases of high stress when necessary. The rigidity of the TTCU is $2.6 \cdot 10^8$ N/m to allow post-limit branching (especially when lateral stresses make sample rigidity smaller than machine rigidity). Under uniaxial compression, only the beginning and ending points (Figure 4.32) of stress decreases are registered. This provides no indication of the behavior of a post-limit branch between these two points. The figure shows complete stress deformation diagrams for sandstones under uniaxial compression. The stones were taken from mines named after A.G. Stakhanov (1) and A.A. Skochinsky (2).

Ultimate and residual strengths of both sandstones were almost equal (residual strength about 20% of the ultimate). Ultimate deformations did not differ much. The biggest difference was in the post-limit deformation (deformation of Stakhanova sandstone was 1.7 times greater than that of the Skochinskogo sandstone).

Figure 4.33 depicts stress deformation for samples from the Skochinskogo mine at different levels of minimal compressive stress σ_3 (constant intermediate stress $\sigma_2 = 10$ MPa). These data indicate this scheme of loading increase of σ_3 leads on one side to coal strengthening and on the other side to changes of sample deformation character. At σ_3 values of 1.8 and 5 MPa, stress decrease and brittle fracture were observed. Ultimate deformation of samples was achieved by continuous power feed from a loading unit. The changes of σ_3 at $\sigma_2 = $ const and σ_1 at $\sigma_3 = $ const led to a change in stress state. Because of

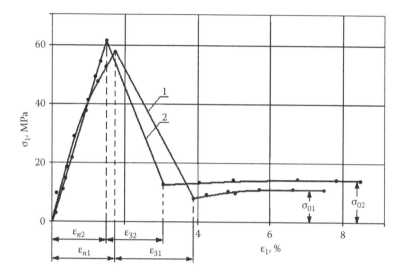

FIGURE 4.32
Complete stress deformation diagrams for sandstones from Stakhanova mine (curve 1) and Skochinskogo mine (curve 2).

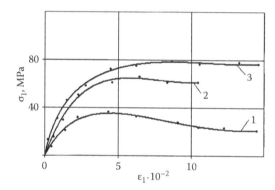

FIGURE 4.33
Complete stress-deformation diagram of coal samples of the seam h_6^1 of Skochinskogo mine at $\sigma2 = 10$ MPa for different values of minimal stress $\sigma3$: 0.2 MPa (1), 1.8 MPa (2), 5 MPa (3).

lack of homogeneity, rock fracturing and plasticity vary considerably under deforming and stress conditions (Figure 4.34) and even under equal stress conditions (for example, $\sigma_1 > \sigma_2 = \sigma_3$). Parameters of post-limit deformation (residual strength and decay) are subject to considerable fluctuations. Depending on lateral stress and type of stress state, a single material can behave as softening and hardening and destruction can be brittle as well as viscous.

The difference between ultimate and residual strengths (stress decrease) is true for coal and sandstone with lateral stress exceeding 30 MPa. The maximal decay connected with the most brittle fracture is observed for

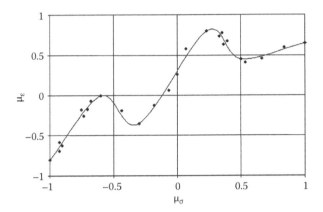

FIGURE 4.34
Dependence of deformation condition on stress state for sandstones in TTCU experiments.

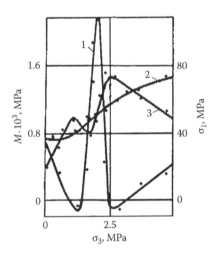

FIGURE 4.35
Dependence of decay module (1), ultimate (2), and residual (3) strengths on minimal compression stress σ_3.

coal (Figure 4.35) at $\sigma_2 = \sigma_3 = 2$ MPa (destruction along crystallites), and for sandstone (Figure 4.36) at $\sigma_2 = \sigma_3 = 0$ and 10 MPa (destruction along cement and flakes of quartz, respectively). Figure 4.37 shows dependence of residual strength and decay modules of coal samples on stress state μ_σ at $\sigma_3 = 9$ MPa. The study proves that in the area of generalized compression, where the parameter of stress state type μ_σ varies from –1 to –0.5, the decay module may vary considerably while ultimate and residual strengths increase.

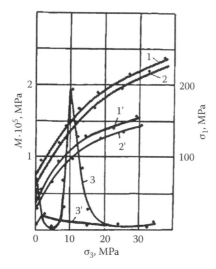

FIGURE 4.36

Dependence of ultimate and residual strengths and decay module on minimal compression stress σ_3 for sandstones (1, 2, and 3) and sandy shale (1′, 2′, and 3′).

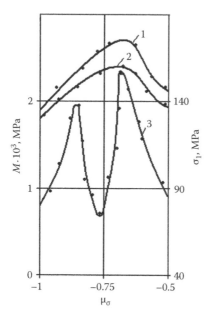

FIGURE 4.37

Dependence of ultimate (1) and residual (2) strengths and decay module (3) on stress state type for coal at $\sigma_3 = 9$ MPa.

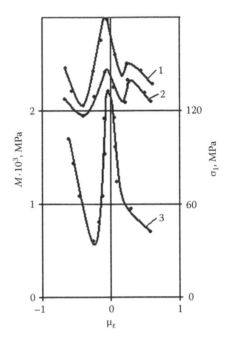

FIGURE 4.38
Dependence of ultimate (1) and residual (2) strengths and decay module (3) on strain state type for coal at $\sigma_3 = 9$ MPa.

Figure 4.38 shows dependences of ultimate and residual strengths and decay modules of coal samples on $\mu\varepsilon$ under conditions of true triaxial compression $(\sigma_1 > \sigma_2 > \sigma_3)$ at $-1 \leq \mu_\sigma \leq -0.5$ and $\sigma_3 = 9$ and 20 MPa. These data prove nonconformity of strain and stress states and also reveal wide ranges of residual strength decay module change. Maximum strength and decay modules were observed under a strain state corresponding to generalized shear ($\mu_\varepsilon = 0$). The maximum decay module for sandstones samples (Figure 4.39) with the growth of σ_3 shifted from generalized tension at $\sigma_1 > \sigma_2 > \sigma_3$ ($\mu_\varepsilon = 0.4$) to generalized shear and compression at σ_3 equal to 9 and 20 MPa ($\mu_\varepsilon = -0.4$ and -0.9, respectively). Underground tunnel support capacity is determined by the residual strength of destroyed rock and the bearing capacity of the lining. The character (brittleness, viscosity) of coal and enclosing rock destruction is based on a decay module.

Under mine conditions brittle fracture and sudden loading decrease are often accompanied by dynamic phenomena. Because the maximum decay module for coal is $\sigma_3 \neq 0$ in a generalized shear area, it is necessary to consider that the focus of a brittle fracture forms in the depth of a massif. Sandstone decreases the capacity for dynamic destruction near working faces and in the depths of massifs.

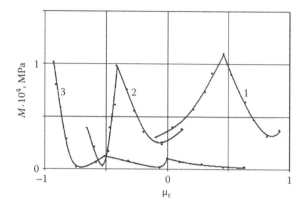

FIGURE 4.39
Dependence of decay modules of sandstone on strain state type $\mu\varepsilon$ at $\sigma_3 = 0$ MPa (1); $\sigma_3 = 9$ MPa (2); and $\sigma_3 = 20$ MPa (3).

To study post-limit branching of a diagram of stress deformations of brittle and gas-saturated minerals with the TTCU, the unloading method was used. After predetermining mine rock strengths under certain stress states by experimentation, samples were loaded at lateral pressures a little higher than the strength limit and unloading along σ_3 continued until the specified value was reached. The accumulated energy in the machine at sample loading scattered before the samples were destroyed so destruction was based only on the energy accumulated in the samples.

Figure 4.40 shows stress deformation for non-gas-saturated and methane-saturated sandstone under 5 MPa pressure for two days. Sandstone strength at $\sigma_2 = 30$ MPa and $\sigma_3 = 5$ MPa was $\sigma_1 = 200$ MPa. For that reason, the sample was loaded at $\sigma_2 = \sigma_3 = 30$ MPa until $\sigma_1 = 200$ to 230 MPa, then σ_3 was unloaded to 5 MPa. After sample destruction, unloading of all three pressures to 0 was realized.

A complete stress deformation diagram may be divided into sections: (1) pre-ultimate deformation (*OA*) in which the space under the curve is the energy accumulated by the sample via deformation; (2) unloading (*Aa*) σ_3 from 30 to 5 MPa (ultimate deformation due to machine energy) in which the space under the curve is the machine energy spent for deformation; (3) sample unloading (*ab*) due to the energy accumulated in the sample; and (4) unloading of stresses to 0 (*bc*) in which the space under the curve is remaining elastic energy (S_{cbf}). The blank area (*OAabc*) is the energy spent on sample destruction. However some part of this energy (*Sodbc*) is from steady defects (precritical cracks and microdefects); the remainder (*SdAab*) is spent directly on new surface formation (material crushing). This system allows us to estimate the dynamics (destruction character) of material under a given stress state as the ratio of destruction energy to total deformation energy

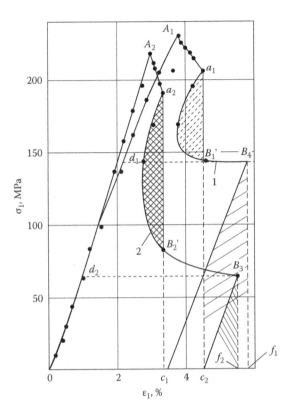

FIGURE 4.40
Complete stress deformation of non-gas-saturated (1) and gas-saturated (2) sandstones.

$K_{br}^{I} = S_{OAabc}/S_{OAabf}$, as the ratio of crushing energy to total deformation energy $K_{br}^{II} = S_{dAab}/S_{OAabf}$, or as the ratio of crushing energy to destruction energy $K_{br}^{III} = S_{dAab}/S_{OAabc}$. The above coefficients of brittleness (dynamics) in contrast to coefficients suggested in [65] have a distinct physical sense. It is not difficult to show that

Total energy of deformation: $A = S_{OAabf} = \dfrac{2\sigma_{ult}\sigma_{sup} \cdot (M+E) - 2E\sigma_{ult}^2 - M\sigma_{sup}^2}{2EM}$

Destruction energy: $A_{destr} = S_{OAabc} = \dfrac{\sigma_{ult}\sigma_{sup}(M+E) - M\sigma_{sup}^2 - E\sigma_{ult}^2}{EM}$

Crushing energy: $A_{rag.} = S_{OAab} = \dfrac{\sigma_{sup}^2(E-M) - E\sigma_{ult}^2}{EM}$

Then:

$$K_{br}^{I} = 1 - \frac{M\sigma_{sup}^2}{2\sigma_{ult}\sigma_{sup}(M+E) - 2E\sigma_{ult}^2 - M\sigma_{sup}^2}$$

$$K_{br}^{II} = 1 - \frac{\sigma_{sup}\left[M\sigma_{sup} + 2(M+E)(\sigma_{ult} - \sigma_{sup})\right]}{2\sigma_{ult}\sigma_{sup}(M+E) - 2E\sigma_{ult}^2 - M\sigma_{sup}^2}$$

$$K_{br}^{III} = \frac{\sigma_{sup}^2(E-M) - E\sigma_{ult}^2}{\sigma_{ult}\sigma_{sup}(M+E) - E\sigma_{ult}^2 - M\sigma_{sup}^2}$$

To estimate mine rock destruction, any coefficient of brittleness may be used. K_{br}^{I} is the easiest to use and understand and for that reason we will use it.

If post-limit strength $\sigma_{sup} = 0$, then all deformation energy is spent for destruction and $K_{br} = 1$. When ultimate and residual strengths are equal, $K_{br} = 0$ (plastic flow). If M and E have the same sign, then $K_{br} > 1$ and destruction arises from elastic energy accumulated in a sample. In all other cases $0 < K_{br} < 1$ is true.

Dynamic destruction of the material is possible in two situations: (1) when $K_{br} > 1$ regardless of rigidity of loading unit and sample; (2) when $0 < K_{br} < 1$, material rigidity is greater than the rigidity of a loading unit. The greater the brittleness, the more dynamic the destruction.

Applying the results obtained for non-homogeneous rocks [66] to conditions in a rock massif, we suggest that outburst- and impact-dangerous situations occur regardless of the rigidity of the destructive layer and enclosing rocks. An outburst occurs when an applied stress exceeds resistibility to the destruction of this layer. For an outburst, the decay module must exceed the rigidity of the enclosing rocks. The destructive layer must be non-homogeneous and characterized by reductions of crack resistance.

Figure 4.40 data show that the ultimate and residual strengths of gas-saturated samples are sufficiently lower than those of unsaturated samples. The energy accumulated in a gas-saturated sample is smaller and the energy spent for destruction is much greater. Thus, a complete diagram of stress deformation can estimate energy capacity, destruction character, residual strength, and bearing capacity of a deposit. Knowing these properties can help solve many mining problems.

Taking into account the results [13] for mine rock dilatancy at destruction in a true triaxial field of compressive stresses, we should note that decay module maxima and dilatancy maxima coincide. That is why we can state that brittle fractures from dynamic effects and considerable stress decreases occur when cracks of in-plane and out-of-plane shear spread and lead to fragmentation and crushing. Conversely, when separation fissures spread in mine rocks, they produce minimum fragmentation and crushing, the least stress decrease, and the least decay.

4.7 Conclusions

We developed a method of strengthening unstable roofs of mine workings by determining the optimal hardening times of carbamide and magnesia solutions. Optimal strengthening time is 3 to 7 days for carbamide solution and 10 to 13 days for magnesia solution. Increases of moisture in strengthened rocks and decreasing the surrounding temperature slow the hardening of the carbamide and magnesia solutions. Increasing glue layer thickness did not affect strength and deformational properties considerably. Results for the carbamide solution strengthening showed marked sample rigidity. Moisture increases led to decreased rigidity in samples hardened by the magnesia solution and a decrease of rigidity in samples bound with the carbamide solution. Increases of moisture improved the plastic properties of all hardened samples.

A temperature decrease in the surroundings led to a decrease of strength of rock samples bound by carbamide solutions; the decrease did not affect the strength of the samples bound by magnesia solution and improved the plastic properties of samples during hardening with one or both solutions.

A complete stress deformation diagram allows analysis of power capacity, destruction character, residual strength, and bearing capacity of a deposit. Such data are important for solving mining problems. During mine rock destruction under a true triaxial field of compressive stresses, the maxima of decay module and dilatancy coincide. Brittle destruction with dynamic effects and considerable stress decreases occur when the fissures of in-plane and out-of-plane shear spread; they are accompanied by considerable material fragmentation and crushing. The spread of normal separation fissures in mine rocks leads to minimum material fragmentation and crushing, least stress decrease, and the lowest value of the decay module.

The principal difference in rock sample behavior during loading and unloading is in the accumulated power and the type of strain state. During unloading, material crushing increases considerably because of the development of in-plane and out-of-plane shear. The ultimate destruction stresses of preloaded rock samples differ little from those of samples that were not preloaded. The results of the tests on mine rock samples without pre-compression are adequate for technical calculations and we can disregard sample loading history.

References

1. Karman Th. 1911. Festigkeitsversuche unter ellseitigen Druck. *Zeitsch. vereins dutgcher Ingen.* 55, 1749–1767.
2. Böker R. 1915. Die Mechanik der bleidenden Förmanderung in kristallinisch aufgebaunten Körpern. *Forschungs. Ing. Weg.* 24, 1–51.
3. Bridgeman P. 1955. *Study of Large Plastic Deformations and Rupture* [Russian translation] Moscow: IL.
4. Vinogradov V.V. 1989. *Geomechanics of Control of Rock Massifs Adjacent to Mine Openings.* Kiev: Naukova Dumka.
5. Bulat A.F., Sofiyskiy K.K., Silin D.P. et al. 2003. *Hydrodynamic Effect on Gas-Saturated Coal Beds.* Dnepropetrovsk: Poligrafist.
6. Sayers C. M., Van Munster J.G., and King M.S. 1990. Stress-induced ultrasonic anisotropy in Berea sandstone. *Int. J. Rock Mech. Min. Sci. Geomech. Abstr.* 27, 429–436.
7. Smart B.G.D. 1995. A true triaxial cell for testing cylindrical rock specimens. *Int. J. Rock Mech. Min. Sci. Geomech. Abstr.* 32, 269–275.
8. King M.S., Chandry N.A., and Chakeel A. 1995. Experimental ultrasonic velocities and permeability for sandstones with aligned cracks. *Int. J. Rock Mech. Sci. Geomech. Abstr.* 32, 155–163.
9. Al-Harthy S.S., Dennis J.W., Jing X.D. et al. 1998. Hysteresis true triaxial stress path and pore pressure effect on permeability. In *Proc. SPE/ISRM Eurock*, Trondheim, Norway, pp. 1–9.
10. Al-Harthy S.S., Jing X.D., Marsden J.R. et al. 1999. Petrophysical properties of sandstones under true triaxial stresses: directional transport characteristics and pore volume change. SPE 57287. Imperial College, London, pp. 1–15.
11. Alekseev A.D., Osyka E.I., and Todoseichuk A.D. 1973. *Setup for triaxial compression testing of samples.* Author Certificate 394692, USSR.
12. Alekseev A.D. and Nedodaev N.V. 1982. *The Limiting State of Rocks.* Kiev: Naukova Dumka.
13. Alekseev A.D., Revva V.N., and Ryazantsev N.A. 1989. *Rock Failure in a Volumetric Field of Compression Stresses.* Kiev: Naukova Dumka.
14. Alekseev A.D., Ryazantsev N.A., and Sukharevskii B.Y. 1985. Structural transformations of rock-forming minerals under the failure of sandstone. *Dokl. Akad. Nauk. SSSR*, 284, 949–952
15. Lapteva A.V., Suljatitskii I.D., Starikov G.P. et al. 1983. *Setup for three-dimensional loading testing of samples.* Author Certificate 989370, USSR.
16. Koifman M.I. 1963. The main scale effect in rocks and coals. In *Problems of Mining Mechanization.* Moscow: AN SSSR, pp. 30–35.
17. Chirkov S.E. 1981. *Methodology of Prediction of Coal Strength.* Moscow: Institut gornogo dela imeni A.A. Skochinskogo.
18. Beron A.I. 1973. *Study of Strength and Deformability of Rocks.* Moscow: Nauka.
19. Ilnitskaja E.I, Teder R.I., Batalin E.S. et al. 1969. *Properties of Rocks and Methods of Their Determination.* Moscow.
20. Sedov L.I. 1920. *Mechanics of Continua.* Moscow: Nauka.
21. Bezuhov N.I. 1961. *Fundamentals of Elasticity, Plasticity and Creep Theory.* Moscow: Vysshaja shkola.

22. Pisarenko G.S. and Lebedev A.A. 1969. *Deformation and Fracture Resistance of Materials under Combined Stress*. Kiev: Naukova Dumka.
23. Alekseev A.D., Revva V.N., Ulyanova E.V. et al. 1989. *Method of determination of degree of strengthening of unstable mine roof*. Author Certificate 1724881 SSSR.
24. Fyfe W.S., Price N.J., and Thompson A.B. 1982. *Fluids in the Earth's Crust*. Moscow: Mir.
25. Kartashov Y.M., Matveev B.V., and Mikheev G.V. 1979. *Strength and Deformability of Rocks*. Moscow: Nedra.
26. Ershov L.V., Liberman L.K., and Neiman I.B. 1987. *Rock Mechanics*. Moscow: Nedra.
27. Sneddon J.N. and Berry D.S. 1961. *Classical Theory of Elasticity*. Moscow: Gos. Izdatelstvo.
28. Norel B.K. 1982. *Variation of Mechanical Strength of Coal Bed in a Massif*. Moscow: Nauka.
29. Alekseev A.D., Norel B.K., and Starikov G.P. 1983. Mechanical tests of coal samples on the triaxial compression set-up. *Fiz. tekhn. probl. razrab. polezn. iskop.* 1, 106–108.
30. Starikov G.P 2001. Peculiarities of coal deformation and fracture under volumetric compression. In *Geotechnologies at the Turn of the XXI Century*, Vol. 1. Donetsk: Donetskii Gosudarstvennyi Tekhnicheskii Universitet, pp P.81–87.
31. Alekseev A.D., Ryazantsev N.A., Starikov G.P. et al. 1989. Effect of adsorbed moist and methane on post-limit properties of rocks and coals. *Fiz. tekhn. vysok. davl.* 30, 48–51.
32. Alekseev A.D., Zhukov A.E., Lunev S.G. et al. 1991. Prediction and control of coal massive state by SAS solutions. *Proc. 24th Int. Conf. Safety of Works in Coal Mining*, Vol. 1, Donetsk, pp. 517–527.
33. Voloshina N.I. and Starikov G.P. 2004. *Experimental Check of Parameters and Technology of Outburst Prevention under Shield Extraction*. Donetsk: Donetskii Gosudarstvennyi Tekhnicheskii Universitet, pp. 66–73.
34. Nakaznaya L.G. 1973. *Flow of Liquids and Gases through Fissured Filters*. Moscow: Nedra.
35. Kerenes E.E. 1975. *Methods of Investigation of Filtration Properties of Rocks*. Moscow: Nauka.
36. Alekseev A.D., Nedodaev N.V., and Starikov G.P. 1990. *Changes in Fissured Porous Structure and Gas Permeability of Coals under Non-Uniform Volumetric Compression*. Moscow: Institut gornogo dela imeni A.A. Skochinskogo, pp 19–23.
37. Vasuchkov Y.V. 1986. *Physico-Chemical Methods of Degassing Coal Beds*. Moscow: Nedra.
38. Alekseev A.D., Starikov G.P., and Filippov A.E. 2003. Numerical simulation of methane escape from coal with the account for unloading wave and porosity opening under change of stresses. *Problemy girnychogo tysku* 9, 120–151.
39. Koifman M.I. 1963. Main scale effect in rocks and coals. In *The Problems of Mining Mechanization*. Moscow: Izdatelstvo Akademii Nauk SSSR, pp. 30–35.
40. Chirkov S.E. 1981. *Prediction Technique for Coal Strength*. Moscow: Institut gornogo dela imeni A.A. Skochinskogo.
41. Ilnitskaya E.I., Teder R.I., Batalin E.S. et al. 1969. *Characteristics of Rocks and Methods of their Determination*. Moscow.

42. Revva V.N., Starikov G.P., and Alekseev A.D. 1981. Variation of mechanical properties of coals with the occurrence depth. *Fiz. tekhn. vysok. davl.* 3, 43–46.
43. Starikov G.P., Streltsov V.A., and Yarembash A.I. 1989. On strength properties of rocks under conditions of high rock pressures. *Fiz. tekhn. vysok. davl.* 32.
44. Starikov G.P. and Grebenkina A.O. 2004. Effect of unloading rate on the change of fissured porous structure of coal. *Visti Donetskogo girnychogo instytutu* 1, 18–21.
45. Alekseev A.D., Ilushenko V.G., and Starikov G.P. 1993. Method of reduction of emission intensity under development workings. *Fiz. tekhn. vysok. davl.* 3, 38–44.
46. Alekseev A.D., Zaidenvarg V.B., Sinolitskii V.V. et al. 1992. *Radiophysics in Coal Industry*. Moscow: Nedra.
47. Fridman Y.B. 1956. Diagram of structural inhomogeneity. *Dokl. akad. nauk. SSSR* 106, 256–261.
48. Fridman Y.B. 1949. *Unified Theory of Strength of Materials*. Moscow: Oborongid.
49. Sneddon J.N. and Berry D.S. 1961. *Classical Theory of Elasticity*. Moscow: Gosudarstvennoe.
50. Genki G. 1948. On the theory of plastic deformations and the resulting residual stresses. In *The Theory of Plasticity*. Moscow: Izdatelstvo, pp. 114–136.
51. Anderson P. 1954. *The Dynamics of Faulting*, 2nd ed. London: Onver and Boyd.
52. Broms-Bengt B.A. 1986. Notes on strength. *Soc Rock. Mech. Jisbon* 2, 146–158.
53. Alekseev A.D., Kogan L.P., and Zhuravlev V.I. 1971. Investigation of conditions of rock failure in a volumetric field of compressing stresses. *Razr. mestoro. polez. iskop.* 26, 62–66.
54. *Atlas of Mechanical Properties of Rocks*. 1968. Leningrad.
55. Doshchinskii G.A. 1957. Theory limiting elastic state. *Izvest. tomckogo politekhn. Inst.* 85, 343–354.
56. Aleksandrov A.P. and Zhurkov S.N. 1933. *Brittle Failure Phenomenon*. Moscow: GITTL.
57. Weibull.V., Ed. 1961. *Fatigue Testing and Analysis of Results*. New York: Pergamon.
58. Kontorova T.A. and Frenkel Y.I. 1941. Static theory of brittle stability of real crystals. *Zhurnal Technicheskoi Fiziki*, 3, 173–183.
59. Volkov S.D. 1960. *Statistical Theory of Strength*. Moscow: Mashgiz.
60. Bolotin V.V. 1965. *Static Methods in Structural Mechanics*. Moscow.
61. Irwin G.R. 1957. Analysis of stresses and strain near end of crack transversing a plate. *J. Appl. Mech.* 24, 361–364.
62. Orowan E.O. 1950. *Fundamentals of Brittle Behavior of Metals: Fatigue and Fracture of Metal*. New York: John Wiley & Sons, pp. 139–167.
63. *Introduction to Mechanics of Rock*. 1983. Moscow: Mir.
64. Stavrogin A.N., Pevzner E.D., and Tarasov B.G. 1981. Post-limit characteristics of brittle rocks. *Fiz. tekhn. probl. razrab. polez. iskop.* 4, 8–15.
65. *Catalogue of Tables of Mechanical Characteristics of Rocks Dangerous in Regard to Dynamical Phenomena*. 1980. Leningrad: VNIMI.
66. Alekseev A.D. and Ryazantsev N.A. 1985. Strength and crack growth resistance of nonuniform rocks. *Razrab. mestor. polez. iskop.* 72, 57–61.

5

Genesis of Natural Gases, Methane Extraction, and Coal Mining Safety

This chapter focuses on the genesis of methane in fossil coal, acceleration of the process of methane extraction from coal, and development of ways to predict and control gas-dynamic phenomena. Some parts of this chapter may be of special interest because they describe techniques for identifying, forecasting and controlling unexpected outbursts of coal, rock and gas developed in our laboratories. This chapter also discusses the first instrument for analyzing the amount and pressure of methane in a coal seam and its practical use for determining safe loads on mining production faces.

5.1 Genesis of Gases in Coal-Bearing Series

The nature of gas genesis in coal-bearing series is not completely known. Most researchers think that methane, the main gas in coal fields (60 to 98%), results from the biochemical processes of vegetal matter decay and the process has not changed for millions of years. Metamorphic coal conversion creates porous structure housing forces that maintain a natural equilibrium of coal and gas [1,2].

The sorption model is a popular tool for analyzing gas genesis in coal seams. Based on the model, intensive flows of mantle gases (hydrogen, methane, and others) were sorbed by pores and cracks of coal-bearing seams during mantle degassing. The sorptive capabilities of coals increased at higher rates of metamorphism during conversion. Saturation of coal-bearing strata by sorption of exogenous gases also occurred in the distant past [3].

Current geological models can be disputed in the light of new scientific achievements. We can assert that the genesis of natural gas in coal-bearing seams varies by time and geological conditions. Field and laboratory research indicates that mine methane has several sources:

1. Methane of metamorphogenic origin forms in coal seams and surrounding rock in situ at different stages of coal deposit formation. It

is the most predictable and indicates background methane saturation of rock massifs.

2. Methane of deep-seated origin migrates along networks of tectonic
fissures in sediments into a developed massif and creates zones of
anomalous saturation. The zones emit methane into mine workings
and dramatically increase the dangers of outbursts, explosions, and
methane poisoning. This type of methane falls into two classes: (a)
methane from deeper coal seams and from oil or gas deposits under
coal deposits [4,5]; and (b) mantle methane that directly penetrates
sediment rock with coal seams through ruptures in crystalline base-
ments [5,6].

3. Methane generated by chemical reactions during coal formation.
This newly formed methane has not been researched extensively.
Its danger arises from its location in local zones of coal seams (up
to several meters) and its presence cannot be predicted by current
techniques.

5.1.1 Methane Isotopic Analysis: Literature Review

Let us examine the generation of deep-seated and newly-formed methane
in more detail. Space–time localization and migration are usually studied
by isotopic analysis. The PDB international standard* is used to analyze the
isotopic compositions of the carbon in coal, methane, and carbon dioxide.
According to the standard, $C^{13}/C^{12} = 0.0112372$. Offset of isotopic composition
relative to the standard ($\delta\,^{13}C$) is determined according to the formula:

$$\delta\,^{13}C = \left[\frac{\left(C^{13}/C^{12}\right)_{sample}}{\left(C^{13}/C^{12}\right)_{PDB}} - 1 \right] \cdot 1000,\ ‰.$$

Isotopic research in many countries [6–13] showed that scattered carbon in
ultrabasic and basic mantle strata has a typical range of values $\delta\,^{13}C$ from –22
to –27‰ [6-8]. A significant range of fluctuations of carbon isotopic composition characterizes hydrocarbon gases. The heaviest are the gases connected
with magmatic activity ($\delta\,^{13}C = -10$ to $-30‰$) and the lightest are the gases
of biochemical genesis ($\delta\,^{13}C = -50$ to $-80‰$) [8,9]. Values for gases of oil–gas
deposits are in the middle ($\delta\,^{13}C = -30$ to $-50‰$) [10]. Methane is considered
abiogenic at $\delta\,^{13}C \geq -20‰$ [11].

Methane under conditions of magmatic melt has the same isotopic composition as the initial graphitic carbon ($\delta\,^{13}C = -3.2$ to $-12.8‰$). Further from
areas of high temperature, due to exchange with CO_2, it is enriched by a

---

* The name of the standard is based on the Peedee formation (PD) in South Carolina and the
belemnites (B) found there. Belemnites were chosen because of their homogeneous isotopic
composition.

light isotope up to the values shown for volcanic gases and thermal springs ($\delta^{13}C = -23‰$) [12,13].

One study [14] shows the role of CO_2 hydrogenation in the formation of natural gas of coal seams in the presence of an iron catalyst. Fe_2O_3 served as a catalyst [15]. CO_2 hydrogenation ($H_2/CO_2 = 4$) reactions [14] were carried out at 167, 180, and 192°C at 1 atm pressure. One hour after the beginning of the reaction (at 192°C), the authors detected methane along with carbon monoxide, ethane, propane, and water. Results indicated that the reaction of CO_2 hydrogenation [14] generates methane and light hydrocarbon with a composition close to natural gas and methane contents more than 90%. The authors calculated the speed of methane generation in an experimental installation and as a result assumed an additional mechanism of natural gas generation in coal seams related to geological epochs [14]. Studies of the artificial metamorphism of kerogens [16–18] show that the compositions of gases generated during thermolysis gases under dry and moist conditions are not similar to natural gas composition. The presence of ferrous minerals and clays during artificial metamorphism confirmed their catalytic effect [19]. Based on these results, we drew a conclusion about the catalytic effects of minerals in gaseous hydrocarbon generation. In addition to open porosities and fractures, coal contains another reservoir for gas accumulation [20–22]. Most gas accumulated in coal seams is concentrated within a solid solution and closed porosities. To confirm this concept, natural coking and fat coals consisting of far fewer open pores than other ranks can contain up to 35 m^3/t of gas.

5.1.2 Methane Isotopic Analysis: Experiment and Discussion

We examined volatile, gaseous, fat, coking, lean coking, lean, and anthracite coals of the Donets coal basin. Table 5.1 lists their characteristics.

Isotopic analyses of methane (CH_4) and carbon dioxide (CO_2) were conducted at the Zasyadko and Krasnolimanskaya Mines. From structural and

TABLE 5.1

Results of Chemical Analysis of Coal Samples

Rank	$V^G\%$	C%	H%	W%	Ash%
Volatile	42.9	81.9	5.6	11.5	2.87
Gaseous	35.6	85.0	5.5	7.4	1.35
Fat	34.1	86.1	5.4	1.0	1.9
Coking	23.7	89.1	5.15	1.0	3.9
Lean baking	21.4	90.0	4.94	1.1	2.79
Lean	11.2	91.8	4.55	1.7	3.5
Anthracite	7.0	95.4	2.2	4.0	1.0

Note: V^G = volatile per gram of coal substance. W = water per gram of coal substance.

tectonic views, the field of the Zasyadko mine (seams l_1 and m_3) is complicated by three thrusts: Vetkovsky (H = 15 to 60 m), Panteleymonovsky (H
= 10 to 20 m), and Grigoryevsky (H = 37 to 55 m). H is the thrust amplitude.
A gravimetric survey on a 1:200,000 scale carried out by the National Mining
University of Ukraine determined that the three thrusts by orientation and
location are close to faults in the crystalline basement.

At the Zasyadko Mine, gas samples were taken from degassing wells
drilled into the roof of coal seam l_1 near small amplitude tectonic faults,
and from holes drilled along coal seam m_3 closer to the Vetkovsky thrust.
Chemical composition of gases was determined on an LHM-8MD chromato-
graph. Isotopic compositions of the carbon of coal, methane, and carbon
dioxide were analyzed on an MI-1201V mass spectrometer and the PDB standard was used.

Analysis of the compositions of gases from two zones of tectonic faults on
seam l_1 revealed increases of heavy hydrocarbon content (up to 8.6%), hydrogen (up to 0.14%), and helium (up to 0.12%). In the first zone, an accident emitted total methane exceeding 100,000 m^3. The registered methane emission
rate in the second zone was 137m^3/min. Another finding was a change of
weighting of methane and carbon dioxide: $\delta^{13}C_{CH4}$ from –31.48 to –29.8‰, δ
$^{13}C_{CO2}$ from –24.9 to –17.2‰, while $\delta^{13}C_{coal}$ was from –26.8 to –26.5‰.

Both zones of high gas-saturation of coal rock massif were characterized
by conformance of isotopic compositions of carbon of methane and carbon
dioxide. Such conformity can result from an increase of coal metamorphism
or fractionation of carbon isotopes between CH_4 and CO_2 at temperatures
above 200°C. Thus the observed conformity of isotopic compositions of CH_4
and CO_2 indicates that the discovered zones formed from gas inflows from
deep sources. Isotopic compositions of the carbon of coal, methane, and carbon dioxide in pairs of samples taken from two corners of the drift bottom
of seam m_3 at different distances from the Vetkovsky thrust are presented in
Table 5.2.

TABLE 5.2

Isotopic Compositions of Carbon of Coal, Methane, and Carbon Dioxide

Sample	Distance from Vetkovsky Thrust (m)	$\delta^{13}C$ (‰)		
		CH_4	CO_2	Coal
1	206	–	–	–24.73
2	206	–42.50	–	–
3	206	–41.65	–	–
4	122	–24.36	–18.55	–
5	122	–35.13	–13.74	–
6	30	–30.30	–21.90	–23.63
7	30	–20.40	–21.35	–

Near to the Vetkovsky thrust, methane carbon became significantly heavier from −42.5 to −30.3 and −20.4‰. This suggests the possibility of heavy methane inflow along the thrust and through its apophyses (minute faults connected with the main thrust) into the developed coal seam. The sample 7 values $\delta^{13}C_{CH4} > \delta^{13}C_{CO2}$ testify to genetically different methane and carbon dioxide in a coal seam. On one side, methane balanced with carbonates at temperatures above 500°C may present a higher $\delta^{13}C$. Its inflow along the Vetkovsky thrust from the depth may increase the $\delta^{13}C$ of total gas composition. From the other side, it is possible that methane formed at deeper layers migrated through the Vetkovsky thrust. In this case, more intensive gas emissions in coal faces may be related to foreign gases.

At the Krasnolimanskaya Mine, analysis of the isotopic composition of the carbon of methane and carbon dioxide was conducted in the drift bottom of seam k_5 closer to the apophysis of the Glubokoyarsky fault (H = 3.5 m). The H of the Glubokoyarsky fault is 20 to 86 m. A network of small amplitude faults complicates coal seam k_5 and surrounding rock in the area of the drift development. At the development, higher gas emissions were registered. The composition of the gas emitted from the rock was analyzed. The results showed that methane content changed from 92.0 to 98.5%, ethane from 0.39 to 0.71%, propane from 0 to 0.36%, and butane from 0 to 0.1%. Helium (0.05 to 0.19%) and hydrogen (0.0 to 0.002%) content increased in the zones of small amplitude faults at Glubokoyarsky, confirming the heterogeneous inflow of these gases from the depth.

The values of $\delta^{13}C$ of methane in gas mixture of coal seam k_5 of Krasnolimanskaya changed between −34.51 and 37.58‰, revealing a slight tendency to increase closer to the Glubokoyarsky fault. Carbon dioxide $\delta^{13}C$ ranged widely from −8.12 to −20.75‰ and decreased closer to the Glubokoyarsky fault. Carbon dioxide enrichment by lighter carbon isotope ^{12}C occurs via greater coal carbonization. The regularity indicates inflow of isotope-light carbon dioxide from deeper coal levels.

The difference $(\delta^{13}C_{CO2} - \delta^{13}C_{CH4})$ in the gas mixture from seam k_5 changed from 16.61 to 28.93‰ and decreased closer to the Glubokoyarsky fault. The variations of $\delta^{13}C_{CH4}$ and $\delta^{13}C_{CO2}$ along the drift indicate that the gas generated from coal of a higher level of metamorphism with $\delta^{13}C_{CH4}$ and $\delta^{13}C_{CO2}$ closer in value to those typical for seam k_5 rose from the depth through tectonic faults near Glubokoyarsky. Figure 5.1 is a comparison of the $\delta^{13}C_{CH4}$ and $\delta^{13}C_{CO2}$ results for the two mines.

The figure shows specifics of distribution of samples from Zasyadko and Krasnolimanskaya. The samples from Zasyadko show a significant range of values of both $\delta^{13}C_{CH4}$ (−42.5 to−−20.4‰) and $\delta^{13}C_{CO2}$ (−24.88 to −6.5‰). Values from Krasnolimanskaya samples covered a wide range ($\delta^{13}C_{CO2}$ from −20.75 to −8.12 ‰) and a small range ($\delta^{13}C_{CH4}$ from −37.58 to −34.51‰). On average, methane at Zasyadko revealed a heavier isotope composition of carbon. In addition, several samples with $\delta^{13}C_{CH4} \geq 25$‰ contained methane with isotope carbon composition characteristic of volcano gases and thermal springs.

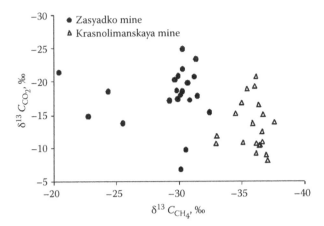

FIGURE 5.1
Distribution of analyzed gas samples in coordinates $\delta^{13}C_{CH4}$ and $\delta^{13}C_{CO2}$ ($\pm 0.5‰$).

Mine carbon dioxide at Krasnolimanskaya was heavier. In the zones of tectonic faults, these differences are more noticeable.

It should be noted that both mines developed seams of the same rank (fat) in different geological–industrial areas and in works at different depths. At Zasyadko, seam l_1 is 1100 m and seam m_3 is 1340 m. At Krasnolimanskata, seam k_5 is 950 m. These conditions most likely determine the differences in genesis and migration of deep methane into mine workings. At Zasyadko, one can find methane close to mantle genesis migrating along the fractures in crystalline foundations and tectonic faults in sedimentary depths from higher mantle layers. The predominant methane at Krasnolimanskaya was formed in coal of higher coalification grade at a deeper stratum than seam k_5 and migrated into the mine via a network of tectonic faults.

5.1.3 Mossbauer Spectroscopy

Supporting the hypothesis that the key moment in the formation of natural gas is the presence of iron catalyst and based on the literature [14–22], we examined the dependence of methane content on iron compound presence. The laboratory conditions in which the work [14] was carried out were relatively free of contaminants. Under natural conditions, minerals are heterogeneous. Their particle sizes, pore water contents, seam thicknesses, and levels of contact with active catalysts can vary widely [23–25].

To attempt to minimize the differences, we analyzed coal samples from fat to anthracite in methane-bearing, outburst-prone, and outburst-proof seams of Yasinovka-Glubokaya, Chaykino, Yuzhnaya, 13-bis, Glubokaya, and 2-2 bis to determine their iron contents. Sample types were (1) a powder with particles smaller than 0.1 mm and (2) plates 3 to 5 mm thick to eliminate the effects of possible heterogeneity in sample iron distribution on spectra and determine texture peculiarities. A Wissel spectrometer produced the Mossbauer spectra of the coals. [57]Co in a chrome matrix with 50 mCi activity served as a radioactive source. Spectra were measured in the mode of uniform acceleration. Mathematical treatment of spectra (fitting) was carried out by the least squares method with the UNIVEM program.

Most Mossbauer coal spectra consist of two or three components—doublets with different quadrupole splitting and chemical shifts. A doublet with small quadrupole splitting and shift is related to pyrite (FeS_2), and a doublet with a high values of quadrupole splitting and shift indicates ferrous iron. The difference in spectra may be seen in the relative intensities of these components. Two orientations of the cubic samples were measured: when planes of preferred layer structure were perpendicular to the gamma quantum beam and when the gamma quantum beam was directed along the seam planes.

For the perpendicular orientation, the component from pyrite was almost 50% larger. Thus, in contrast to ferrous iron, pyrite is located on planes of layer structures. We also noted a strong correlation between the combinations of ferrous and ferric iron and methane content as well as outburst potential. Figures 5.2 through 5.4 present the results.

The spectra of lean baking coal (Yasinovataya Glubokaya, seam l_4), coking coal (13-bis, seam l_1), and fat coal (Chaykino, seam m_3) revealed small intensities of components connected with ferrous iron. Most iron compounds were trivalent pyrite and marcasite that differ from pyrite only in the type of crystalline lattice (Figure 5.2). Two components in the spectrum are connected

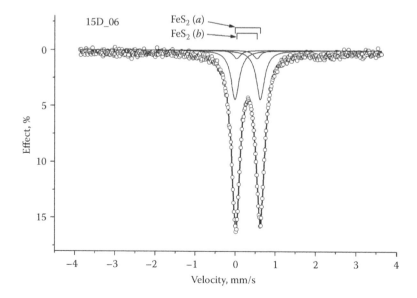

FIGURE 5.2
Mossbauer spectrum of outburst-proof fat coal from Chaykino seam m_3.

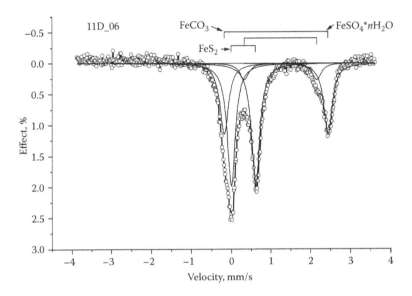

FIGURE 5.3
Mossbauer spectrum of methane-saturated, outburst-prone lean baking coal from Yuzhnaya seam h_{10}.

with two crystallographic positions occupied by iron atoms. These coal seams are outburst-proof and contain large amounts of methane (12 to 30 m^3/t).

The spectrum of outburst-prone lean-baking coal from Yuzhnaya, seam h_{10} (Figure 5.3) suggests a high iron content that correlates with the methane-bearing capacity of this seam. Ferrous iron appeared in two forms: $FeCO_3$ (8.55% of the whole iron spectrum area), a component resembling $FeSO_4 \cdot 7H_2O$ (about 34.55%), and the remainder as pyrite FeS_2. This combination of iron compounds and their proportions shown on Mossbauer spectra allow us to conclude that the samples contain large amounts of methane and that the increase of outburst activity of seam h_{10} is correlated with the intensive ferrous iron content shown in its spectrum.

Spectra of lean coal from long wall 14 of Glubokaya (unlike samples from other long walls of the mine), anthracite from 2-2 bis, and anthracite from Kommunist contained only one component of ferrous iron (Figure 5.4). These coals are outburst-prone and methane-bearing.

Table 5.3 shows hyperfine structure parameters of Mossbauer spectra in Figures 5.2 through 5.4.

Based on analyses of Mossbauer spectra, we surmise that methane-bearing capacities of coals correlate with total iron and the methane-bearing ability and outburst potential correlate only with ferrous iron.

Joint analysis of x-ray and Mossbauer data [26] confirmed iron in coals in at least three compounds: trivalent (pyrite and marcasite), bivalent as siderite ($FeCO_3$), and sulphates containing crystalline hydrates of $FeSO_4 \cdot nH_2O$ type. Ferric iron is mainly concentrated in flat precipitates in interlayer spaces;

TABLE 5.3

Parameters of Hyperfine Structures of Mossbauer Spectra

Rank	Spectrum Component	Chemical Shift (mm/sec)	Quadrupole Splitting (mm/sec)	Relative Intensity (%)
Fat, outburst-proof	$Fe^{+3}S_2$ (1)	0.3094	0.6270	91.73
	$Fe^{+3}S_2$ (2)	0.2887	0.4986	8.27
Lean baking, outburst-prone	Fe^{+2} (1)	1.1394	2.6194	34.55
	Fe^{+2} (2)	1.2545	1.8249	8.55
	$Fe^{+3}S_2$	0.3133	0.6139	56.90
Anthracite, outburst-prone	Fe^{+2}	1.2350	1.7858	100.00

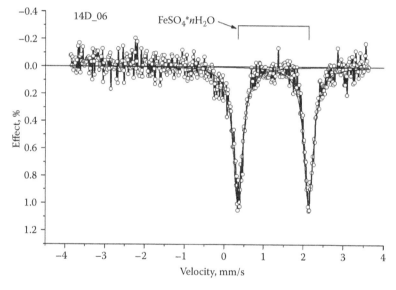

FIGURE 5.4

Mossbauer spectrum of outburst-prone anthracite from 2-2 bis, seam h_8.

ferrous iron concentrates in clusters that are more or less uniformly distributed throughout coals. The presence of ferrous iron suggests that methane was generated in the distant past and the formation continues by Fischer–Tropsch (FT) synthesis on iron catalysts [27,28].

5.1.4 Stoichiometry of Reaction

According to FT synthesis, methane forms via the following reactions:

$$CO + 3H_2 \rightarrow CH_4 + H_2O \ (\Delta H^0 = -206.4 \text{ kJ}) \tag{5.1}$$

Water formed over the iron catalyst in the presence of CO is converted into $CO_2 + H_2$:

$$CO + H_2O \rightarrow CO_2 + H_2 \ (\Delta H^0 = -39.8 \ kJ) \tag{5.2}$$

The overall reaction of methane formation is

$$2CO + 2H_2 \rightarrow CH_4 + CO_2 \ (\Delta H^0 = -254.1 \ kJ) \tag{5.3}$$

Stoichiometric methane yield is 178.6 g/m^3 of CO + H_2. Also methane is formed by CO_2 hydration at 192°C [10] according to

$$CO_2 + 4H_2 \rightarrow CH_4 + 2H_2O \ (\Delta H^0 = -164.9 \ kJ) \tag{5.4}$$

The presence of iron disulfide suggests that a portion of iron atoms that combine with sulfur during methane generation are gradually excluded from "young" methane generation or excluded from synthesis earlier during the interaction of iron-containing pyroxene and olivine with hydrogen disulfide outbursts from the mantle. Thus, the free hydrogen formed ensures reactions of all hydrocarbon formation.

It is useful to explain the principles of FT synthesis kinetics. Synthesis speed (along with product yield from unit of catalyst volume per time unit) increases with increased pressure and temperature of the gas mixture. Because synthesis reactions differ in activation energy, changing the temperature can influence the selectivity of FT synthesis. The rise of temperature accelerates methane formation. All the efforts of the developers of artificial fuel technologies focused on finding a process optimized for gasoline and oil synthesis. Methane synthesis was considered an unwanted side reaction. For coal scientists, the formation of "young" methane directly in a coal deposit via FT synthesis is an important concept.

Tannenbaum and Kaplan et al. [19] found that the activities of minerals during the formation of light hydrocarbons decreases in the presence of water although other work [14] made no note of catalyst deactivation as methane formation continued. To examine the influence of moisture early in the reaction, water equal to the number of micromoles of CO_2 was introduced. This water increased the amount of generated methane while the catalyst retained its activity over the entire 24-h experimental period. This may be explained by the important role of water in a complex FT synthesis reaction [14]. We conclude that coal seams contain all the necessary components to allow these processes. FT synthesis in a coal massif requires iron compounds and a mixture of CO, CO_2, and H_2 as initial synthesis products. The hydrogen and carbon dioxide necessary for these reactions are always present in a massif [14].

The gases are emitted during mantle degassing. Hydrogen, carbon monoxide, and carbon dioxide are always present in coal seams. The deeper the production working, the higher the average temperatures in seams (as high as 40 to 50°C at 1000 to 1500 m depths). At these temperatures, hydrocarbon synthesis reactions via catalysts become possible.

To verify this concept, we heated volatile (almost iron-free) coal in a mixture containing 1% ferrous iron ($FeCO_3$). Heating above 50° C led to a significant increase of hydrogen, methane, and CO yield. We concluded that very intensive coal hydrogenation and methane generation via FT synthesis is possible using catalysts containing ferrous iron compounds. The exothermic character of most reactions of FT synthesis may increase temperatures in small local areas containing iron atoms As a result, these areas emit gaseous methane. Synthesis may be intensified by using promoting agents such as Al_2O_3, CaO, CuO, MgO, and other metallic oxides in clays found in coals, as determined by roentgen diffraction experiments [26].

Superfine ferriferous minerals are found in coal strata at concentrations ~1% and Fe concentrations up to 2.67% have been noted [29]. The most common ferriferous mineral is pyrite FeS_2 but iron is also found as oxides and carbonates of siderite $FeCO_3$ [30], rosenite $FeSO_4 \cdot 4H_2O$, melanterite $FeSO_4 \cdot 7H_2O$ [31] , and micaceous clays such as illite. The so-called organically bound iron is found in coal as porphyrins and protein-like structures [32], iron acetate $Fe(C_2H_3O_2)_2$ [31], and iron bound with carboxyl groups [33]. Thus, coals contain sufficient materials capable of reduction to metallic iron at 200°C.

5.1.5 Conclusions

Based on the variations in isotope compositions of methane and carbon dioxide disclosed for several areas of the Donets coal basin, we conclude that areas of methane and carbon dioxide form by the constant intake of deep and mantle gases. These areas are near zones of tectonic faults that allow gases from deep sources to flow into the seams. Thus, anomalous areas of gas saturation form and lead to outbursts, outbreaks, and explosions. We can utilize the differences in isotopic and geochemical characteristics of coal seam gases and deep gases to detect zones of long-term methane production within coal fields of closed and working mines.

Mossbauer investigation revealed another source of high gas content in coal seams. The correlation of the amounts of iron and methane accumulated in coals led us to suggest the model of "young" methane generation. According to this model, generation of "young" methane and other hydrocarbons in coal seams is constant at the expense of non-standard reactions of FT synthesis on catalysts from ferrous iron compounds. Generation speed increases with the rise of average seam temperature. This condition leads us to conclude that outbursts in deep (more than 800 m) seams are connected to "young" methane. Because iron atoms are intrinsic to coal structures, generation of these methane molecules around a catalyst explains the generation of large amounts of methane in solid solutions and closed porosities.

5.2 Dependence of Outburst Proneness on Mineral Inclusions

During coal extraction, an outburst-prone zone represents less than 3 to 6% of the total seam area. To reduce production costs (decrease outburst prevention measures), it is necessary to increase the accuracy of predicting outburst-prone areas [34–48]. Coal seams in natural bedding are complex systems that are thermodynamically balanced. Coal structures and the fluids and admixtures they contain greatly influence seam behavior during mining. Section 5.1 showed that mineral inclusions containing ferrous iron serve as catalysts to generate "young" methane. It is logical that the locations of these inclusions indicate outburst-prone zones [49–52].

5.2.1 Experiments and Discussion

We investigated iron content in coking coal in seam h'_6 at the Skochinskogo Mine. The seam has a complex structure and is subject to unexpected and dangerous coal and gas outbursts (density = 1.44 t/m^3, ash A^{daf} = 19.6%, moisture W = 1.9%, sulfur S^{daf} = 1.1%, and volatiles V^{daf} =31.0%). Natural gas content is 20 to 30 m^3/t of dry ash-free mass. Coal samples in the form of lumps (0.5 to 0.8 kg) were taken every 4 m during well drilling as the drift face advanced. Successive sampling during the development of outburst-prone seams was necessary to analyze the changes in structure depending on natural methane content and zones of gas-dynamic phenomena (GDP).

Sampling continued for one month. Two GDPs were noted. All coal samples were examined on an Elva X roentgen fluorescent (RF) spectrometer to determine iron content. RF analysis showed that the samples from safe areas contained very small percentages of iron (Figure 5.5) [52].

A big halo at left (beginning at 16 keV) is background from the instrument. The less the contribution from iron (0.065 a.u.), the greater the contribution of the instrument (0.4 a.u.). The iron doublet is on the left of the spectrum (6 to 7 keV). The doublet on the right indicates a small amount of strontium.

Based on the spectrum in Figure 5.6, the iron quantity increased one order in an outburst-prone area. Samples 34 and 36 taken directly from outbursts revealed even more iron (Figure 5.7). Since the background intensity decreased thrice, we can surmise that iron content increased three times as well. Other elements found in samples from outburst-prone areas were titanium (4 keV), zinc (9 keV), arsenic (11 keV), a lot of rubidium (13 keV), strontium (14 keV), and zirconium (16 keV); see Figures 5.6 and 5.7. A yttrium doublet (15 keV) also appeared in samples 34 and 36. Samples 33 and 35 (GDP, March 19, 2009) revealed copper and bromine.

The advantage of RF analysis is the rapid yield of information about iron in a sample and its comparative quantity (the time interval of a spectrum record is 5 min). However, RF analysis does not reveal which compounds contain iron. To determine the iron valence that characterizes a compound,

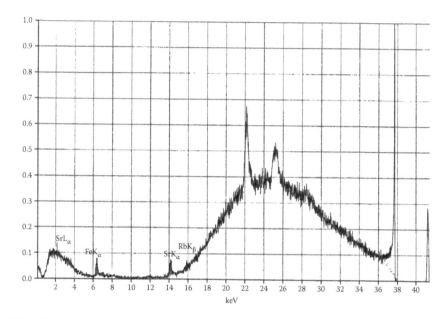

FIGURE 5.5
Roentgen fluorescent spectrum of sample 6 from safe area.

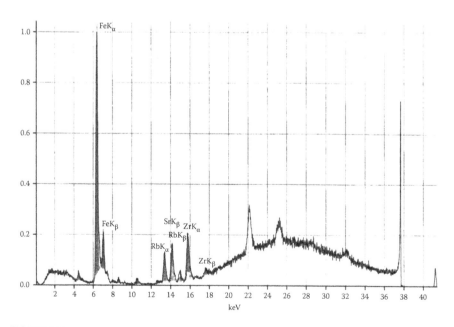

FIGURE 5.6
Roentgen fluorescent spectrum of sample 25 from GDP on March 14, 2009.

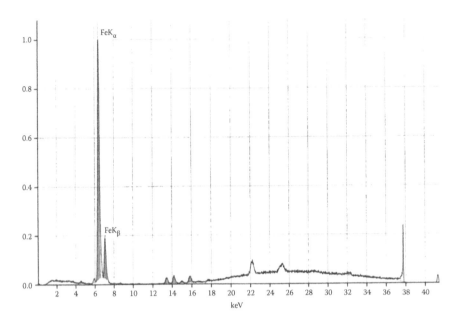

FIGURE 5.7
Roentgen fluorescent spectra of samples 34 and 36.

investigations were carried out on a WISSEL Mossbauer spectrometer at the G.V. Kurdyumov Institute of Physics of Metals of the National Academy of Sciences of Ukraine. Most of the iron found was ferrous. Ferric iron was in the form of pyrite (FeS_2) or marcasite; the ferric compounds differ by type of crystalline lattice. Trivalent iron was minimal in all samples. Only in sample 33 did pyrite represent 83% of all iron content.

We also monitored iron content and valences while working in an outburst-prone seam of coking coal (Praskovievsky) at the Sovetskoi Ukrainy 60 Mine [51,53]. The samples were selected every 5 m during seam development. Mossbauer spectra showed pyrite and two types of ferrous iron compounds. Table 5.4 shows the changes of intensity correlations of three components of ferrous and ferric iron in the samples.

The spectra for outburst-proof areas of the Smolyaninivsky seam and the Praskovievsky seam are not significantly different. In coal selected 14 h before an outburst, the total ferrous iron increased 1.5 times in comparison with ferric iron. Ferric iron in outburst coal increased approximately three times. Moreover, in outburst coal and coals of outburst-prone areas, the correlations of different types of ferrous iron changed but the chemical content of iron was fixed.

We suggested [50–55] that a change of S_{II}/S_I in a Mossbauer probe indicates a higher amount of local distortion during an outburst. Assume that a coal before an outburst contains regular polymer structures and iron in organic compounds. Under influence of high-pressure during an outburst,

TABLE 5.4

Parameters of Hyperfine Structures of NGR Spectra of Samples from Praskovievsky Seam

Sample Area	Fe (+2)/Fe(+3) $[(S_I + S_{II})/S_{III}]$	S_{II}/S_I
Outburst-proof	1.83	2.05
Outburst-prone	2.67	0.26
Outburst	4.06	0.38
Outburst	5.53	0.18

discontinuity of polymer chains occurs and monomerization increases. The value of quadrupole splitting must rise as the number of local distortions increases. A similar analogy may be based on Mossbauer research of polymer metal organic compounds of polyferrocene type [56]. As a result of quadrupole splitting, the number of monomer compounds of ferrocene is smaller than the number of polyferrocene compounds with lattice-like structures. Discontinuity of polymer chains due to high mine pressure must be accompanied by gaseous hydrocarbon emissions that create a reducing atmosphere. This factor may cause the increase of ferrous iron during an outburst as revealed by Mossbauer spectra.

Suppose that local temperature in a coal seam increases (for example due to generation of geological faults). As described earlier, synthesis reactions near a catalyst occur. The reactions are later supported by the exothermic reactions of FT synthesis. Copper makes iron reduction easier and decreases the process temperature to 150°C and the speed of FT synthesis increases.

The synthesis of heavy hydrocarbons at low temperatures (50 to 200°C) may change a coal rank via the addition of aliphatic components at FT synthesis locations. These processes are important because catalysts usually concentrate in coal bands. That is why the presence of heavy hydrocarbons (propane, benzene, paraffin, and other fatty acids) at low temperatures of synthesis near catalyst concentrations radically decreases coal shear strength and increases outburst-proneness without increasing intrastratal methane concentration and thus without increasing gas pressure in a seam.

A number of observations prove this hypothesis. In outburst-prone areas, the percentage of volatiles increases [57–61]. This concept is known as coal reduction. Coal types that display similar petrographic compositions and ranges have other identical indices. Two types of coal exhibit several different characteristics [23,57]. The first type has darker orange organic mass and contains less hydrogen and sulphur and more carbon and nitrogen. The H_{at}/O_{at} and H_{at}/C_{at} ratios are lower and this coal type exhibits less volatile emission, reactive capability, solubility, and heating capacity. It also has greater microhardness and more paramagnetic centers. The mineral content includes a fine-grained pyrite, and base oxides dominate over acids in the ash [59]. This coal type is designated *r* (reduced) because it contains more hydrogen and less oxygen. According to some authors, *r* coals are predominant in outburst-prone areas [59–61] and contain large numbers of hydrocarbons in aliphatic rows. FT synthesis reactions sufficiently explain the presence of *r* coal in outburst-prone areas.

5.2.2 Conclusions

Analysis of Mossbauer spectra of coal suggest that outburst-prone zones in coal seams are connected to the presence of ferrous iron.

5.3 Electromagnetic Method of Seam Degassing

The need to develop methane utilization technology arises from environmental problems. Releases of gas into the atmosphere aggravate global warming. The Kyoto agreement requires industrially developed countries to reduce production when environmentally unfriendly chemicals contributing to the greenhouse effect are released into the atmosphere.

Coal extraction worldwide is carried out at great depths; in Ukraine, depths can now extend more than 1300 m. Methane concentrations in the atmosphere of such mines threatens miners' lives; 89% of mines contain gas and 60% of mines are dangerous because of explosions of coal dust and gas. Many miners die as results of accidents at coal mines. Ukraine statistics reveal that about half the accidents with serious and lethal outcomes occur in the coal mining industry from explosions of methane-containing mixtures and sudden coal and gas outbursts.

Preliminary methane extraction (before coal extraction) for reasons of safety and utilization represents only a partial solution. Most methane (up to 70%) remains in coal seams and extraction is a slow process because of solid state diffusion.

Experiments using electromagnetic radiation [62] on coal revealed strong fields with amplitude $E > 10^4$ V/m at a frequency interval of 1 to 7 MHz. More intensive fracture formation makes gas filtration of coal samples easier. Results described below explain the influence of high-frequency electromagnetic vibration in accelerating methane desorption without changing coal structure.

However, questions on the mechanism and necessary energy consumption to achieve efficient porous coal degasification in production conditions remain. It is necessary to activate methane sorbed in interlayer spaces and solid solutions without breaking coal and causing formation of additional fissures. One technique is to generate low frequency vibrations and impulses of electromagnetic force (EMF) of high frequency. Impulse generation must lead to pressure pulsations in the medium (reverse seismic and electromagnetic effect). That is why it is necessary to associate electromagnetic impulse amplitude with pressure impulse. Effects of such wave-like influences on gas-bearing coal substances were studied [63,64].

One method to initiate methane emission from a gas-saturated sample involves the influence of resonance on the molecules embedded into the coal structure. Let us assume that a resonance frequency $\Omega_0 \approx 70$ MHz and it is necessary to generate shock waves (or weak discontinuity) with the same frequency. As shown in experiments on conducting mediums [65], the most efficient variant is when the interval between sequential impulses T and impulse time τ is $q = T/\tau = 5$ or 10. For instance, consider the variant when $\tau = 10^{-7}$ sec, and $T = 10^{-6}$ sec. Within 10^{-6} sec, a seismic impulse will traverse path $R = T \cdot v_p = 2 \cdot 10^{-3}$ m ($v_p = 2 \cdot 10^3$ m/sec is longitudinal seismic wave velocity).

Thus, if a generator is 10 m from a working face, we must provide sufficient power and time resolution so that a following impulse does not interfere with the earlier impulse. When $R = 10$ m, this interval Δt must be more than $R/v_p = 10/2 \cdot 10^3 = 5 \cdot 10^{-3}$ sec. In a seismic influence mode, it is impossible to produce a frequency f more than $1/\Delta t = 200$ Hz. Therefore, the only alternative for additional methane extraction activation from a coal seam at a frequency $\Omega_0 \approx 70$ MHz is electromagnetic impact. Wavelength $\lambda = C \cdot T = C/f = 3 \cdot 10^8/10^7 \approx 30$ m in open air. For coals, where phase velocity is $v_{ph} = C/\sqrt{\varepsilon\mu}$ and $\varepsilon = 3$ to 5, we can get wave length value $\lambda = 15$ m. Therefore, with this interval of frequency, we work in the nearest zone where electromagnetic vibrations in the first instance are defined by electric component $\dot{E}$. Since the distances are rather small, we have no problem with wave attenuation because the typical distance of wave attenuation is much more than $R = 10$ m.

One study [66] indicates that impulsive electromagnetic radiation observed during earthquakes has a limited frequency $\sim 10^7$ Hz. This frequency was determined by charge relaxation time, $\tau = 1/\omega_{max}$, and corresponds to wave length $\lambda \approx 30$ m. Thus, there is no point to using frequencies higher than the value specified if a technique is based on seismic and electromagnetic mechanism [67–69] of methane emission activation.

One way to generate electromagnetic radiation (EMR) is radiation of an elementary Hertz dipole transmitter. In a nearer field zone $r \ll \lambda$, the solution is found in a spherical coordinate system for an azimuthal component of magnetic field $\dot{H}_{m\theta} = \dot{H}_{mR}$ and two components of dielectric field intensity $\dot{E}_{m\theta}$, $\dot{E}_{mR}$ in a complex form [70]:

$$\dot{E}_{mR} = \frac{\dot{I}_m le^{\frac{\dot{u}R}{c}} e^{-j\frac{\pi}{2}}}{2\pi\varepsilon\omega R^3}\cos\theta, \tag{5.5}$$

$$\dot{E}_{m\theta} = \frac{\dot{I}_m l e^{-j\frac{\omega R}{c}} e^{-j\frac{\pi}{2}}}{4\pi\varepsilon\omega R^3} \sin\theta, \tag{5.6}$$

$$\dot{H}_m = \frac{\dot{I}_m l e^{-j\frac{\omega R}{c}}}{4\pi R^2} \sin\theta, \tag{5.7}$$

where $i = I_m \sin\omega t$ is direct current in a conductor with length l. Value $\varepsilon = \varepsilon_r\varepsilon_0$ is an absolute dielectric permeability value. Let us determine impedance $Z = E/H$ for efficient values of the EMF components $E = E_m/\sqrt{2}$ and $H = H_m/\sqrt{2}$. Consider a point in the plane (x, y) that is $\theta = 90°$. For real components of dielectric field intensity we have

$$H_{eff} \equiv H = \frac{I_A \cdot l}{4\sqrt{2}\pi R^2}, \tag{5.8}$$

$$E_{eff} \equiv E = \frac{I_A \cdot l}{4\sqrt{2}\pi\varepsilon\omega R^3}. \tag{5.9}$$

Then,

$$Z = \frac{1}{\varepsilon\omega R}. \tag{5.10}$$

For the frequency $\omega = 70$ MHz, $\varepsilon_r = 4$, and $R = 1$ m, we have $Z = 404$ Ohm, and for $R = 10$ m we have $Z = 40.4$ Ohm, whereas impedance in the plane electromagnetic wave is $Z = 377$ Ohm.

Let us estimate energy to be used in a coal seam to make methane molecules more active. While coal has a complex internal structure, its molecular bond energy distribution depends on the molecular arrangement of its porous system. Keep in mind that methane in fossil coal is found in three states: (1) as a gas in transporting channels and pores; (2) sorbed on the surfaces of channels and closed pores and emitted via solid-state diffusion; and (3) dissolved in organic coal substance; this methane requires maximal energy activation for emission from coal substance.

Initiating methane emission from a solid solution requires energy of ~45 kJ/mole or $u_0 = 0.5$ eV per molecule. If coal density $\rho = 1.5 \cdot 10^3$ kg/m³, 1 m³ contains $N_c = \rho/m_c = 0.75 \cdot 10^{29}$ carbon atoms. The 0.01 to 0-.1 fraction of methane molecules can be found in interplane spaces of coal structures. To estimate the energy necessary to active methane emission, we assume that the number of such molecules is $N \approx 0.1 \cdot N_c = 0.75 \cdot 10^{28}$. Total energy needed to activate methane molecules can be taken from the equation $W = 0.75 \cdot 10^{28} \cdot u_0 = 0.6 \cdot 10^{9J}/m^3$.

To estimate the dielectric field intensity required, note that all Joule losses in dielectric media are determined by

$$W_L = \varepsilon_0 \varepsilon_r E^2 \omega \cdot \mathrm{tg}\,\delta \; W/m^3, \tag{5.11}$$

In a broader interval in metamorphism, coals exhibit electrical conduction $\gamma \approx 10^{-7}\,(\mathrm{Ohm \cdot m})^{-1}$. Thus, for dielectric losses:

$$\mathrm{tg}\,\delta = \frac{\gamma}{\omega \varepsilon_r \varepsilon_0}, \tag{5.12}$$

$$W_L \approx \gamma E^2, \tag{5.13}$$

Then

$$W = W_L \cdot t. \tag{5.14}$$

where t is impact time. At intensity E = 100 V/m, the required time should be $6 \cdot 10^4$ sec = 16.7 h.

It is necessary for a resonance exposure method to make methane molecules embedded in coal structures more active. In the presence of an electric field, polarization of methane occurs, that is, an electric dipole moment P = αE where polarizability $\alpha = 2.6 \cdot 10^{-30}$ m^3 develops. Relative dielectric methane permeability ε in normal conditions is 1.00804. The relation of α, ε, and molecular dipole moment may be defined by the Clausius-Mossotti equation that takes the following form for non-polar molecules:

$$\frac{\varepsilon - 1}{\varepsilon + 2} = \frac{4}{3}\pi n \alpha_e, \tag{5.15}$$

where n is the number of molecules in a volumetric unit, ε is dielectric permeability, and α_e is electronic polarizability. Methane embedded in a coal structure as a result of adsorption is polarized by an efficient electric field with E_0 intensity and may be defined by the expression for the energy of molecule interaction for a molecule having polarizability α:

$$\vec{W}_p = -\vec{p}\vec{E}_0, \tag{5.16}$$

where P = $\alpha \varepsilon_0 E_0$ is a dipole moment and ε_0 is a dielectric constant. If $W_p = 0.5$ eV = $0.8 \cdot 10^{-19}$ J, we get $E_0 = 5.9 \cdot 10^{10}$ V/m. We also have a dipole molecular moment P = $1.36 \cdot 10^{-32}$. While considering A_{work}, which is fulfilled by carrying over an electron between lamellae in coal substance, average electric field intensity between layers is defined by

$$E = -\frac{\Delta U}{\Delta L} = 2.06 \cdot 10^8 \tag{5.17}$$

where distance between lamellae is $\Delta L = 4 \cdot 10^{-10}$ m and $\Delta U = -A$. In this case a dipole moment of methane molecule will be P = $4.74 \cdot 10^{-33}$ (SI system). It is absolutely clear that it is necessary to create high electromagnetic field intensities in the medium to make methane molecules in a sorbed state more active. To calculate essential EMR generator parameters let us start with the need for a pressure impulse ~0.01MPa to create coal seam deformation leading to activation of methane emission. In classical treatment [70], dipole radiation generates mechanical vibrations with the force (in volumetric units):

$$f = -\frac{\varepsilon_0(\varepsilon_r - 1)}{2} = \text{grad } E^2. \tag{5.18}$$

In this case, for efficient electric field intensity E_{eff} [Equation (5.5)], recall that pressure can be determined by the expression:

$$p = \frac{\partial f}{\partial R} = \frac{21(I_A l)^2}{64\pi^2} \cdot \frac{(\varepsilon_r - 1)}{\varepsilon_0 \varepsilon_r^2 \omega^2} R^{-8}, \tag{5.19}$$

We can see that the product $I_A l$, that is, a feature of radiation generator power, is expressed through pressure change in this way:

$$(I_A l)^2 = 64\pi^2 \cdot \varepsilon_0 \varepsilon_r^2 \omega^2 \cdot pR^8 / 21 \cdot (\varepsilon_r - 1). \tag{5.20}$$

For $\varepsilon_r = 5$, $R = 10$ m and $\omega = 7 \cdot 10^7$ Hz, we get $I_A l = 2.8 \cdot 10^9$. $I_A l = 2.8 \cdot 10^5$ is sufficient to provide the same intensity at distance $R = 1$ m. When current $I_A = 100$ A, we have $l = 2.8$ km, which is absolutely impossible to obtain by means of radiation source construction. The value $E_{eff} = 7.3 \cdot 10^6$ V/m is unachievable technically.

Let us define some deformation values and the intensities required to activate methane emission from interlayer voids under electrostrictive effects. In this case, homogeneous and isotropic medium relative volume change is calculated by

$$\frac{\Delta V}{V} = A \cdot E^2 \tag{5.21}$$

where A=($\beta/2\pi$) ρ ($\partial\varepsilon/\partial\rho$), β is compressibility, ρ is density, and ε is dielectric permeability. For organic liquids, coefficient $A = 10^{-12}$ in CGSE. To calculate coefficient A for coals, we use Equation (5.15) in the following form for coals with different degrees of metamorphism:

$$\frac{\varepsilon - 1}{\varepsilon + 2} \cdot \frac{1}{d} = 0.3, \tag{5.22}$$

where d is relative density. For density derivative we will get

$$\frac{\partial\varepsilon}{\partial\rho} = \frac{0.3(\varepsilon + 2)}{1 - 0.3 \cdot d} \cdot g \tag{5.23}$$

Assuming that $d = 1.4$ g/sm^3, $\varepsilon = 4$, and $g = 981$cm/s^2, we get $\partial\varepsilon/\partial\rho = 3.10^3$. Compressibility β can be calculated according to the definition:

$$\beta = \frac{1}{\rho}\frac{\partial\rho}{\partial\rho} = \frac{1}{\rho \cdot c_p},$$

where longitudinal velocity C_p of elastic waves is in the denominator. The velocity of sound in coals can be expressed by modulus of elasticity:

$$c_p = \sqrt{\frac{E}{\rho}}. \tag{5.24}$$

Taking on the values $E = 5 \cdot 10^8$ Pa and $\rho = 1.373 \cdot 10^3$ kg/m^3, we get the value $C_p = 603.5$ m/sec. Then, compressibility $\beta = 2 \cdot 10^{-10}$ (CGS) and coefficient A in Equation (5.10) is $1.3 \cdot 10^{-7}$ (CGSE). If we take $\Delta V/V = 10^{-6}$ for relative volume change, we get electric field intensity $E = 2.76$ V/cm ≈ 300V/m.

Let us estimate generator parameters assuming that at a distance of radiation source $R_1 = 1$ m it is necessary to achieve $E = 300$ V/m intensity. In the atmosphere, this value corresponds to $E_0 = 300$ $\varepsilon_r = 1.2$ kV/m derived by considering the continuity of the tangential field component at the air–coal seam interface. From the expression for $E_0 = 300$ $\varepsilon_r = 1.2$ kV/m, we get $I_A \cdot l = 13.2$ ($R = 1$ m) leading to dipole length $l = 0.26$ m if current $I_A = 50$ A. These parameters allow us to achieve an efficient field E_{eff} value of $300/10^3 = 0.3$ V/m at $R = 10$ m from the radiation source and it produces coal deformation $\Delta V/V = 10^{-8}$ due to electrostrictive effect.

What should be the medium deformation under electromagnetic vibration? Connection of deformation and intensity in the lithosphere obeys universal dependence, which is linear in logarithmic coordinates [68] and covers relaxation processes from slight earthquakes to rock bumps at stresses ranging from 10 M/Pa to 100 Pa. Knowing the dependence of relaxation of stress from differential adjustment movements u [70], we obtain the dependence:

$$\lg(\Delta\sigma) = 6.2 + 0.75\lg(u) \tag{5.25}$$

That allows us to calculate dumps of stress occurring at rock bumps $\sigma\rho = 280$Pa at adjustment movement $u = 10^{-5}$ cm. For slight earthquake vibrations, we have $\sigma\rho = 0.320$Pa ($u = 10^{-9}$ cm). From Equation (5.25), it follows that adjustment of movements about 10^{-4} cm results in stress release $\sigma\rho = 1.6 \cdot 10^3$ Pa. As for the homogeneous isotropic medium, Hooke's law applies to relative volumetric change:

$$\varepsilon = \frac{\Delta V}{V} = \frac{12}{E(1-v)} \cdot P, \tag{5.26}$$

where E is modulus of elasticity, v is Poisson coefficient, and $P = \sigma/3$. At deformation along one direction, we can write:

$$\varepsilon_x = \frac{2}{E(1-v)}\sigma, \tag{5.27}$$

which conforms to experimental data regarding uniaxial deformation of coals [71]. Equation (5.27) enables us to calculate relative deformations for processes of differing energies. Thus, for rock bumps starting with (5.25), we get $\varepsilon_x = 1.7 \cdot 10^{-6}$, and for $\Delta\sigma = 1.6 \cdot 10^3$ Pa, we get $\varepsilon_x = 10^{-5}$. The $E = 5 \cdot 10^8$ Pa and $v = 0.35$ values used are typical for coals.

When a coal seam is under electromagnetic influence, deformations and stresses cannot exceed the values typical for rock bumps and they should approach the values typical for slight earthquakes. That is why we start with the following values as the most rational: maximal stresses $= 0.5$ kPa and $\Delta\sigma = 100$ Pa. For $\Delta\sigma$, $\varepsilon_x = 0.5 \cdot 10^{-6}$ and $\Delta V/V = 10^{-6}$. That is why relative medium deformations created by a generator during electrostriction should be restricted by the interval $10^{-8} < \Delta V/V < 10^{-6}$.

We now describe studies of the influences of electromagnetic vibrations that lead to acceleration of methane desorption without changing coal structure based on nonthermal mechanisms of electromagnetic vibrations in the radio-frequency range. The intent was to devise a method for degassing coal using electromagnetic influence of radiofrequency range [72,73]. We sought to

1. Find optimal conditions of electromagnetic field influence based on metamorphism stage.
2. Study influences on coal samples under laboratory conditions.
3. Design a generator appropriate for use under production conditions.
4. Provide a rational technological scheme for coal seam degassing.

At a frequency bandwidth of electromagnetic radiation $F = 10$ to 60 MHz in the zone near the radiator, it is possible to provide electromagnetic field power and amplitude of electric component E in a coal seam; this amplitude is sufficient to accelerate methane diffusion. The method is based on the physical concept stating that diffusion of methane molecules in coal substance is determined by chemical potential gradient. Generally, the flow is described by Fick's law and has the form:

$$\vec{j} = -D\,\nabla c + c\,\langle v \rangle_F + c\,V_k, \tag{5.28}$$

where D the is coefficient of diffusion and C is the concentration of the component diffused. The second part of the equation represents substance flow affected by different chemical, electrical, and gravitational forces. The final segment is defined by the medium flow as a whole. Therefore, if an electromagnetic field becomes available in the medium, polarizing effects occur and

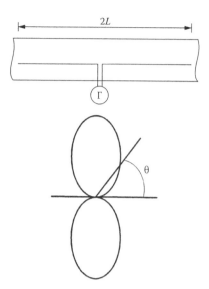

FIGURE 5.8
Elementary dipole (a) and direction diagram of half-wave electric dipole (b).

additional flow of methane molecules takes place. Two variants of elementary radiators create alternating electromagnetic fields in media: the Hertz dipole and the magnetic dipole.

The Hertz dipole is symmetrical (Figure 5.8); an alternating current going through a linear conductor is in a unitary phase. Direction is defined by ratio of dipole length $2L$ to wavelength λ. Figure 5.8 b depicts a half-wave dipole $2L = \lambda/2$. In a nearer zone where the distance from radiation source is less than the wavelength $R \ll \lambda$, a solution in a spherical coordinate system for the azimuthal component of the magnetic field and two components of field intensity [71] can be found:

$$H_\varphi \sim 1/R^2, \; E_\theta \sim 1/R^3, \; E_R \sim 1/R^3 \tag{5.29}$$

Power radiated by the antenna will be defined by the electric moment of antenna $I_A L$, where I_A is an amplitude current value under harmonic vibrations.

A magnetic dipole is a current loop (coil) with alternating electric current having one phase and amplitude at any point. From a theoretical view, the similarity and structure of a field of electric and magnetic dipoles follows the principle of permutation duality of Maxwell's equations. For a nearer zone we generate expressions for electromagnetic field components if we use the current moment notion of a magnetic dipole $I^M \times L$, which is equivalent to the moment of a magnetic coil:

$$H_\theta^M = \frac{I^M \cdot L}{4\pi\mu_a\omega} \cdot \frac{1}{R^3} \sin\theta \cdot j \cdot \exp(-jkR)$$

$$H_R^M = -\frac{I^M \cdot L}{2\pi\mu_a\omega}\frac{1}{R^3}\cos\theta \cdot j \cdot \exp(-jkR) \tag{5.30}$$

$$E_\varphi^M = -\frac{I^M \cdot L}{4\pi}\frac{1}{R^2}\cdot\sin\theta \cdot j \cdot \exp(-jkR),$$

where I^M is magnetic dipole current; μ_a is magnetic permeability of the medium where the transmitting coil can be found; L is a magnetic dipole length equivalent to the transmitting coil, ω is frequency, and k is a wave vector [71]. Using the measured values and transformations:

$$H_\theta = -\frac{I_p d^2}{8R^3}\cdot\sin\theta\cdot\exp(-jkR)$$

$$H_r = -\frac{I_p d^2}{4R^3}\cdot\cos\theta\cdot\exp(-jkR) \tag{5.31}$$

$$E_\varphi = \frac{\mu_a\omega I_p d^2}{8R^2}\cdot\sin\theta\cdot\exp(-jkR),$$

where I_p is current in transmitting coil and d is diameter of a current loop.

The field and direction diagrams of both transmitters correlate, but their radiation fields differ because electric field vectors (E) and magnetic field vectors (H) exchange their positions in space. Thus, a horizontal transmitting coil can be treated as a dummy vertical elementary dipole meant to be fairly small in comparison with the wavelength element of a linear magnetic current with invariable length, amplitude, and phase.

To achieve effective coal seam degassing [70], it is necessary to create an electric field intensity $300 < E < 3$ V/m near a down hole of a radius of 10 m. That is why it is important to create electromagnetic vibrations that are equivalent to those that should generate prospective antenna arrangements placed in a coal bearing mass to accelerate degassing.

During the first stage of the experiments, a high frequency transmitter created electromagnetic vibrations in the interelectrode space of a plane capacitor [70,71]. The amplitude of the electric field intensity was 40 V/m at the voltage output of the generator of 1 volt. The electromagnetic vibration power was W = 11 μW/m³.

The samples were coal fractions about 3 to 4 mm saturated with methane up to 10 MPa in a high pressure chamber. We studied the effect of exposure to an electromagnetic field on methane desorption at generator operating frequencies of 10, 30, and 60 MHz for 7 h. The most efficient effect was achieved at 30 MHz where we could approach a "nearer zone" for electromagnetic radiation and ignore the offset current. Size of this zone is determined by $\omega \times \varepsilon = \sigma$, where σ is the electric conductivity of the medium.

Methane content was registered by a laboratory device that allowed simultaneous measurements. Methane content was measured in a closed chamber with accuracy up to 2 ml. We could calculate the velocity of methane emission at any moment of electromagnetic influence and determine the entire methane emitted from samples through the end of the experiment. Based on rank, we noted an increase of desorption coefficient at the final stage of electromagnetic influence from 26 to 130% in comparison with the check sample. This result was sufficient to decrease the time of methane emission; the time required by the check variant without electromagnetic influence was 1.5 to 2 times as long.

The second stage used the same methodology as the first stage above, but a time solenoid to provide electromagnetic vibrations was placed in the container with the sample. The solenoid was supplied with simple harmonic current from a high frequency lamp transmitter 6C5D. The transmitter could change anodal voltage from 30 to 150 V at anodal current $\approx$100 mA and served as the source of radiation. We measured relations between the magnetic and electrical components of electromagnetic vibrations. The amplitude of vibrations of the electrical component reached 900 V/m (effective value 645 V/m). Energy density of the electromagnetic field inside the solenoid was 3.7 μW/m^3 if frequency f = 28.5 MHz. However, significant inconsistency of the electromagnetic field within the solenoid prevented a significant increase of the degassing efficiency achieved in the first stage of the experiment. At the final stage of electromagnetic treatment, methane emission from experimental sample increased only by 39% as compared to the check sample.

To achieve an effective degassing project in the conditions common to modern coal mines, it is necessary to know required generator power and the power of the electromagnetic field in a coal seam. Using the electrostrictive mechanism that is responsible for accelerating methane emission via electromagnetic influence, we can get an electric field intensity E $\approx$300 V/m for relative volume change value $\Delta V/V$ = 10^{-6}. From the expression for E$_{eff}$ (5.32), we can get value of the electric dipole moment $I_A \times l$ = 13.2 (R = 1 m). If current I_A is 50 A, a dipole length l = 0.26 m is fully realizable. With these parameters, we can calculate an efficient field with a vibrator (Hertz dipole) at a distance of R = 10 m from the radiation source; E$_{eff}$ = 300/10^3 = 0.3 V/m. Such a field can provide coal deformation by electrostrictive effect to $\Delta V/V$ = 10^{-8}.

5.3.1 Electric Dipole

Let us determine an impedance value $Z = E/H$ for effective electromagnetic field components $E = E_A/\sqrt{2}$ and $H = H_A/\sqrt{2}$. We are going to consider a point in the plane (x,y), that is, θ = 90 degrees. For components of electromagnetic field intensities we have

$$H_{eff} \equiv H = \frac{I_A \cdot l}{4\sqrt{2}\pi R^2}, \quad E_{eff} \equiv E = \frac{I_A \cdot l}{4\sqrt{2}\pi\varepsilon\omega R^3}, \quad (5.32)$$

where ε is absolute dielectric permeability. Then,

$$Z = \frac{1}{\varepsilon \omega R}. \tag{5.33}$$

For relative coal dielectric permeability we have $\varepsilon_r = 4$. At a distance from antenna $R = 1$ m, Z is 404 Ohm. However, if we have distance $R = 10$ m, then $Z = 40.4$ Ohm; meanwhile in the plane, electromagnetic wave impedance $Z = 377$ Ohm. Let us compare energy densities of electric and magnetic components $W_E = \varepsilon E^2/2$ and $W_H = \mu H^2/2$. Their ratio can be expressed as

$$\frac{W_E}{W_H} = \frac{\varepsilon_r \varepsilon_0}{\mu_r \mu_0} \left(\frac{E}{H} \right)^2. \tag{5.34}$$

Considering values $R = 1$ m and $R = 10$ m, we get $W_E/W_H = 4.6$ and 4.6×10^{-2}, respectively.

5.3.2 Magnetic Dipole

We get the expression for impedance $Z = E/H$ at any point of r space in a nearer zone of the radiation source. From the above expressions for the field components, we have

$$H^2 = \left(\frac{I_p d^2}{4 \cdot r^3} \right)^2 \left[\frac{1}{4} \sin^2 \theta + \cos^2 \theta \right]. \tag{5.35}$$

In the plane of the orthogonal coil axis when $\theta = 90$ degrees:

$$H^2 = \left(\frac{I_p d^2}{8} \right)^2 \cdot \frac{1}{r^6}. \tag{5.36}$$

Then, we get the following expression for impedance:

$$\frac{E}{H} = \mu_a \omega \cdot r. \tag{5.37}$$

Taking $\omega = 3 \times 10^7$ Hz, $\mu = \mu_0 = 4\pi \times 10^{-7}$, and $R = 1$ m, we have $E/H = 37.7$. This relation will grow 10 times in the distance of 10 m. At the same time an electric component voltage will become 100 times lower due to quadratic dependence on the distance. We now determine the relation of electric and magnetic field energy density:

$$\frac{W_E}{W_H} = \frac{\varepsilon_a E^2}{\mu_a H^2}. \tag{5.38}$$

This relation will equal 0.01 when the distance from the source of radiation $R = 1$ m. If the distance is 10 m, the relation will grow to 0.1. The experiment simulates two types of transmitters: the elementary electric dipole and magnetic dipole [72] that can serve as antennas during degassing in coal mine

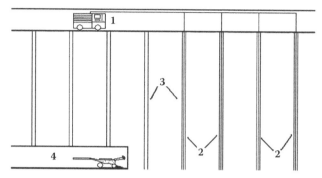

FIGURE 5.9
Degassing under electromagnetic influence using vertical pilot holes. (1) Generator of electromagnetic radiation. (2) Antennas (electric vibrators). (3) Holes. (4) Mine working.

conditions. Our experiments and the experience gained while using transmitters in underwater and underground conditions [73] indicate that a Hertz dipole vibrator is the most efficient antenna. A generator of high frequency vibrations under conditions of mining production should meet ecological and electrical safety requirements. That is why antenna-transmitting tools should be placed directly in holes. The electromagnetic vibrations that scatter through the thickness of rock mass are not dangerous.

One method of generating electromagnetic influence (Figure 5.9) suggests using traditional advance boring and placing symmetrical vibrators as electromagnetic vibration sources to activate degassing. A generator must provide radiation power not less than 2 kW at $f = 20MHz$.

5.3.3 Conclusions

Electromagnetic influence increases the velocity of methane degassing as it decreases methane content in a coal mass within the time correlated with coal extraction. A transmitter in the form of a symmetric electric vibrator is preferable. For coals of the Donets field (anthracite, coking and fat), the optimal operating mode of the generator was $f = 30$ MHz. The generator should provide radiation of 0.4 to 2.0 kW, depending on the frequency at which intensity of the EMF will exceed 0.3 V/m at a distance 10 m from a hole.

5.4 Physics of Fissured Porous Coal Structure Transformation under Influence of Unloading Wave

Some theoretical material is necessary to explain the concepts and calculations covered in this section. We will consider the stability conditions of coal–gas systems and discuss the results of computer modelling of porous system stability during unloading wave movement and transformation of pores into fissures.

5.4.1 Influence of Gas Content on Stability of Coal–Gas System

Based on the results of the analysis of the existing views of the structures of fossil coals, the modern view is that coal substance is a complex polymeric system whose main characteristics arise from three-dimensional configurations and atomic groups of polymer chains i.e., so-called macromolecule conformations. Coal also contains some rather large monomeric components. The main method of theoretical examination of conformational properties is based on classic static thermodynamics [74,75].

The examination of coal substance within a polymeric model must consider the gases (CH_4, CO_2, etc.) that saturate coal during its development. Moreover, the fluid concentration in a coal constitutes one of the main controlling parameters of the system. According to contemporary views [76,77], methane can be found in fossil coals in a free, adsorbed, or absorbed (solid solution) state. Methane solution mainly occurs in the aliphatic (polymeric) component of coal; in the arranged part (crystallite), it can be found in the intercorrelational state between carbonic lattices.

According to Juntgen and Karweil [78], 200 liters of methane are produced from 1 kg of coal by coalification. Static pressure of rock does not contribute to chemical changes during metamorphism; it slows chemical changes because gas emission is made difficult [79].

The pressure contributes to physical and structural properties, whereas a temperature rise accelerates chemical coalification.

Thus, it is quite possible that when conditions prevent methane from leaving coal, the potential methane content can be as much as 150 to 200 m^3/t. When methane dissolves in a coal seam, two changes occur: (1) the formation of voids in the coal substance by methane molecules that move and expand the coal structure [38] (unlike a laboratory situation in which methane achieves volumetric filling of the existing micro pores); and (2) increased absorbed methane while the temperature under which metamorphism took place rises.

We used Flory's polymeric model as a basis for studying a solid solution of gas and coal [75]. Flory's model is based on self-coordinated fields formed within the lattices of polymer chains.

Flory's results were later found inaccurate, especially for the semi-dissolved solution. The theory did not consider fluctuations in the polymeric solutions. At present, the most complete model of polymeric alloys and solutions is the scaling (fluctuating) theory of polymeric solutions [80].

However, as polymeric solution concentration takes place, the role of fluctuations decreases. This was discovered in an experiment comparing Flory's theory with the scaling theory. Many of Flory's predictions related to concentrated solutions and polymeric melts were proven accurate.

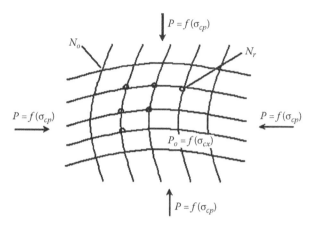

FIGURE 5.10
Model of gas and coal molecule.

In examinations of coal–gas systems, we usually work with concentrated polymeric solutions. The "polymeric solution" phrase generally specifies a situation in which molecules of a polymeric solution are dissolved in a monomeric solvent. In reality, the situation is converse; the polymeric molecules of a coal substance act as molecules of solvent and the monomeric molecules of gas act as a dissolved substance. Thus, a weak polymeric gas and coal solid solution is a concentrated polymeric solution in a traditional view.

Let a coal–gas system consist of N polymeric molecules of coal substance, each of which contains as many as r elements and N_r monomeric molecules of gas. Thus, in a hypothetical lattice that models a mix of coal and gas (Figure 5.10), N_0 units are engaged:

$$N_0 = N_r + rN. \tag{5.39}$$

Each molecule of the dissolved substance has one position in the lattice; each molecule of polymer has r places. We assume the polymeric chain is flexible and consists of r movable segments, so if there is one attached segment of the molecule, the following segment takes any vacant neighboring unit of the lattice. A flexible molecule of a polymer can have different configurations and a configuration may be oriented differently in a lattice.

According to general thermodynamic theory [74], the state of methane dissolved in coal can be described on the basis of the available energy of the system, stated as a function of sub-allocated density $\psi = \psi(r)$. For density $\psi(r)$, we usually consider a proportion of the share of the states filled with methane molecules in a given part of the volume $N_r = N_r(r)$ to the complete number of states N_0, that can be potentially filled by them:

$$\psi(r) = N_r(r) / N_0. \tag{5.40}$$

The available energy consists of the inner energy of the system E and entropy contribution S_{conf} that can be determined by the configuration number $W(\psi)$, by which the state with data ψ can be formed:

$$S_{conf} = \ln [W(\psi)]k.$$

$$S = \ln [W(\psi)]. \tag{5.41}$$

The exact calculation S_{conf} is a difficult separate problem. However, while developing the theory, one can focus on the mean field for ΔS_{conf} entropy. That change is normally calculated in statistical physics [74,81,82] as follows:

$$\Delta S_{conf} = -k N_0 [\psi \ln (\psi) + (1-\psi) \ln (1-\psi)/r], \tag{5.42}$$

where k is Boltzmann's constant and r equals the number of equivalent states that can be filled by each molecule in each local state and as a result defines the proportion between the energies of the free and filled states. The change of the inner energy of a gas can be given as series according to the degrees ψ and in the approximation of pair-wise interaction can be given in simple quadratic form [82,83]:

$$\Delta E = E - E_0 = N_0 T \chi \psi (1-\psi). \tag{5.43}$$

where T is the temperature of the system and constant χ (Flory and Higgins parameter) [80] fixes the proportion between energy and entropy contribution to the full energy of the system. The required free energy has the following view:

$$\Delta F = \Delta E - T \Delta S_{conf} = N_0 T \{\chi \psi(1-\psi) + k[\psi\ln(\psi)+(1-\psi)\ln(1-\psi)/r]\}. \tag{5.44}$$

The χ can be defined by the interaction of methane molecules and coal molecules. The intensity of the interaction depends on pressure P in the coal–gas system in the proximity of pressure P_0 realized in the coal seam. The dependence $\chi(P)$ has no peculiarities and can be given as follows:

$$\chi(P) = \chi(P_0) + \beta (P - P_0) \approx const - \kappa (1 - P/P_0). \tag{5.45}$$

The value of the free energy of a coal–gas system may be determined by the concentration of methane in coal and proportion of inner and outer pressures. Equilibrium values of concentration ψ can be determined by the state of minima of free energy. The corresponding states of thermodynamic equilibrium depend on outer conditions where the system is found. Thus, if methane emission from a seam is blocked, bear in mind that $(1/V)\int\psi dV = \psi_0$, where ψ_0 is the specified mean concentration. In some cases, additional conditions

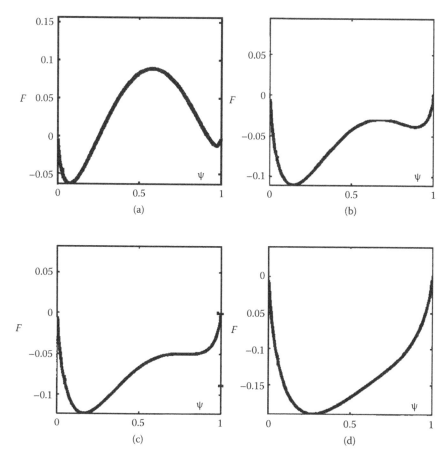

FIGURE 5.11
Change of free energy $\Delta F(\psi)$ at different pressure values P/P_0: (a) 1, (b) 0.9, (c) 0.88, and (d) 0.8.

will not greatly change the state of the minima that can be found according to the equation:

$$\partial \Delta F/\partial \psi = N_0 T \{\chi (1-2\psi) + k (1-1/r)+[\ln (\psi)- \ln (1-\psi)/r]\} = 0 \qquad (5.46)$$

One can see from Equation (5.46) that both equilibrium concentration ψ and one or two different minima of the energy depend considerably on the value of χ. Figure 5.11 shows typical dependences of energy $\Delta F(\psi)$ calculated with different χ values. The state filled with gas equals the minimum of energy (~1) if concentration is ψ, which corresponds to the metastable state of the energy.

If the pressure is rather high (χ is large), this state is divided from a stable global minimum with a high potential barrier if $\psi \ll 1$. As a result, in a high gas content, the state remains stable almost indefinitely. When the barrier is lowered, diffusion is intensified and the system steadily leaves this minimum and transfers to C $\psi \ll 1$. It is obvious that this process should be described using a variable concentration (both in space and in time) $\psi(r,t)$. Evolution $\psi(r,t)$ with variable total quantity of gas in the system $M = \int\psi(r,t)$ dV is given by Cahn–Hilliard's equation, whose solutions, although weak and slowly changeable in the space ψ, can be in proximity to the answers to Cahn–Allen's equation:

$$\partial\psi(r,t)/\partial t = -\gamma\, \delta\Delta F[\psi(r,t)]/\delta\psi(r,t), \tag{5.47}$$

This equation including a particular form of free energy:

$$F[\psi(r,t)] = \int \{D_{eff}(\nabla\psi(r,t))^2/2 + \Delta F(\psi(r,t))\}dV = \int \{D_{eff}\,(\nabla\psi(r,t))^2/2$$

$$+ N_0 T \{\chi\, \psi\,(1\text{-}\psi) + k\,[\psi\,\ln\,(\psi) + (1\text{-}\psi)\,\ln\,(1\text{-}\psi)/r]\}\}dV, \tag{5.48}$$

can be given as

$$\partial\Delta\psi/\partial t = D_{eff}\Delta\psi(r,t)\text{-}\gamma\, N_0 T\, \{\chi(1\text{-}\,2\psi)+k(1\text{-}1/r)+[\ln(\psi)\text{-}\ln(1\text{-}\psi)/r]\}, \tag{5.49}$$

where γ is a relaxation constant that determines the time of the process and D_{eff} is the effective coefficient of diffusion. Thus, the kinetics of free energy of a coal–gas system are determined by gas diffusion depending on the pore volume.

5.4.2 Modeling of Porous Medium

Analyzing gas emission from a system involving a change of total concentration $M = \int\psi(r,t)dV$ while maintaining constant exterior conditions is a complicated task. First, the process takes place in a semilimited medium, and the equation of partial derivatives should be solved using corresponding boundary conditions. In other words, it is necessary to create a condition in which concentration $\psi(r,t) = 0$ on the boundary and its convergence to initial (equilibrium) value into the depth of the seam. Consider a boundary plane $x = 0$; the direction into the depth of the seam is $x > 0$. The described boundary conditions can be given as

$$\psi(r,t)|_{x=0} = 0;$$

$$\psi(r,t)|_{x\to\infty} = \psi_0. \tag{5.50}$$

The solution is considerably complicated because the reaction occurs in a medium of complex and inconsistent porosity. From a phenomenological view, the porosity in Equation (5.49) can be described by means of an unequally spread diffusion coefficient $D = D(r,t)$. Function $D(r,t)$ represents the characteristics of the massif and can be found in experimental data. However, it can be modeled satisfactorily only while retaining the characteristics of the medium such as (1) mean porosity of the system $\rho_0 = \int \rho\,(r,t)dV/V$, (2) characteristic pore size (i.e., medium size $<R>$ of areas where diffusion coefficient significantly differs from 0 $[D(r,t)>0]|_{r<R}$), and (3) relative dispersion of their sizes $<\delta\rho>$ that determine the pore spread along radii. These requirements can be met with an incidentally spread medium with a Gauss correlation function:

$$G(r-r',t) = <\rho\,(r,t)\,\rho\,(r',t)> \tag{5.51}$$

given as

$$G(r-r',t) = G_0 \exp[-\,(r-r')^2/2\sigma]. \tag{5.52}$$

Such a medium can be naturally achieved by the kinetics of phase separation (segregation) that can be described by a simple kinetic equation:

$$\partial\rho\,(r,t)/\partial t = \Delta\rho\,(r,t) + \rho\,(r,t)(1-\rho\,(r,t)) \tag{5.53}$$

when fixed total density is $\int\rho(r,t)dV = \rho_0$. If one uses δ correlation density distribution as an initial condition correlative density distribution:

$$<\rho\,(r,0)\,\rho\,(r',0)> = \delta(r-r'), \tag{5.54}$$

the process generates the desired distribution with correlation function close to $G(r-r',t) = G_0 \exp[-\,(r-r')^2/2\sigma]$; half width depends on time $\sigma = \sigma(t)$. By fixing the process at some instant t^*, one can model the incident medium with the structural porosity determined previously (corresponding to the real one) and thus the unequally spread diffusion coefficient:

$$D(r,t) = D_0 + \text{const } \rho(r,t^*). \tag{5.55}$$

Each initiation of the process creates a new incidental realization of the medium fragment. The information is useful for single numerical experiments for studying specific characteristics of the process in limited space (multiple repeats of numeric implementations) to gather statistics and calculate general system parameters. Figure 5.12 depicts the structure. The sequence *a* through *d* illustrates changes of the distribution of system density with the change of scale. The areas of closed porosity are darkened; the areas of open porosity are light. The inset shows a three-dimensional correlation function calculated for the same density distribution $\rho\,(r)$ as in the figures (*a - d*).

Figure 5.13 and Figure 5.14 illustrate diffusion in the system both with and without open pores in the cluster. The area of the space where gas diffusion started (or continues) is shown in gray. As in Figure 5.12, the areas of mainly open porosity are light; closed porosity areas are dark. For a convenient comparison, the early stages of the process are light gray. The last two configurations coincide although they correspond to different moments. This means the system starts the stationary density distribution within a limited cluster of open porosity (because of the absence of other ways to flow). In summary, the kinetics of forming a coal–gas system at a given density distribution

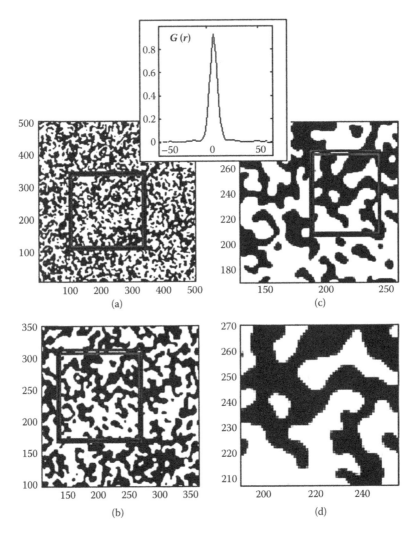

FIGURE 5.12
Distribution of modeling density of system ρ(r). Scale = angstrom units.

(differential porosity) leads to the formation of a system of closed pores that prevents gas emission from the coal.

5.4.3 Movement of Wave of Unloading

During coal excavation, a wave of unloading is formed. The wave can be described and numerically modeled according to the mechanics of continua. The general theory [82] of the equation of the movement of arbitrary elements of the continua volume starts with total force that influences the emitted volume and equals the integral of the pressure $p(r,t)$ over the surface

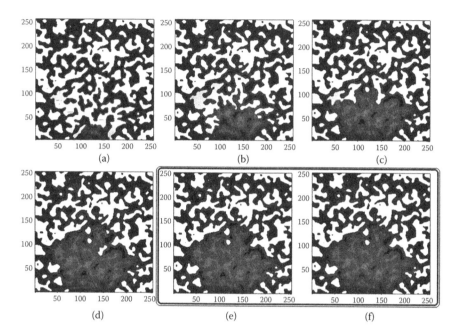

FIGURE 5.13
Diffusion development in absence of flow on cluster of open pores.

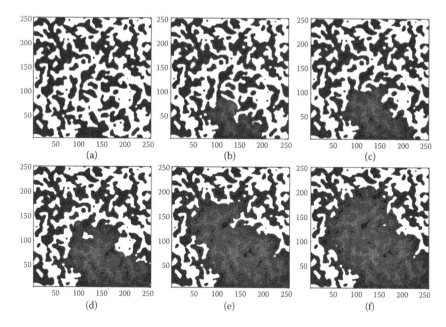

FIGURE 5.14
Diffusion in the presence of flow through cluster of open pores.

of the examined volume. While transposing the force into the integral over the volume and making it equal to the change of the impulse of this medium volume, we get:

$$\rho(r,t)\, dv(r,t)/dt = - grad\ p(r,t) \equiv -\nabla\ p(r,t). \tag{5.56}$$

Here $v\ (r,t)$ is velocity vector and $\rho(r,t)$ is density of the substance that fills the volume at the proximity of point r at moment t. The time derivative determines the velocity of movable medium particles. To adapt the equation of movement using fixed coordinates, one should make a change:

$$dv(r,t)/dt = \partial v(r,t)/\ \partial t + (v(r,t)\ \nabla)\ v(r,t). \tag{5.57}$$

Equation (5.56) is transformed to the Euler equation:

$$\rho(r,t)[\partial v(r,t)/\partial t + (v(r,t)\nabla)v(r,t)]=-grad p(r,t) \tag{5.58}$$

that cannot be used for modeling the wave of unloading in a real system in its unmodified state; it can be used only when no energy is dissipated. The movement of the wave of unloading in a real medium is accompanied by inner friction that naturally leads to additional velocity loss of volume element that is noted as an additional summand in the left part of Equation (5.58). The inner friction processes are characterized by effective viscosity and their corresponding power additive can be given as follows:

$$\delta f = \eta \Delta v(r,t) + (\xi +\eta/3)\ \nabla\ (\nabla v(r,t)), \tag{5.59}$$

where ξ and η are viscosity coefficients of the first and second sort. If we assume the contractibility of the medium is slight and omit the summand representing velocity divergence, we get the equation of motion as follows:

$$\rho(r,t)[\partial v(r,t)/\partial t + (v(r,t)\nabla)v(r,t)] = - grad\ p(r,t) + \eta\Delta v(r,t). \tag{5.60}$$

Equation (5.60) should be related to the equation determining pressure $p(r,t)$ and velocity $v(r,t)$:

$$\Delta\ p(r,t) = - \rho(r,t)(\nabla v(r,t))^2. \tag{5.60}$$

If we disregard small anisotropy corrections for non-contractable medium $\rho(r,t) \approx \rho$, this equation can include the approximate ratio of Bernoulli $p(r,t) = p_0 - \rho\ v(r,t)^2$ for a change of local pressure (according to initial pressure at standstill p_0) to local kinetic energy of the medium volume.

For common non-homogeneous porous media, the Navier-Stokes equation (5.60) describes the motion of viscous non-contractable liquid. We think it acceptable to consider a model in which partially loosened coal together with its contained fluids (at proper viscosity and density) is a liquid of a special sort. Solution of the many-component vector equation of motion (5.60) is difficult and time consuming, but the equation may be simplified by suggesting a motion (cylindrically symmetrical in a flat coal seam) of the wave of unloading from the initial center of destruction.

Let us suggest that the medium velocity (except for a small area near the boundary) is radial almost everywhere $v(r,t) = v(r,t)\, r/r$. In this case, the operators of Laplace and gradient calculated in spherical coordinates are reduced to the following simple and short forms:

$$\Delta v(r,t) = \partial^2 v(r,t)/\partial r^2 + (d - 1)\, \partial v(r,t)/\partial r,\ \text{где}\ (\nabla v(r,t)) = \partial v(r,t)/\partial r. \quad (5.62)$$

As a result the equation of motion (5.60) can be given as follows:

$$\rho(r,t)[\partial v(r,t)/\partial t + v(r,t)\partial v(r,t)/\partial r] = -grad\ p(r,t) + \eta \partial^2 v(r,t)/\partial r^2 + (d-1)\partial v(r,t)/\partial r, \quad (5.63)$$

Equation (5.63) allows easy numerical integration. Here d is a space dimension that may equal 1 to 3 in practical calculation. The results of solving Equation (5.63) may be applied to different configurations of a system. For example, when the process develops in a thin seam, the system can be considered quasi-two-dimensional, described by $d = 2$. If the wave travels in a thin channel (or initially generated as almost plane), the process can be described as $d = 1$. We performed integration of Equation (5.63) using $d = 3$ and following boundary and initial conditions. We surmised that virgin coal comes in contact with the free half space $x<0$ along the plane boundary and the following borderline condition is met:

$$\rho(r,t)|_{x<0} = 0. \quad (5.64)$$

For the initial condition, we assumed that at $t = 0$ a small volume of the substance is almost instantly emitted from a small spherical cavity of some radius r_0. It causes substance flow directed inside the resulting void along with a release of pressure and initiates the wave of unloading that enters the half-space $x > 0$. Because Equation (5.63) is given for velocity V as the independent variable, we should choose as the initial condition the burst V directed toward the center of destruction (and negative coordinates x at the open boundary). As a result we have

$$V(r,t = 0)|_{r < r_0} = V0 < 0. \quad (5.67)$$

Figure 5.15 illustrates the results of numerical integration (5.63) at $d = 3$ at the specified boundary and initial conditions. It is difficult to depict velocity $V(r,t)$ and pressure $p(r,t)$ distributions in a wave as three-dimensional. Therefore, they are shown in the figure in a two-dimensional section above the plane (x,y) taking symmetry into account. The two configurations in the figure are for two typical moments. The first shows the onset of wave development almost immediately after its appearance $(t = 0.1)$; the second is at $(t = 30)$ when the wave has developed form and stresses generated by pressure release approximate to the ultimate stress limit of coal. At higher values, the destruction stops and the wave dies without significantly changing a system.

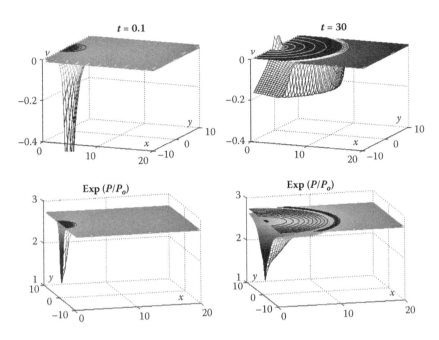

FIGURE 5.15
Two typical configurations of the wave of unloading at two moments of time.

Figure 5.16 shows pressure distribution $p(r,t)$ in the form of curves for different initial velocity V_0 of the wave of unloading. Because the system is open toward the half space $x < 0$, the coal destruction should continue until the stress limit for stretching is reached [84,85]. Figure 5.17 shows the distribution of instant pressure configurations in seam depths at different viscosities of a coal–gas system.

The ultimate strength σ^* should be considered in the solution of Equation (5.63) as an additional intrinsic characteristic of the medium. As a wave widens, the $R \gg R_0$ stresses become less intensive, so that with some r values the condition $\sigma = \sigma^*$ or $\tau_{C_{II}} = \tau_{st}$ can be met. Destruction of coal structure almost stops. In particular, no opening of closed porosity is connected with that irreversible methane emission. As a result, despite large r after the wave $p(r,t)$, the system relaxes to its initial condition and pressure p_0 is restored.

As noted earlier, methane desorption from coal depends on an external parameter (pressure). It is natural to expect parameter changes during the wave of unloading. Using dependence (5.45) of parameter χ on pressure $\chi(P)$ $\approx \chi(P_0)-\kappa(1-P/P_0)$, one can restore the potential of free energy with different pressure values, hence the dependence on radius R at every moment t while the front of the wave moves inside the seam. This process may be partial (Figure 5.18).

The top left of the inset in Figure 5.18 shows the disappearance of the barrier that separates high and low concentrations at a fixed distance R^* from the

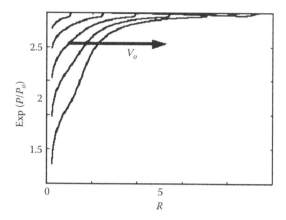

FIGURE 5.16

Distribution of instant pressure along radii with different initial velocities (V_0) of seam selvedge destruction. (1) 1. (2) 0.1. (3) 0.01. (4) 10. (5) 100.

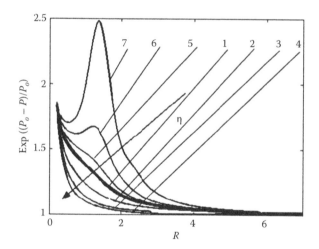

FIGURE 5.17

Distribution of instant pressure configurations along radii at different viscosities. (1) η. (2) 2η. (3) 4η. (4) 8η. (5) 0.5η. (6) 0.25η. (7) 0.125η.

center of the wave. The bottom segment illustrates the behavior of extremes of free energy depending on R. The black curve indicates maximum energy; gray curves show two different minimums of energy. In the area of small R, all three extremes join, indicating disappearance of the potential barrier which separates a metastable condition with high concentration from a stable one with small concentration. Figure 5.18 clearly shows that a quick decrease of external pressure (unloading) in a coal–gas system instantly transforms the system to $\psi \ll 1$ and generates free energy of the phase transfer. This process usually occurs during drilling and blasting operations when GDP exceeds 90%.

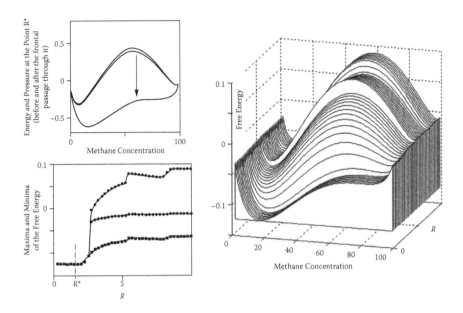

FIGURE 5.18
Transformation of free energy of methane during wave of unloading in radius-concentration variables.

5.4.4 Transformation of Pores into Fissures

A physical result of the change of the form of free energy as a wave of unloading flows is the loss of stability from high gas content in closed pores and its transport into fissures. At this point, the possibility of gas emission from the system, as we have seen, depends completely on its flow over a system of interdependent open pores.

It is natural to expect that simultaneously with the loss of stability, the wave of unloading may produce other effects, namely an increase of open porosity areas at the expense of pore transformation into fissures and the development of an interdependent fissure system. This results because the release of pressure on the front of the wave causes stress in the pores that can exceed the critical level necessary for their spontaneous transition into fissures. According to the general theory of elasticity [86,87] exceeding critical pore size causes instability, determined by a pressure change $\Delta P = (1 - P/P_0)$ under the following condition:

$$L_{cr} = L_0/\Delta P^2 \tag{5.68}$$

Here L_0 is a constant of the length dimension that can be determined by characteristics of the medium and includes the elastic modules and surface tension [74,86,87]. However, as a real system under examination is difficult to analyze, it may be more consistent to keep only the general dependence L_{cr}

on ΔP. Even in this form, the use of the condition (5.68) looks rather extraordinary as real medium contains many pores of different diameters. Thus, in the presence of the wave of unloading, the condition is met only for the biggest pores.

In these circumstances, two opposing processes occur simultaneously. The number of fissures increases and the mean distances between them decrease accordingly. That should contribute to merger of pores into a single cluster. However, as the largest pores transform into fissures first, the medium size of the rest of the pores decreases, thus decreasing the probability of their intercrossing with cavities and thus generating clusters. To solve this issue, we suggest that the medium has a specific structure—porosity—according to Equation (5.51) by a correlation function $G(r-r',t) = <\rho(r,t)\,\rho(r',t)>$. The given porosity corresponds to a distribution function according to $\rho_{pore} = \rho_{pore}(r,t)$, which is also characterized by medium pore radius:

$$<r_{pore}(t)> = \int_{r_1}^{r_2} r\rho_{pore}(r,t)dr \tag{5.69}$$

Taking into account the change of pore distribution depending on size as they transform into fissures, we will determine whether the mean distance between the cavities will decrease or equal the mean pore radius. In other words, we will learn under what pressure (or distance from the center of destruction) the opening of closed porosity takes place at the expense of fissure system increase. If we suggest that coal density $\rho(r,t)$ approximately equals 1 inside the substance or is ~0 inside a pore, we may think that correlation is $G(r,t) \approx \int_r^\infty \rho_{pore}(r,t)dr$ and $\rho_{pore}(r,t) \approx -\partial G(r,t)/\partial r$ accordingly. Taking into consideration the form shown in the inset of Figure 5.12, we find the final function of pore distribution along the radius $\rho_{pore}(r,t)$ (Figure 5.19a). Using $L_{cr} = L_0/\Delta P^2$ and suggesting that at a given ΔP each pore whose diameter exceeds L_{cr} transforms into a fissure, integrating $\rho(r,t)$ with respect to r from $L_0/\Delta P^2$ ad infinitum, we calculate the dependence of the number of fissures on pressure:

$$\rho_{fis}(\Delta P,t) = \int_{L_0/\Delta P^2}^{\yen} \rho(r,t)dr. \tag{5.70}$$

The corresponding curve is shown in Figure 5.19b. Finally, if we assume the fissures are parallel to each other (in the one-dimensional case and with $d = 2$ and $d = 3$), we calculate the mean distance between them

$$L_{fis}(\Delta P,t) = 1/\rho_{fis}(\Delta P,t). \tag{5.71}$$

This value decreases with the intensification of ΔP. In turn, the mean pore radius $<r_{pore}(t)> = \int r\rho_{pore}(r,t)dr$ decreases as the largest pores transform into fissures. The flow along the system occurs only when the distance between

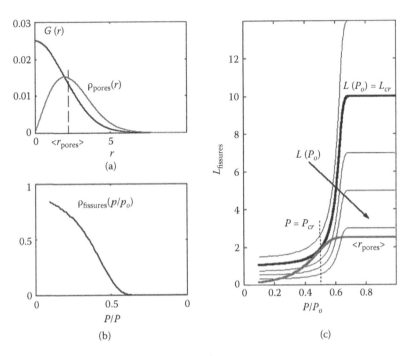

FIGURE 5.19
Calculation of mean distance between fissures as pressure function. (a) Correlation function of porous medium and pore distribution along radius $\rho_{pore}(r)I$. (b) Dependence of cavity density on pressure $\rho_{fis}(\Delta P)$. (c) Comparison of mean distance between cavities $L_{fis}(\Delta P)$ and mean pore radius $<r_{pore}(\Delta P,t)>$ depending on pressure.

the fissures is shorter than the typical pore scale $L_{fis}(\Delta P,t) \ll r_{pore}(\Delta P,t)>$. As pressure $L_{fis}(\Delta P,t)$ is determined by the initial value L_0, for $L_{fis}(\Delta P,t)$, certain curves cross the line only with small L_0, i.e., starting with L_{cr} (Figure 5.19c). The thick black line that touches the (gray) curve for $r_{pore}(\Delta P,t)$ at the single point $P = P_{cr}$ indicates that all methane in the system of closed pores will go into fissures from where it will be emitted by intensive filtration [88].

We can see that wave distribution contributes to porosity opening and methane emission [88]. This effect can be significantly reduced only by the reduction of the number of open pores in the initial system. If density distribution is fixed $\rho_{pore}(r,t)I$, it can only be reached by applying additional (external) factors.

5.5 Classification of Gas-Dynamic Phenomenon Type

Geological and gas-dynamic phenomena (GDP) are important issues, particularly under steep pitch conditions [89–99]. As a rule, determination of GDP type (namely sudden outbursts, gas emissions, or sudden coal falls) is the job of an expert commission that applies the results of studies of a site in accordance with a protocol [100]. At depths of 800 m and more (where the main contributor to the GDP power balance is rock pressure) the sizes and forms of potential outbursts are determined by methane volume and concentration [20]. After a determination is made, a set of measures to prevent GDP is required. As a rule, such measures concern changes of coal mechanical features and decreases of massif elastic parameters. Coal that is prone to sudden falls before mining works reveals low values of such properties [101]. That is why poor mining practices may initiate a GDP.

NMR analyses of coal samples taken after sudden coal outbursts and gas emissions and from pillars [102] showed structural changes of outburst coal. The decrease of intensity of the maximum spectrum line in the aromatic area of coal is obviously connected with the breaking of light hydrocarbons from aromatics. The greatest decrease of maximum intensity of a spectrum line was noted in aliphatics, i.e., coal and gas outbursts are accompanied by breaking of aliphatic chains that in turn leads to increased methane. The destruction of aliphatic bonds during sudden outbursts must be accompanied by adsorbent capability changes of coal due to the decreases of adsorbent centers. To confirm this conclusion, we studied methane-saturated coal taken after outbursts and from pillars [50]. The width of a spectrum line of adsorbed methane in coal samples after an outburst is narrower than it was earlier. This confirms the conclusion that at the time of outburst, a break of finite groups at adsorbent centers takes place and mobility of methane-adsorbed molecules increases.

Nevertheless, the question of methane formation during a GDP remains controversial. More widely believed are studies that show that unloading part of a gas-saturated coal massif causes stress in the porous system of coal and transforms pores into cracks that emit methane via filtration. As shown in subsection 5.4.4, the critical pore size (greater than the size when instability develops) is determined as pressure changes, $\Delta P = (1-P/P_0)$:

$$L_{aver} = L_0/\Delta P^2.$$

The analysis of the formula shows that increasing numbers of shallow closed pores are involved in formation of new fissures as ΔP increases. The number of cracks increases and the average distance between them declines. The decreased distance aids binding of pores into clusters. However, because the largest biggest pores transform into fissures, the average size of the remaining pores decreases, i.e., closed porosities break during outbursts.

The mechanism of sudden coal falls with simultaneous methane emissions is connected with stability loss due to a transition into generalized tension. In contrast to coal and gas emissions at lower values of unloading wave amplitude, sudden falls do not cause porous system openings. Thus, depending on GDP type, the degree of coal structure damage varies. One of the most promising methods of estimating structural changes in coal types prone to different GDP is based on sorbent property recognition, particularly, the kinetic energy of activation and methane desorption.

5.5.1 Criteria of GDP

To reveal the destruction mechanism impact on methane desorption kinetics, experiments to define the diffusion coefficient and processes of power activation were conducted on coal samples of layer m_3 from the Zasyadko Mine [103,104]. The sample size was 6.0 × 6.0 × 6.0 cm, deformed under three-axial loading according to three different schemes. The first scheme modeled generalized compression ($\mu_\sigma = -1$) $\sigma_1 = \sigma_2 = \sigma_3 = 25.0 - 50.0$ MPa with simultaneous unloading along the three facets. Sample fracture did not occur. The second scheme modeled the conditions of generalized shift ($\mu_\sigma = 0$); loading parameters were $\sigma_1 > \sigma_2 > \sigma_3$ where $\sigma_1 = \sigma_{st}$, $\sigma_2 = \sigma_1/2$, and $\sigma_3 = \sigma_{comp}$. The third scheme utilized generalized tension ($\mu_\sigma = 1$); parameters were $\sigma_1 = \sigma_2 > \sigma_3$, $\sigma_1 = \sigma_{st}$, and $\sigma_3 = \gamma H$ with decrease of σ_3 to 0. In the second and third cases the samples were destroyed.

Two fractions were selected from every sample. Particles were 0.25 to 0.5 and 2 to 2.5 mm and weighed at least 10 g in accordance with the method cited in Section 3.3.4 in Chapter 3. Methane emission activation energy was determined. Figure 5.20 shows typical dependences of methane desorption from two fractions at 25°C. Along the ordinate axis, the relative decrease of methane in a coal sample over time is fixed and calculated by

$$\Theta = 1 - \frac{Q_M - Q_C}{Q_M},$$ (5.72)

where Q_M is the maximum methane in a coal sample at initiation ($t = 0$) and Q_C is the current value of methane ($t > 0$). The analogical curves were obtained

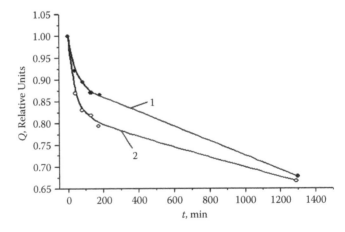

FIGURE 5.20
Desorption of methane from coal particles: (1) $R = 2$ to 2.5 mm. (2) $R = 0.25$ to 0.5 mm.

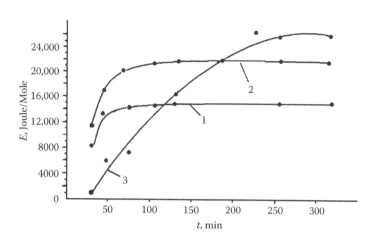

FIGURE 5.21
Transforming of methane emission activation energy in time depending on coal deformation scheme. 1. Hydrostatic compression. 2. Deformation by shear. 3. Deformation by cleavage.

at 50°C. The only difference was in the initial methane amount in coal fractions and speed of mass loss. Applying the results of methane desorption, we calculated diffusion coefficients for each scheme of loading and determined the activation power of methane emission (Figure 5.21).

The results indicated that the initial 35 min for all schemes of loading activation energy was minimal, taking into account the dependences in Figure 5.21. This shows that methane is emitted from open pores whose volume is more than 20% of total porosity. Between 50 and 350 min, the activation energy for hydrostatical compression is smaller than energy from deformation by shear. A decrease of activation energy (1.5 to 1.7 times) in comparison with hydrostatic loading is connected with a greater amount of closed pore system deformation during coal fall. The greatest transformations of activation energy were observed when samples were deformed by cleavage. This is connected to an increase of open porosities due to the transformation of a considerable number of pores into fissures (Figure 5.21, curve 3). During the initial 70 min, activation energy was minimal based on methane filtration from the formed system of fissures. The activation energy later increased and exceeded the energy of samples deformed by shear and compression. This proves the decrease of average size of pores not transformed into fissures.

Figure 5.22 shows the integral dependence $E = f(\mu_\sigma)$ that characterizes the transformation of activation energy of methane emission from coal based on stress condition. A coal massif under conditions of generalized tension ($\mu_\sigma = 1$) demands on average only one fourth the power needed for methane desorption from a porous system. The regularity of methane diffusion serves as a basis for predicting the geomechanical condition of a face working space.

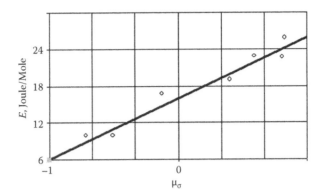

FIGURE 5.22
Transformation of methane emission activation energy depending on stress condition during loading of TTCU.

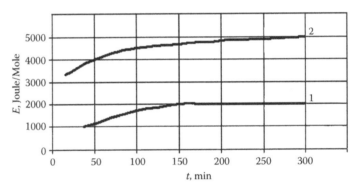

FIGURE 5.23
Methane emission activation energy transformation of coal samples from outburst coal (1) and virgin rock mass (2).

As methane emission activation energy depends on stress condition (one of the main parameters determining outburst hazard), it may also serve as a predictor of outburst danger. To prove this possibility, we studied methane emission activation energy of samples of outburst-prone coal (intensity of 515 tons) from seam h_{10} at Kalinin Mine. The results (Figure 5.23) indicate that coal after a coal and gas outburst shows 2.5 to 3.0 times less methane emission activation energy than coal from an outburst-safe massif. We now know that methane emission activation energy from coal characterizes its porosity and outburst danger and depends on sorbent centers. Thus the degree of destruction at the molecular level (sorbent centers) will be different in outburst- and eruption-prone coal. NMR proved this by calculating methane absorbed (desorbed) by coal prone to GDP.

To determine GDP type via NMR spectroscopy data, the coal under analysis was reduced to fine particles of 0.3 to 0.4 mm, dried, degassed, and put into

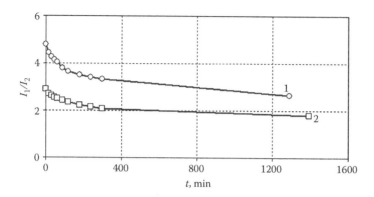

FIGURE 5.24
Dependence of NMR spectra lines of coal from GDP area and virgin rock mass over time. (1) Coal from pillar. (2) Outburst-prone coal.

test tubes where it was saturated under 10.0MPa pressure for 120 h. A wide line NMR spectrometer measured the decrease of adsorbed methane while the sample was degassed every 120 min for 24 h. Using I_1/I_2 (I_1 = intensity of the narrow line from the virgin rock mass area and I_2 represents coal from a GDP area) over time t, graphs of spectra intensity decreases were built.

Figure 5.24 presents the results. Lines 1 and 2 show the transformation of NMR spectra line value for gas-saturated coal samples from virgin rock mass and gas-saturated samples from GDP areas (seam h_{10} of the Kalinin Mine), respectively. At the initial moment of time ($t = 0$) the ratio of intensities I_1/I_2 of coal spectra lines from the pillar was 4.9. For coal from the GDP area, the ratio was 2.9. At 400 min, the ratios were 3.0 for the pillar coal and 1.35 for the outburst coal; at 0.5 day, $I_1 = 2.6$ and $I_2 = 0.9$, respectively.

Figure 5.25 shows results obtained for a GDP in coal seam h_8 'Proskoviyevsky' of the 60 Soviet Ukraine Mine when the seventh eastern haulage drift advanced at an intensity of 318 t. Thus, essential differences exist in NMR spectra for coals from outburst-proof and outburst-dangerous areas. For this reason such analysis was conducted on coals in dangerous steep seams subject to sudden falls. During exploratory studies, two GDP were under consideration (1) coal seam m_3 'Tonkiy' at the Kochegarka Mine at 1080 m depth and (2) coal seam K_4^1 at the Rumyantseva Mine at 910 m depth. Expert commissions classified them as sudden falls. Fallen coal from each GDP was compared with coal from a pillar 5 m from the GDP site. Using the methods described above, the coal fractions were prepared and the gas emission kinetics from methane-saturated samples from the GDP area and the outburst-proof area determined. Figure 5.26 shows results for the Tonkiy seam as the dependences $I_1/I_2 = f(t)$. The same result was obtained for seam K_4. Analysis of the results indicates that spectra lines from the pillar and coal fall area samples differed not more than 1.05 to 1.1 times (for coal gas outbursts, 2.7 to 3.2). Small differences in graphs 1 and 2 prove that

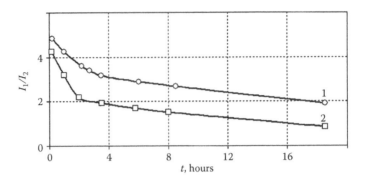

FIGURE 5.25
Dependence of NMR spectra lines of coal from virgin rock mass (1) and GDP area (2).

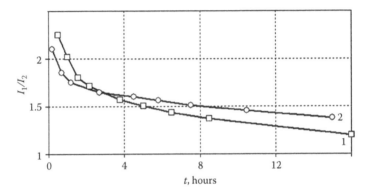

FIGURE 5.26
Kinetics of methane desorption in coal NMR spectra intensity of Tonkiy seam from pillar (1) and coal fall zone (2).

during a sudden coal fall caused by loss of stability of a gas-saturated coal massif, the transition from generalized shift to generalized tension via a gravitational component of mine pressure from an unloading wave does not extend the limit in coal or open the porous system to turn pores into cracks. Thus coal deformation at the molecular level is not efficient. Considering the ambiguity of results, further testing related to GDP classification method was conducted.

5.5.2 Parameters and Technology of Method

The first step was selection of samples from a virgin rock mass (pillar) and erupted samples from an outburst. The samples were treated by breakage, drying, vacuuming, and saturation with methane. Intensity of narrow and wide lines of NMR spectra were measured and we then calculated an integral index to characterize GDP type. Pillar samples were taken 0.2 to 0.3 m below the face surface or drilled from 1.0 to 1.5 depths.

If a GDP occurs at a shield face, selection of samples must be done 5 to 10 m from the GDP site. Samples taken after a GDP at a working area or face entry must be taken at least 3 to 5 m from a development working area or face entry. In a seam with clustered structures, a sample must be taken from every cluster and weigh not less than 0.1 to 0.2 g. In development workings and shield long wall faces, sample selection from outburst (erupted) coal is carried out from the slope at a distance of 2 to 4 m from the side of the GDP site. At least three samples are required. In overhand stopes and face entries, at least four samples are scraped from cavity walls. If sample selection from a GDP cavity is not possible, coal rubble as close as possible to the GDP area is chosen. A diagram detailing sample selection sites is then prepared.

Coal samples must be placed into hermetically closed tubes of 20 to 30 cm^3 capacity or into polyethylene bags. Sizes of the fines or outburst (erupted) coal pieces must be at least 2 mm. The time interval between sample selection and delivery to a laboratory cannot exceed 24 h. Before drying, the samples must be milled to fractions 1 to $3 \cdot 10^{-4}$ m. The fractions must be dried to eliminate all adsorbed moisture. Control of the degree of drying is conducted according to NMR spectra. Coal is considered completely dried when no Lorentz line is present in a spectrum.

Granulated and drained samples weighing 20 to 25 g are placed in a methane saturation system and kept at 8.0 MPa pressure for 120 h. Before methane saturation, samples are vacuumed for 3 h at 10^{-3} MPa pressure (Figure 5.27).

A total of 8 sudden coal falls and 13 sudden coal and gas outbursts were analysed. Figure 5.28 shows the gas emission kinetics curves for samples from a coal and gas outburst, from a sudden coal fall, and from a distance of 5 to 7 m from a GDP site.

NMR spectrometry determined the transformation value of methane in coal in relation to a reference standard sample (etalon). The methane in samples was not calculated. Only the correlation of the intensities of a Lorentz

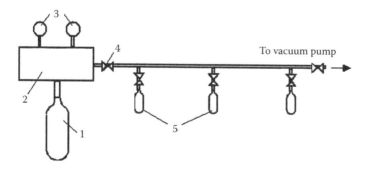

FIGURE 5.27
System for vacuuming and saturating coal probes with methane. 1. Methane cylinder. 2. Speed reducer/gearbox. 3. Manometers. 4. High-pressure valve. 5. Containers.

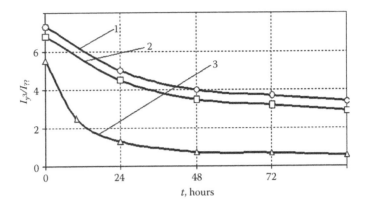

FIGURE 5.28
Dependences of methane emission of coal probes showing different degrees of damage. Samples were selected from safe massif (1), site of coal fall (2), and (3) site of coal and gas outburst.

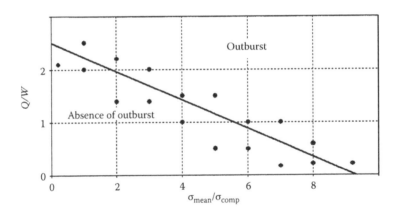

FIGURE 5.29
Regularities of coal deformation while modeling outbursts.

line and a Gaussian line of methane-saturated samples chosen from virgin rock mass with methane-saturated coal samples from a GDP cavity were compared. These studies enabled us to classify GDP types. We suggest the following criterion:

$$B = \Psi_V / \Psi^{-1}_{GDP},$$ (5.73)

where ψ_V and ψ_{DGP} are NMR spectra intensities of the coal from virgin rock mass and a GDP site, respectively. Based on statistical analysis [105], if B ≥ 0.6, a GDP is related to coal and gas outbursts; if B < 0.3 the GDP is a coal fall. To summarize the results of tests, a National branch document titled "The method for differentiating sudden rock fall (eruption) of coal from outburst for expert estimation of its GDP type" was published.

5.6 Predicting Coal Seam and Sandstone Outburst Danger

During coal seam development, local and area outbursts occur and prevention measures are expensive. It is important from both safety and economic views to be able to forecast outbursts accurately at seam breaks, working faces, and development workings. Coal, gas, and sandstone are all outburst-dangerous materials.

5.6.1 Outburst Dangers of Coal Seams

5.6.1.1 *Criteria for Outburst Danger*

We used the data on the impact of sorbent gas and unloading speed on coal mass deformation to define the parameters that determine outburst danger in the form $B = (\sigma_{mean}/\sigma_{comp}; Q/W)$ [92–95] where Q is the methane in coal and W is the bound water in coal. To establish a quantitative criterion, the investigations [106] utilized a high pressure chamber set between the rods of three hydraulic cylinders on a desk-top triaxial loading unit (see Chapter 4 for details). To model an outburst in a loaded sample, a sample facet 3.14 cm^2 in area was unloaded at 5 m/sec. After a complete cycle of investigations, the chamber was opened and the sample was studied to define outburst cavity depth and sizes of the formed fractions. Table 5.5 presents the results.

Table data indicate that the cavity depth in a coal sample and the reduced radii of fractions are determined by the extent of gas and moisture saturation of coal and the speed of tension release from the loaded sample facet. The maximum cavity depth was $l \geq 7$ to 7.9 mm and the minimum fraction sizes were $r \leq 0.25$ to 0.5 mm—values representative of coal containing methane $Q \leq 20$ to 22 m^3/t, physically bound water $W \leq 0.5\%$, and unloading speed $V \geq 4$ m/sec. The fixed properties of fluid (gas, water) influence and unloading speed on cavity parameters in a sample are realized under the conditions of a stress deformed state typical of a face working seam area. To determine the power accumulated in the massif, we included mean stress σ_{mean} and coal compression strength σ_{comp} as criteria. Thus, the dependence $Q/W = f(\sigma_{mean}/\sigma_{comp})$ allowed us to devise criteria for predicting the outburst danger of a coal seam.

Figure 5.30 presents the results of the investigations of coal outburst modeling at $V_p \geq 4$ m/sec. It reveals the formation of two areas with different mechanisms of deformation. Approximation of the dividing line yields a qualitative index of outburst potential during breaking of a coal seam. See Equation (5.75).

5.6.1.2 *Criteria for Outburst Estimation*

Criteria to classify outburst coal seams and sites by potential for outburst (very hazardous, hazardous, or outburst-proof) [96–98] and monitor activities

TABLE 5.5

Influence of Unloading Speed on Depth of Formed Cavity

$Q, \dfrac{m^3}{t}$ / $W, \%$	Unloading Speed (V_p, m/sec)	Cavity Depth (l, mm)	Reduced Particle Radius r (mm)
$\dfrac{20-22}{0.5}$	0.05	–	–
	0.5	2.0 to 3.0	3.5 to 3.8
	2	3.0 to 3.5	2.4 to 2.8
	4	4.1 to 5.2	3.5 to 3.8
	5	7.0 to 7.9	0.25 to 0.5
$\dfrac{10-12}{0.5}$	0.05	–	–
	0.5	–	–
	2	1.0 to 1.5	5 to 7
	4	2.1 to 2.5	3 to 4
	5	3.1 to 3.3	3.5 to 4.2
$\dfrac{10-12}{2.0}$	0.05	–	–
	0.5	–	–
	2	–	–
	4	1 to 1.3	9 to 10.5
	5	1.8 to 2.1	8.1 to 8.8
$\dfrac{3-5}{2.0}$	0.05	–	–
	0.5	–	–
	2	–	–
	4	–	–
	5	1 to 2	12 to 15

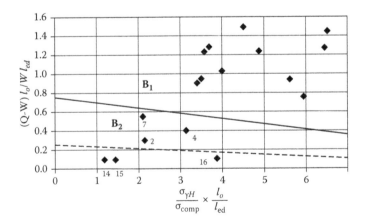

FIGURE 5.30

Dependence $((Q - W)/W) \cdot (l_0/l_{ed}) = f\,(\sigma_{\gamma H}/\sigma_{comp}) \cdot (l_0/l_{ed})$ characterizing degree of coal seam outburst danger.

to prevent GDP are essential. Categorizing outburst potential of coal seams is based on a physical model that reflects the unstable conditions of a working space based on water–methane content in porous volume, resistance, stress deformed condition, and expected transformation. As noted earlier, the construction requires the connection of three non-dimensional parameters:

$$B = f(K \frac{Q-W}{W}, \frac{\sigma_{\gamma H}}{\sigma_{comp}}, \frac{l_0}{l_{ed}}). \tag{5.74.}$$

where $\sigma_{\gamma H}$ is geostatic stress at a specific depth; the other parameters are described above. The connection between the first two non-dimensional indices is determined while modeling partial unloading of a sample facet:

$$B = Q/W - 0.27 \sigma_{mean} / \sigma_{comp} - 2.5 = 0. \tag{5.75}$$

Replacement of Q/W with $(Q–W)/W$ increases the accuracy of an outburst prediction criterion. Non-dimensional index l_0/l_{ed} [99] represents the tension transformation gradient and gas recovery factor in the face working space of a seam. Depending on the unloading technology used in driving a mine working, l_0 is the shift face advance and l_{ed} is the distance from a seam edge to maximal tension concentration; $l_0/l_{ed} = 0.1 \div 1$.

When unloading speed $V_{unl} < 1.0$ MPa/sec $(l_0/l_{ed} \leq 0.1)$ and deformation state is characterized as generalized shear, deformation of crack-porous coal according to gas-permeability parameters is minimal and in a resulting gas emission, only free methane in free pores and cracks will be involved; thus the probability of a coal and gas outburst is small. During a GDP, most free and sorbed methane is emitted. At unloading speed $V_{unl} \geq 1.0$ MPA/sec $(l_0/l_\kappa = 0.9$ to $1.0)$ and high methane capacity, a working face area subjected to generalized tension forms a secondary system of cracks that intensify the opening of most of the closed pores and actively desorb the sorbed gas.

We devised a method for using numerical criteria to predict coal seam outbursts [Equation (5.74)] by instrument measurements and calculating dependence values. Further investigations and tests to determine acceptance of the method were conducted in the Central and Donetsk–Makeyevka regions of the Donbass, namely, seams m_3 "Tolstiy", h_{11} "Bezymyanniy", m_2 "Tonkiy", k_8 "Kamenka", and m_4 "Peschanka", all from 790 to 970 m in depth; and flat seams h_8 "Praskovievsky", h_{10} "Livensky", and h_6 "Smolyanovsky" all from 810 to 1144 m in depth. The measurements at steep seams were taken from flat sections in four faces while driving development workings and in a face working. The face advance was 620 m during investigation and acceptance tests. Figure 5.31 presents the results as dependence $B = (Q–W)/W; \sigma_{\gamma H}/\sigma_{comp}; l_0/l_{ed})$ of the coordinates $((Q–W)/W) \cdot (l_0/l_{ed})$ along axis y and $(\sigma_{\gamma H}/\sigma_{comp}) \cdot (l_0/l_{ed})$ along axis x. The data point labels correspond to the number of measurements made.

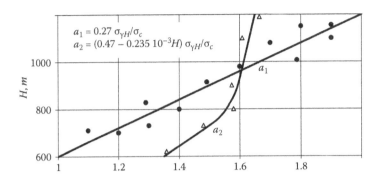

FIGURE 5.31
Dependence of coefficient a on mining depth H.

A coal seam is considered extremely dangerous when its outburst criterion is

$$B_1 = \left[(Q-W)/W + 0,06\sigma_{\gamma H}/\sigma_{comp} \right] \times l_0/l_{comp} \geq 0,75. \qquad (5.76)$$

A seam may be considered outburst-dangerous or outburst-proof when the criterion equals:

$$B_1 = \left[(Q-W)/W + 0,025\sigma_{\gamma H}/\sigma_{comp} \right] \times l_0/l_{comp} \leq 0,25. \qquad (5.77)$$

Based on exploratory testing, when mine layers were classified by Equation (5.76) or (5.77) as extremely outburst-dangerous or outburst-proof, errors of the first type that characterize refusal to accept the prognosis of outburst degree were not available. Errors of the second type connected with consistent prediction of a higher degree of danger that did not correspond to reality constituted 14%. The results of predicting the outburst danger of seam h_8 while driving a track during a regime of shock blasting at 1090 m at Donbass confirmed the productivity of the B_1 criterion. Three outbursts forecast by index B_1 occurred.

Qualitative analysis of the explosion danger of seam h_8 based on Equation (5.76) showed that the biggest factor of explosion danger is the non-dimensional index $(Q - W)/W$ representing the water–methane fluid contribution. We reiterate that the suggested criterion characterizing the deformation of a face working space of a seam based on fluid saturation and mine pressure can be regarded only as a generalized predictor of a GDP.

Physically bound water content W, coal strength σ_{comp}, and unloading area value before a face l_0 must be determined to classify a coal seam. In a working face, these parameters are measured every two cycles of an advance, and in a development working, at every cycle. In stoping faces during a face advance of a haulage roadway of a working face, line measurements and sample collection are conducted in a face at a distance of 15 to 25 m from the roadway. For a system long wall roadway, measurements and selection are conducted 10 m from stope toes at pitching and at the first bench slope at stope toes in development workings. Determination of gas saturation (Q) for a potential or definite coal seam site is conducted once according to the methods described in Chapter 3.

Determination of physically bound water content in a seam during face and development working advance is conducted and involves sampling from drilling rubble. A hole is made to measure initial speed dynamics of gas emission from a depth of 1.0 to 1.5 m. At least three rubble subsamples are placed into metal hermetic capsules and taken to the surface. Direct qualitative dampness determination is conducted by a ^{1}H NMR unit using the above methods. Mean dampness is calculated from three determinations. The time interval between sample collection from coal mass and dampness determination must not exceed 24 h to ensure that initial moisture is maintained.

Ultimate strength in a face is determined by an "express" method using a strength meter [34]. Coal compressive strength is measured using all the patches of coal, power m that exceed 0.2 m. Five individual measurements are made for every measurement cycle. The mean value peak or spear implementation depth h_ψ is determined and the compressive strength index q_i is calculated. The distance between the two "shot spots" is 0.25 m:

$$q_i = 100 - h_\psi. \tag{5.78}$$

The average resistance of a seam that has a complex structure is determined as the weight-average value of resistance of the patches that comprise a seam:

$$q_{seam} = \frac{q_{i_1} m_1 + q_{i_2} m_2 + q_{i_n} m_n}{m_{i_1} + m_{i_2} + \cdots m_{i_n}}. \tag{5.79}$$

According to resistance index data, seam resistance is calculated [90]:

$$\sigma_{comp} = \frac{800}{95 - q_{seam}}. \tag{5.80}$$

Determination of unloading area before the face of a seam is conducted by the standard method based on the dynamics of the initial speed of gas emission [100]. As an unloading value, the minimal distance from the face breast to the middle of the interval where the maximal initial gas emission speed has been measured is accepted. To classify coal seams as outburst-dangerous according to criterion B_1, the complex of observations in the stope face covering more than 20 m of a face advance (in a development working, 10 to 15 cycles over a length of more than 30 m). The number of test holes, their parameters, schemes of location for stope faces and development working conditions, and anomalies (GDP areas) are accepted in accordance with the standard [100]. The periodicity of classification of a very dangerous coal seam is conducted every 100 m of a face advance in a stope face or development working. If an outburst coal seam has geological faults, a GDP outburst-danger classification is set up.

5.6.1.3 Grounding and Testing of Outburst Prediction Method at Openings of Steep Seams

Theoretical studies established that the kinetics of free energy of a coal–gas system is defined by the effective diffusivity of methane from coal in the transport channels in the form of gas adsorbed on the surface and dissolved in the coal substance. The only way to intensify gassing and the transition from diffusion to filtration is to examine the porous system in the coal at high speed unloading by electromagnetic effect (see Section 5.3). Speed load is always used since the opening of coal seams by crosscuts is carried out by drilling and blasting operations (DBO). The lower the toughness of a coal which is proportional to the shear modulus, the greater the penetration depth of the unloading wave. To predict the outburst danger at the opening, we used Equation (5.75) to determine the loss of stability of the coal.

The functional test was carried out on heavily pitched seams with varying degrees of outburst dangers at their openings by shafting and crosscutting at depths from 530 to 1180 m. The results indicated that the method reliably described the outburst hazardous states of the coal seams. However, it is impossible to verify whether the outbursts will occur at the opening because when outburst danger is determined in the usual way, outburst control measures are introduced at the face.

Results of modeling a coal massif to depths of 3000 m [91] indicate that share of the elastic energy recovery initiated during unloading in the balance of power emerging from rock pressure tends to increase with depth. For example at 800 m, 27% of the elastic energy is released; at 2000 to 3000 m, recovery is 12 to 15%. Therefore, determining the coefficient a_1 of the dimensionless parameter $\sigma_{\gamma H}/\sigma_{Comp}$ must consider the depth of the work at the opening of the seam. Figure 5.31 shows the function $H = f(a_1)$ for this regularity. The final criterion for outburst danger prediction at the opening of a coal seam has the form [102]:

$$B_l^! = Q/W + (0.47 - 0.235 10^{-3} H)\sigma_{\gamma H}/\sigma_{comp} 2.5. \qquad (5.81)$$

The functional areas for analysis may be heavily pitched seams subject to outburst danger at depths from 600 to 1200 m. Predicting outburst danger of such seams includes preliminary well drilling, the selection of coal samples and placing them in airtight containers to determine water content and coal resistance, and calculating the degree of danger. The drilling of prognostic wells is performed in the opening pit. Samples (Figure 5.32) are taken from the working face of the shield long wall 10 to 15 m along the dip of the field haulage roadway.

The main parameters determining the stability of a coal–gas system are the concentration of methane in the open and closed pores [101] and the rate of change of gradient stress. Therefore, these parameters may serve as predictors of outburst danger in coal seams. One of the most difficult problems

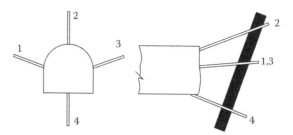

FIGURE 5.32
Recommended scheme of outburst prediction at openings of heavily pitched seams.

at present is prediction of outburst-dangerous seams during blasting of preparatory workings. This is not well covered in the literature or regulations.

The criteria for classifying gas-dynamic phenomena are based on results cited in Sections 4.1 and 5.4. The degree of change in free energy ΔF characterizes the "hidden" burst state of a coal–gas system by the density of closed and open pores bridging Ψ with methane in the plane of the seam X and velocity of the unloading wave that causes tension in the pores, During mining of an outburst-dangerous seam, we can assess the free energy of a coal–gas system by developing workings in areas not subject to GDP (ΔF_b -= baseline) and compare the energy value with the current ΔF_c, to predict GDP location according to the ratio:

$$B \approx \left(\Delta F_c - \Delta F_b\right)/\Delta F_b \geq 1.$$

However, the calculation of free energy in a moving face is not possible. The energy can be calculated if density (methane in open and closed pores) is used as a parameter [101]. Based on the foregoing, the formula to determine locations of GDP is

$$B_2\left(\Delta\psi\right) = \frac{L}{\psi_m}\frac{d\psi_m}{dx} > B_n, \tag{5.82}$$

where $\psi = I_n/I_w$ is the amount of methane in samples determined by NMR spectra; B_n is the index characterizing nonoccurrence of GDP; x is the current sampling coordinate in meters; and L is the face advance value of development working for one cycle.

Index testing using Equation (5.82) was carried out on two very outburst-dangerous seams (h_{10}, Glubokaya, 1190 m and h_4, Sixty Years of Soviet Ukraine, 770 m) during the drifts and well advance. Samples were taken from the left and right wells drilled from the faces of development workings in the direction advance to a depth of 1.8 m (Figure 5.33). Processing and calculation of B was conducted according to Equation (5.82). The concentration of methane in the samples was determined by NMR using the methods described in Chapter 3.

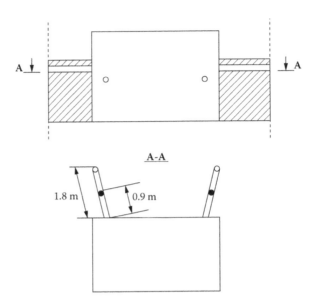

FIGURE 5.33
Location of sampling wells.

Nine outbursts (average of two outbursts per 10 m of well) occurred during drilling in seam h_4 before the experiment started. The background value of index B was determined first (average = 0.41). Table 5.6 illustrates results for the outburst location based on samples selected from a 60 m ventilation well drilled in the seam. Figure 5.34 shows the B calculation data for samples from the left hole and outburst locations.

In general, the results indicated that prior to each of the six outbursts in development workings (concussion blasting over a length of 60 m), the minimum value of B_1 = 0.6. The distance of gas dynamic phenomenon fixing did not exceed one or two cycles of blasting (1.5 to 3.0 m). Extensive tests confirmed the physical sufficiency of the established mechanism for GDP forecasting based on measuring the packing density of a coal pore structure with methane and a standard was thus developed [101]

Theoretical studies led to development of a physical model, explaining the methane density mechanisms that confer stability on a coal–gas system, and defining the terms of methane emission (coal outburst and gas emission) from the pore structures of coal by unloading waves and analyzing diffusion and filtration processes. The main findings were verified experimentally by bulk loading of unequal components, in particular, to study the effect of methane saturation on unloading speed and the toughness of coal based on unloading distribution depth. We established for the first time that unloading of gas-saturated coal samples leads to the formation of generalized tension consisting of both stresses and strains oriented perpendicularly at the unloaded stress of the system of cracks at high filtration

TABLE 5.6

Results from Samples from Seam h_4 and Calculation of B_1 Criterion

	Left Well				Right Well			
	Parameters of NMR Spectra				Parameters of NMR Spectra			
Sampling Place along (m)	I_{nar} a.u.	I_w a.u.	ψ_m	B_1	I_{nar} a.u.	I_w a.u.	ψ_m	B_1
40	0.323	0.855	0.378	−0.055	0.256	0.854	0.3	−0.250
41.8	0.493	1.2	0.411	0.027	0.649	1.037	0.410	0.622
46	0.706	1.438	0.490	0.225	0.791	1.48	0.534	0.335
48	0.482	0.768	0.490	0.247	0.29	0.867	0.334	−0.165
48.7	0.936	1.042	0.898	1.245	0.479	1.159	0.413	0.032
50.5	1.212	0.77	1.570	2.35	0.379	0.95	0.399	−0.002
51	0.402	0.801	0.502	0.255	0.375	0.737	0.509	0.272
52	0.523	0.781	0.670	0.675	0.948	0.492	1.927	2.25
53	0.631	0.774	0.815	1.037	0.672	0.79	0.851	1.127
54	0.758	0.82	0.924	1.310	0.633	0.792	0.799	0.997
63.2	3.323	3.219	1.032	1.580	1.637	3.104	0.527	0.317
73.6	1.792	2.419	0.741	0.850	1.396	2.096	0.666	0.665
81.6	1.207	1.158	1.042	1.605	1.344	1.227	1.095	1.737
84	0.603	1.343	0.449	0.122	1.203	1.059	1.136	1.840
85.6	1.149	1.127	1.020	1.550	1.317	1.205	1.093	1.732
87.2	0.488	1.065	0.458	0.145	0.727	1.022	0.711	0.7775
92.8	0.619	1.237	0.501	0.25	0.874	1.592	0.549	0.34
94.4	0.377	1.149	0.382	−0.035	0.551	1.356	0.406	0.03
96.8	0.489	1.08	0.453	0.125	0.433	1.292	0.335	−0.15
98.4	0.677	1.346	0.503	0.264	0.655	1.087	0.603	0.38
100	0.399	1.267	0.315	−0.02	0.548	1.127	0.486	0.2

rates, thus aiding the development of a GDP. After simulating high speed unloading on the faces of the samples, we found a significant impact of fluid saturation (gas and liquid), level of stress and strain, and the elastic properties of coal on the mechanisms of destruction based on coal grade. Coal characteristics and numerical values serve as the bases for predicting coal seam outbursts.

5.6.2 Rock and Gas Outbursts

The earliest data on rock and gas outbursts in the Donbass were obtained at a depth of 800 m at the Novo Tsentralnaya Mine. The vertical shaft was driven by a new explosive method that removed sandstone and gas. Outbursts have occurred at half of the mines in the Donbass. A monograph [106] contains a detailed bibliography on the rock outbursts.

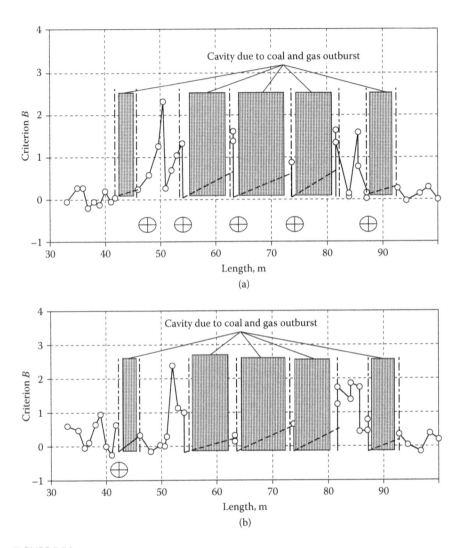

FIGURE 5.34

Gas-dynamic phenomenon prediction using index B. (a) Left well. (b) Right well. $\oplus$ = outburst locations.

Natural rock and gas outbursts result from a complex interaction of three factors: the stress of a rock massif, gas contained in rocks, and the physical, mechanical, and structural properties of the massif. No single factor should be exaggerated or underestimated. The most important feature of an outburst is the release of large reserves of energy consisting of the energy of elastic deformation of the material being destroyed, the energy of elastic deformation of bearing rocks, and the energy of compressed gas. In contrast to sudden coal and gas outbursts, the energy of compressed rocks is more than the energy contained in the gases during a rock outburst.

The present methods of predicting sandstone outbursts provide the necessary security of works during the excavation but often forecast dangerous conditions in non-hazardous areas. Incorrect predictions of sudden coal and gas outbursts cause serious personal injuries; sandstone outbursts normally increase expenses. Treating outburst-hazardous sandstone as dangerous leads to widespread use of standing explosives and large cost overruns for labor and materials. That is why, despite already established methods, a procedure for predicting rock outbursts based on effective surface energy (ESE) was developed and is currently approved as a standard [107].

Rock and gas outbursts result from breaking of rocks outside a massif being exploded or drilled. The duration of the outburst is usually 1.2 to 2.6 sec. During dynamic breaking, the energies of the various elements of the rock massif change and redistribute. All changes during transformation take place in accordance with the energy conservation law. From a practical view, it is not necessary to trace in detail the transformations of energy during an outburst. It is enough to consider the system state immediately before destruction and immediately after the outburst ends. In this case, the energy balance during the outburst can be described by the equation:

$$W_r + W_M + W_n = W_p + W_\kappa + W_s, \tag{5.83}$$

where W_r is the energy of gas expansion; W_M is the energy of elastic deformation of material being destroyed; W_n is the energy of elastic deformation of the surrounding rocks; W_p is the energy spent on forming new surfaces; W_κ is the kinetic energy of expansion of the destroyed material; and W_s is the energy of seismic vibration loss (10%). Neglecting the scattering of energy one can state that, for pressure bump occurrence, the release of energy accumulated in a rock mass should exceed the energy of rock destruction and the outburst. The total energy of elastic deformation [108] of the material is

$$W_M = \frac{1}{6K(1-2v)}\left[\sigma_1^2 + \sigma_2^2 + \sigma_3^2 - 2v(\sigma_1\sigma_2 + \sigma_2\sigma_3 + \sigma_1\sigma_3)\right], \tag{5.84}$$

where σ_1 through σ_3 are the principal stresses; K is the modulus of dilatancy; and v is the transverse strain coefficient. The energy density of the elastic deformation of sandstone is 1 to 10 mJ/m³. The kinetic energy of expanding gas at the onset of destruction (isothermal process) can be estimated by

$$W_T^l = \frac{P_0 T}{T_0}(\mu - a)\ln\frac{p}{p_0}, \tag{5.85}$$

where $P_0 = 0.1$ MPa; $T = 273$ K; T is the absolute temperature of the rock–gas system; P is gas pressure; and $(\mu - A)$ is the amount of free gas. The final period of gas expansion is considered adiabatic. For the conditions of an adiabatic process:

$$W^a{}_{0r} = \frac{p_0 T}{T_0(k-1)}(\mu - a)\left[1 - \left(\frac{p_0}{p}\right)^{\frac{k-1}{k}}\right], \qquad (5.86)$$

where $k = 1.32$ is the adiabatic index for methane. The energy expansion of the gas is 10^4 J/m^3. The energy intensity of the destruction of sandstone [109] is $10^6 - 10^7$ J/m^3.

The energy of elastic deformation of the material being destroyed is almost entirely spent on destruction. An additional source of energy is needed to compensate for the speed of the particles of the destroyed material. These sources may be the energy of elastic recovery of the surrounding rock or the energy of the compressed gas expansion. The velocity $V = 20$ m/sec, the kinetic energy of expansion of the material being destroyed $W_K = \rho\, V^2/2$ mJ/m^3, and the energy of compressed gas is two orders of magnitude smaller. Thus the main source of energy required for material removal should be the compressed energy accumulated in the surrounding rocks.

Unequal components of stress field and the heterogeneity of crack resistance (fracture toughness) of rocks are important factors of destruction [109]. During the development of a heterogeneous rock massif in a hard layer, horizontal or vertical stress can be destructive in the nearby areas because of stress concentration arising from the development.

In a face zone where minimum compressive stress is caused by the residual strength of rocks or approaches 0, a stress state close to a generalized displacement is formed and may well lead to destruction. If the same heterogeneity of elastic properties is observed in a layer between blocks, the vertical stresses will break. The destruction of a hard and abnormal stress layer, according to the laws of destruction of heterogeneous materials, can cause an outburst.

Outburst-pone conditions start when the heterogeneity of rock layers (or blocks) causes an inhomogeneous state in virgin rock. If a more rigid and abnormally stressed layer is subject to impacts from working, stress concentration from the heterogeneity of rock properties is superimposed on the stress concentration peak from the influence of the working and the more rigid layer is destroyed. If this layer is crack-resistant, the less crack-resistant neighboring layers are also destroyed. The flush of retraction energy of the bearing rocks and the energy from compressed gas expansion are consumed by the removal of damaged material. Thus, the model of inhomogeneous rock mass in which a hard layer (block) is also crack-resistant (Figure 5.35) is a model outburst situation.

Because the destruction in the depth of a massif when $\sigma_3 \neq 0$ is energetically efficient only near the generalized shift, that is, at $\sigma_2 = (\sigma_1 + \sigma_3)/2$ and

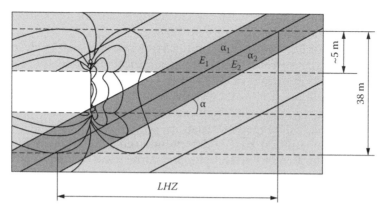

FIGURE 5.35
Outburst-dangerous situation model. LHZ = length of hazardous zone.

the fracture toughness of inhomogeneous rock $\gamma = \sigma_2 (\gamma_1/\gamma_2)^{1/2}$, the condition for the destruction of the strongest seam [109] is

$$\frac{\gamma_2(2-v)^2}{l(1-v)^n} = \frac{1}{\mu}\sigma_2 \frac{\left(4tg^2\alpha + 3\lambda + 1\right)\left[\sqrt{3(1-\lambda)} - p_0\sqrt{4tg^2\alpha + 3\alpha + 1}\right]}{16\left(1+tg^2\alpha\right)^2}, \quad (5.87)$$

where $n = (\gamma_1/\gamma_2)^{1/2}$ is relative viscosity of the rock seam boundary and μ is the movement of inhomogeneous strata:

$$\mu = \frac{(1-v)^2 E_1 E_2 (E_1 + E_2)}{(1+v)\left[E_1 + E_2(3-4v)\right]\left[E_2 + E_1(3-4v)\right]} \quad (5.88)$$

Assuming that the stress state at infinity is created only by the weight of rocks lying above $p_0 H$, in the zone of impact of working we have $\sigma_1 = k\rho H$. Denoting

$$A = \frac{16\left(1+tg^2\alpha\right)^2(2-v)^2}{l(1-v)(4tg^2\alpha + 3\lambda + 1)\left[\sqrt{3(1-\lambda)} - p_0\sqrt{4tg^2\alpha + 3\lambda + 1}\right]}, \quad (5.89)$$

let us show the condition of destruction as

$$A\gamma_2 = \frac{1}{\mu}n(k\rho H)^2, \quad (5.90)$$

where ρ is rock density and k is the stress concentration coefficient for the material destroyed and depends on the lateral pressure, crack size, and the coefficient of friction between the crack edges. For the bulk stress state parameter, A varies within 10^5 to 10^6 m^{-1} [109].

The left sides of Equations (5.83) and (5.84) denote the energy destroying more durable layers of rock; the right sides indicate the density of energy stored in a massif in a state of stress. If the right side exceeds the left, the destruction condition (5.83) and (5.84) becomes a condition of outburst because the difference between right and left sides represents the energy expanded to remove damaged material.

Location of an outburst zone is determined reliably by the frequency of selection and definition of ESE samples and correctness of the hypothesis of monoclinal layers of sandstone. If you use this method during driving, corning of a pilot hole to determine the ESE should be made at 1 to 2 m. The methods for determining the effective surface energies of rocks are described in detail in a monograph [106]. Predicting outbursts at the stage of exploration does not require the coordinates of outburst zones. Establishing the variability of ESE rocks within the mine field, the layer along the strike, and across and down the strike is sufficient.

The density of well grids determines the prediction reliability of rock outburst location in this situation. The distance between the holes in the line or profile is generated, depending on the specific characteristics of the deposit and prediction problems. According to the model set out in the rock massif, there are no outburst-dangerous or safe seams (layers). It is more correct to talk about the outburst conditions if two or more rock layers show definite differences of ESE values. After these layers contact with working zone, they are destroyed. The hotbed of destruction (outburst cavity) will be in the direction toward the working output that has a bed boundary (in the soil, the roof, the sides, or the working). Since layers will change their positions based on working contours during the face advance, the direction of the outburst cavity will change. For the case shown in Figure 5.36, the cavity will move from the floor over the course and continue to the roof, with the working moving from the opposite side, from the roof to the floor.

Thus, in contrast to methods that assess only a certain probability of outburst (it may or may not happen) or potential danger of sandstone thickness, prediction based on ESE differences enables us to assess outburst danger and also establish the coordinates of outburst-dangerous zones along pilot holes and the direction of spread of an outburst cavity.

Industrial tests were conducted at the Skochinskogo and Stakhanov Mines to assess the advance of obviously outburst-dangerous sandstone before blasting and check the reliability of rock outburst forecasts. Before the tests, 13 outbursts of various intensities occurred.

A pilot hole (70 mm diameter, 110 m length) was drilled for sampling the sandstone and predicting outbursts along the axis of future development of a conveyor roadway. The hole was driven in a well location with slashing. Sampling (at least to 2.5 m depth) to predict outbursts according to ESE was carried out from each sandstone interlayer. Fourteen outbursts shown in

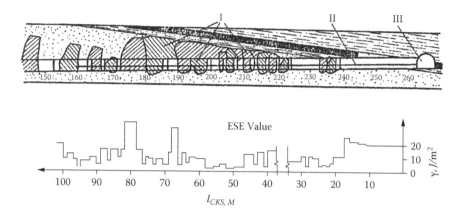

FIGURE 5.36
Working of Skochinskogo mine. (I) Outburst-destroyed rock. (II) Pilot hole. (III) Haulage gate of fifth long wall faces of center panel.

Figure 5.36 occurred during the advance of the working. All outbursts were predicted within 1 to 2 m accuracy. The directions of the outburst cavities coincided with the directions predicted.

Results indicated that all outbursts correlated with the boundaries between the layers of rocks with different surface energies. Outbursts occurred as long as this boundary was in the zone of working. When the distance between the boundaries was small, several layers of rocks were involved in outbursts. In cases of only one boundary, two adjacent layers of rocks were destroyed and the outburst cavity was directed toward the layer with greater ESE. Outbursts occurred when the working was driven from rocks with low ESE to rocks with greater ESE. Outburst prediction for sandstones based on ESE is the only way to establish the site of an expected outburst and the direction of its cavity.

5.7 Safe Extraction of Coal

This section focuses on the regional way of dealing with sudden coal and gas outbursts, allowing a large load (speed of advance) at the working face and the calculation of methane released during coal extraction.

5.7.1 Outburst Control Measures

Worldwide experience mining coal seams led to a variety of methods for fighting sudden outbursts in a working face [110–135]. They are classified as regional and local. Regional methods allow removal of dangerous conditions in a coal massif before stoping. With local methods, coal is not extracted during layer stimulation directly in the zone of excavation. The present level of coal extraction mechanization allows large loads on working faces, but local outburst control measures deter face advances.

Regional outburst prevention measures based on adding water solutions of surface-active substances (SASs) to coal mass developed by the Institute of Physics of Coal and Mining Processes SU have been used in mines for a long time [89]. We covered the conditions that turn pores into fissures in Section 5.4. Crack development leads to the weakening of flat areas filled with gas parallel to a face [136] and the areas then become outburst-dangerous. The input of a number of SASs into a massif will cause fissures to develop due to rock pressures, based on the Rehbinder effect. Because the fissures in a virgin coal seam are in disorder and no single bearing pressure is dominant, fissure development is likely in all directions.

As stoping work moves to a surfactant-treated zone, the stronger sorptive power of water in coal (in comparison to the power of methane in coal) displaces (drains) the methane in branched and locked channels. The addition of aqueous surfactant solutions plasticizes coal seam zones and enlarges the water removal zones, decreasing the potential for outburst-dangerous situations. Since the penetration of fissures and changes in physical and mechanical properties of a seam affect only the adsorbed (physically bound) moisture, this quantity can serve as a criterion for assessing outburst-hazard. The only current method of determining adsorbed moisture is NMR. To meet this need, portable moisture meters based on NMR technology are now used at mining operations.

Most coal seams are heavily soaked with water because coal is hydrophobic and water has high surface tension of $K = 7.3 \times 10^{-3}$ J m^{-2} that significantly increases the degree of penetration of water. Water penetration into coal massifs started in the early 1960s but the method does not completely solve the problem. No single type of surfactant is suitable for all grades of coal or the surfactant may produce different effects in samples of the same coal rank taken from different mines. Furthermore, surfactants are expensive and increase the costs of outburst control measures [137].

Certain surfactants give good results at low concentrations. The most effective and least expensive is sulfanol. A 0.5% sulfanol solution increases physically adsorbed moisture content to 2.5% a day after the beginning of wetting (natural moisture = 0.95 to 1.3%). The radicals of surfactant molecules are grouped around the adsorption centers of pores and then they become adsorption centers for water molecules. The activation energy of new adsorption centers is higher than the original energy [138,139]. This is clearly illustrated by NMR spectra. For example, the line width of an NMR spectrum of *H* (see Chapter 3) indicates that moisture level after adding sulfanol to coal grade K is nearly twice the level before the application of surfactant.

We know that fluids, in particular methane and water in pore spaces, change the behavior of coal in a triaxial stress state based on the load of the rock massif during its development, the character of fracture (brittle, ductile), and resulting gas-dynamic activity. Laboratory and field experiments confirmed this principle [140].

Studies of triaxial stress compression (see Chapter 4) determined the effects of methane and aqueous surfactant solutions on the strength, elastic, and deformation properties of coal samples subjected to gas and water saturation in a high-pressure chamber. Saturation time T_s varied from 5 to 25 days; the pressure of saturation P_s = 8 to 10 MPa. Before saturation, coal samples were under vacuum for 5 to 10 h. The first phase studied the strength, elastic, and deformation properties of samples (1) moisturized without gas, (2) saturated with gas and dried, and (3) saturated with gas and left moist; the samples were subjected first to uniaxial compression and tension. The presence of methane and adsorbed water in pore spaces significantly affected the behavior under load. In comparison to gas-saturated samples, the elastic modulus of moisturized samples measured 4 to 5 times smaller; moist gas-saturated sample modulus values were 3 to 4 times less. Although the rupture stress value in gas-saturated samples is reduced by 15 to 20% in comparison to naturally moist samples, the Young's modulus increased 1.3 to 1.4 times, indicating embrittlement in the presence of methane. A 2 to 3% increase of physically bound moisture decreased the strength of degassed coal samples 2.5 to 4 times; a further increase produced little effect on strength.

Another study focused on the impact of adsorbed moisture on changes of elastic energy accumulated gas-saturated coals of various grades. Cubic coal samples were loaded to voltages of 80 to 85% of destructive levels and then unloaded. The graphs of the results of loading and unloading allowed us to estimate the energy released during unloading. The amount of elastic energy released at unloading after triaxial stress and increase of moisture from 1 to 2.5% decreased 3.5 times for anthracite coal and 6 times for coking and lean coals.

The true triaxial compression unit (TTCU) can create conditions of loading in coal samples and model conditions in virgin rocks and face zones of coal seams where outburst-dangerous situations usually form. Stress measurements (σ) were carried out in steep and flat-lying coal beds and analysis indicated that the edges of seams are under unequal stress conditions. For this reason, the program of loading in laboratory conditions corresponded to the natural conditions: $\sigma_1 \neq \sigma_2 \neq \sigma_3$. A sample placed in a TTCU chamber was stressed at regular intervals ($\sigma_1 = \sigma_2 = \sigma_3$) to loads normal for virgin rock. Then stress σ_1 was raised to $k\gamma H$, and σ_2 to $\lambda k\gamma H$, and simultaneously σ_3 was reduced until destruction. Here k is the coefficient of stress concentration (for steep layers k = 1.8 to 2.5) and λ is the coefficient of lateral thrust. The gas-saturated, gas-saturated–moistened, and gas-free–moistened samples of coking, lean and anthracite coal were loaded and unloaded. The increase of the adsorbed moisture to 2.5% (for example, for lean coal) increased the plastic properties (average deformation) of gas-saturated coals 2 to 2.5 times and reduced the volume 2.5 to 3 times. Three-dimensional stress studies showed that full unloading caused a 2 to 3% volume increase of gas-saturated samples. The volume of moistened samples was almost restored. These results confirm the irreversibility of physical processes in coal structures based on increases of adsorbed moisture content.

In conclusion, a specific amount of adsorbed moisture in each rank of gas-saturated coal is capable of reducing its elastic and effective superficial energy, thus creating a non-outburst condition. This was confirmed by mine experiments at Donbass (Removsky layer of Uglegorskaya mine and Almazny layer of Stozhkovskaya mine) and also by industrial tests on the outburst-dangerous Bezymyanny seam at the Krasny Oktyabr Mine.

The stress study at Bezymyanny was conducted by means of hydraulic sensors placed in holes drilled from a push-down to a depth of 6 to 8 m. Self-recording manometers compiled pressure data. Results showed that in a massif with adsorbed moisture W of 0.6 to 0.9%, the stress increased. The strip of raised pressure started 8 m from a face and decreased 1.2 to 1.8 m from a line of stoped excavation. The coefficient of stress concentration changed within 1.5 to 1.8 m. In the moistened coal massif, within the limits of measurement of a 12 m strip from a face, no maximum of stress approaching the line of a working face was noted by the gauges. This indicates the absence of increased stress or the movement of increased stress in the depth of the massif. At $W = 1.93$ to 2.5 the change of stress occurred at regular intervals, confirming the positive influence of moistening on stress in critical areas of coal formations. Also, a test hole defined the exit of coal slack before and after moistening. The exit of slack is an indirect indication of stress in a coal massif. At Bezymyanny, the exit of rubble per meter of hole decreased 2 to 3 times as humidity increased 2.05 to 2.2%.

Gas emission from holes characterizes changes of massif stress conditions, natural gas content, permeability to gas, speed of gas recovery, and physical properties of coal. To find the influence of physically bound moisture on emission of methane within a coal massif in and out of zones of moistening, we studied the dynamics of gas emission. Based on the results of interval measurements of gas emission and determination of physically bound moisture, we determined the dependence $g = f(W_{av})$. Measurements of gas dynamics at drilling holes 2.5 m long were analyzed. This allowed us to develop an equation describing the connection of gas emission speed change with physically bound moisture in a coal massif:

$$g = 4,24 - 3,25W_{ph} + 0,64W_{ph}^2 \qquad (5.91)$$

The level of physically bound moisture in a coal massif within 2 to 2.5% can serve as a criterion to ensure that conditions are safe from sudden coal and gas outbursts. Studies showed that to prevent sudden coal and gas outbursts, adsorbed (physically bound) moisture must be maintained at levels of 2.5% for coking and fat coals, 2% for lean and lean baking coals, and 3% for anthracites. After these levels were determined, the technological parameters related to aqueous surfactants and moisture control via NMR were developed.

The basic indicator of the efficiency of coal seam processing is the content of physically bound moisture. To control moistening efficiency after the termination of this process, test holes are drilled between heating holes. The lengths of the test holes are equal to the lengths of injection holes (for steep beds). Sampling is carried out every 5 m along the length. Test holes in flat-lying seams are drilled from long wall faces along the seam course to depths equal to the distance between two injection holes every 5 to 6 m, with 2 to 3 m sampling intervals. The test holes can be situated as they are for steep seams but control using EPR is required [141]. As in steep seams, holes must be drilled with flushing because it is not possible with NMR testing to separate moisture in coal after infusion from moisture added during flushing. Before sampling, the holes are cleared of rubble. The drill fines intended for sampling are drawn from the holes by natural flow (steep seams) or by ribbon spiral conveyers (flat-lying seams). Coal samples are placed in glass or metal test tubes 20 to 30 sm^3 and delivered to a laboratory. The time between sampling and testing should not exceed 50 h. A sample is placed in an ampoule and inserted in a high frequency circuit of an NMR device. The basic characteristic is the percentage of physically bound moisture defined after infusion during drilling of test holes.

To define moisture via electron paramagnetic resonance (EPR), a spin probe representing a stable nitrate radical is added to water infused into a coal seam. The spin probe content defines the amount of additional moisture in a sampling place. Device calibration is conducted before every test. A water solution of the nitrate radical is inserted into dried (dehydrated) coal. The solution initial concentration should not exceed 10^{-3} mole/l to prevent weakening of the spin–spin interaction. Figure 5.37 presents the calibration curve of dependence of the relation of EPR signal line intensity of a radical-probe 2,2,5,5-tetramethyl-1-carbamide pyrrolidine-1-oxyl, dissolved to 10^{-3} mole/l to the intensity of EPR signal line of the etalon on humidity W.

The spectrum of electronic paramagnetic resonance of a spin probe represents the first derivative of three lines with a 1:1:1 ratio of intensities (Figure 5.38). As the paramagnetic centers (PMC) of coals partially block two lines of a spin probe by their absorption line, the calculation of spin probe concentration in a coal sample simply involves comparison of the internal etalon with one of the EPR spectrum lines of a spin probe outside the limits of the EPR PMC spectrum. The comparison of intensity of a spin probe line with the etalon is best presented as the relation intensities I_x/I_E, representing a standardized dimensionless value proportional to the concentration of a spin probe. Concentration may be more exactly defined by

$$c_x = (I_x/I_E)\frac{(\Delta H_x)^2}{(\Delta H_E)^2}c_E,$$

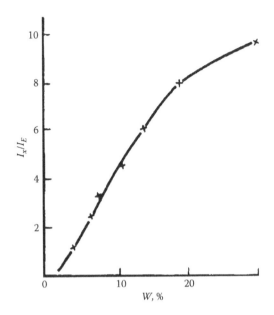

FIGURE 5.37
Dependence of the relation of signal EPR line intensity of radical-probe 2,2,5,5- tetramethyl-1
-carbamide pyrrolidine-1-oxyl, dissolved to 10^{-3} mole/l to line intensity of etalon I_x/I_E on coal
moisture W (calibration curve).

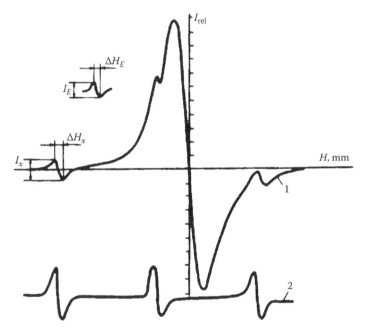

FIGURE 5.38
EPR spectra. (1) Spectrum of coal containing solution of nitrate radical. (2) Etalon spectrum.

where I_x and I_E are the intensities of the first derivative line of the spin probe EPR absorption not overlapping with the EPR PMC signal of coal and the etalon intensity, respectively, ΔH_x and ΔH_E are the respective widths between maxima of the first derivative line of EPR signal absorption of a spin probe and the etalon in millimeters; and C_E is the etalon concentration in moles per liter. Since the $(\Delta H_x)^2/(\Delta H_E)^2$ ratio is a constant for the concentration of a spin probe below 10^{-3} mole/l and approximates unity, it is acceptable to measure W by the ratio of intensities I_x/I_E without a need to define quantitatively the concentration of a spin probe in a sample.

The best internal etalon for EPR was the third line Mn^{2+} in a MgO etalon made in VNIIFTRI and fixed immovably in the microwave resonator of an EPR radio spectrometer of 3-cm range. For this test, an EPR radio spectrometer of any type in a mode of continuous microwave radiation, with sensitivity better than 10^{11} spin/gauss and resolution at least $5 \cdot 10^{-1}$ gauss/mm is acceptable.

Unlike other methods, the suggested technique for defining coal moisture analyzes the additional amount of moisture added to a coal bed. In the event the air tightness of a sample container fails during transportation to the laboratory and drying results, it is still possible to define initial humidity as the amount of spin probe in a sample proportional to the amount of moisture that was in it; drying and dilution do not change it.

Mine samples contain methane and water. Because the NMR spectrum of methane (created by hydrogen protons) is imposed on the spectrum of water, it is necessary to divide their contributions. The simplest and most reliable method based on water and methane sorption characteristics is used. The heat of water adsorption by coal reaches $7.1 \cdot 10^4$ J/mole, but the heat of methane sorption does not exceed $2.5 \cdot 10^4$ J/mole. Therefore several times less energy is required to activate methane molecules than is required for water molecules. Thus, by creating underpressure around the coal particles, the detachment of methane molecules can be made much easier than detaching water molecules. This is the concept behind the method. After the spectrum of a damp gas-saturated sample is recorded, vacuum degassing is carried out for 3 min. The amount of methane left in the coal is beyond the sensitivity of installation and does not exceed 0.16 m^3/t.

The experience of mining in outburst-dangerous coal seams has shown that the most sudden coal and gas outbursts occur in zones of geological failures and increased rock pressure. As a rule, small-amplitude plicate and disjunctive faults (such as thrusts) are characterized by various degrees of coal preparation connected with gas and hydrodynamic features of coal seams. At the approach of a zone of limit equilibrium to such failures, the plateau sectors with the prepared coal are under small rock pressure. The areas of not-destroyed and over-compacted coal will transform into a limiting state and accumulate considerable elastic energy. Areas of sufficient methane saturation will exhibit brittle fracture-outburst. The same features found in zones of over-compacted coal characterize zones of increased rock pressure (IRP). Prevention of coal and gas outbursts in such seams is complicated by geological faults and IRP should be carried out beforehand, at the approaches of face zones, to decrease elasticity and increase deformation properties. One method that meets these needs is preliminary processing of the specified zones by water solutions of surface-active substances through upward holes (injection and test) while using NMR to control the amount of adsorbed moisture.

The basis for treating zones of geological faults and IRP by aqueous SASs is humidifying the coal consolidated by high rock pressures in the surrounding fault and IRP zones by utilizing the high filtration ability of the aqueous solution. The results are plasticization and relaxation of high mechanical stress. Furthermore, replacement and partial blocking of methane in pores and cracks are achieved, but the possibilities of desorption and outburst development decrease sharply. Efficient processing in geological fault zones is carried out by drilling samples from test holes and analyzing them with NMR equipment. The parameters of coal seam processing are orientation and depth of injection and test holes, hole diameter, concentration of aqueous SAS, and the pressure and the supply rate of SAS injected (per ton of coal). Efficient processing in outburst-dangerous zones depends on the choice of optimum values of the parameters.

The depth of the injection holes is defined by the distance of geological faults from the mine working where drilling is carried out, and also the sizes and directions of faults in a seam. In cases where zones of increased rock pressure extend along the strike of a seam in the form of a strip over a haulage gate, the depth of injection holes should be 3 to 5 m less than the width of a zone in a direction of seam rise. If a coal seam contains small-amplitude plicate faults scattered along the height of a lift, injection hole depth should be 8 to 10 m less than the inclined height of a lift to avoid the break of injected fluid in a ventilating drift. The depths of test and injection holds should be equal. The diameters of injection and test holes depend on the orientation requirements and reliability of well head sealing. The rational diameter sizes of 8 to 120 mm holes are determined experimentally.

The distance between the holes depends on the radius of effective action (moistening). At optimum distance, sufficiently deep moistening of a coal seam zone between two neighboring holes should be provided. As the humidity further from the hole is lower than the humidity nearer the hole, an acceptable distance between the holes is slightly less than two radii of the effective action, that is, $H_{hole} = (1.6 \text{ to } 1.8) R_{ef}$, where H_{hole} is the distance between injection holes and R_{ef} is the radius of the effective action (moistening). The R_{ef} is not a constant; it depends on the filtration and sorptive properties of coal in the processed zone that in turn depend on rock jointing and coal metamorphism (rank). R_{ef} has been defined experimentally.

A moistening hole is drilled in a zone of increased rock pressure over the intake entry. After injecting an aqueous SAS for at least seven days, test holes are drilled parallel to the moistening hold. The first hole is at a distance of 3 m, the second at 5 m, and the third at 6 m. The content of adsorbed

TABLE 5.7

SAS Content in Mixture (%)

| | Coal Class | | | | | |
SAS Rank	Gaseous	Fat	Coking	Lean Baking	Lean	Anthracite
Sulfanole	0.2	0.3	0.2	0.1	0.1	0.1
Teepol	0.2	0.3	0.3	0.2	0.2	0.3

(physically bound) water in samples from the holes is analyzed. The R_{ef} of the moistening hole is the distance at which the humidity of coal increases 2% or more, depending on the rank. On average, the R_{ef} from injection holes is 6 to 7 m in natural humidity below 1% for seams of coking, fat, lean baking, and lean ranks.

The SAS content in water injected in a coal seam is assumed from data in Table 5.7. The discharge pressure should exceed the pressure of gas in a coal seam, but not cause hydraulic fracturing. Liquid consumption and pressure are quality indicators of coal moistening. Despite possible losses, SAS content should be 20 to 25 liters per ton of coal to increase humidity of a coal massif by 1%. The consumption and pressure should be monitored constantly to detect possible leaks of solution into cavities and neighboring holes. The amount of aqueous SAS required to moisten zones of geological faults and IRP is calculated according to

$$Q_{hole} = L_{hole} h_{hole} m \gamma (W_{ads} - W_{nat}) q \qquad (5.92)$$

where L_{hole} is the depth of the injection holes in meters; h_{hole} is the distance in meters between holes (in faults lacking continuity of coal seams and corner deviations 10 to 15 degrees from seam pitch, the width of the fault zone is entered); m is seam thickness in meters; γ is coal density in tons per cubic meter; W_{ads} is the percentage of adsorbed moisture; and q is the rate of water consumption to reach 1% humidity (0.025 m^3/t).

The preventive treatment of outburst-dangerous coal seams in zones of geological faults and IRP involves drilling of injection holes, well head sealing, injecting aqueous SAS into holes, drilling of test holes and coal sampling, analysis based on content of adsorbed (physically bound) water, and determining decrease or full elimination of coal outburst danger. Technological schemes for positioning holes and competing tests based on the parameters and the characteristics of geological faults and IRP zones are found in the literature [89].

5.7.2 Safe Load on Working Face

Coal mine engineers and technicians use standards published more than 15 years ago to calculate the amount of air necessary for ventilation of working faces. The use of highly efficient testing equipment for analyzing the behavior of gas-bearing coal rock massifs under difficult geological and technical conditions present in mines requires better criteria for determining loads on working faces. One of the basic factors for calculating optimum load on a working face is the methane-bearing capacity of the working area. Several studies in Russia and Ukraine focused on improving the calculation of optimum load under a variety of conditions in gas-saturated coal seams [142–145].

Studies of methane phases in coal, and the filtration and diffusion concepts of desorption (Chapter 2) led to invention of a desorption meter (DS) to measure methane quantities and pressures in coal seams. Based on the desorptive activity in a seam (Section 2.4.1), DS equipment yields exact calculations of times of maximum permissible methane concentration in working spaces during the coal extraction. The DS-01 has an independent power supply and simultaneously measures the amounts and pressures of methane in three samples. Integral electronics, intercommunication capability, and control of the sequence of operations allow display of results, storage in long-term memory, and automatic transfer of TTCU data from the mine to the surface in real time. Some technical characteristics of DS-01 are

Range of measured pressure of methane in coal bed	2 to 10MPa
Error of measurement of pressure without individual calibration of pressure sensor	+5 %
Temperature range	40 to 50°C
Preparation time	Maximum 900 sec
Measurement time	Maximum 660 sec

The use of the DS-01 in mines helps optimize loads on working faces and also promotes safe working conditions. Information about methane characteristics of a seam can be used to calculate the kinetics of gas emissions in stoping or development working. The calculation of dangerous emission levels requires data for a number of geological, physical, and technological factors that apply to a specific working face. The initial physical and technological parameters for calculating the onset of dangerous methane concentration formations (based on a real example from the Zasyadko mine) are

Seam thickness, m	2.05
Porosity of coal, m^3/m^3	0.05
Air consumption in stope, m^3/min	2100
Effective factor of diffusion of methane in coal, m^2/sec	$1.8 \cdot 10^{-6}$
Sectional area of working face, m^2	5.74
Coverage of operating element, m	0.8
Length of working face, m	300
Speed of working face advance, m/day	5.5
Factor of extracting equipment loading	0.7

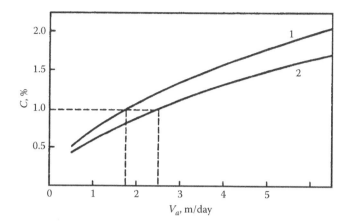

FIGURE 5.39

Dependence of methane concentration on face advance rate at maximum (1) and minimum (2) values of methane diffusion.

Even at constant pressure and methane level measured by a DS, the speed of advance of a working face leading to the formation of unacceptable methane concentration (more than 1%) depends on the effective diffusion K (Figure 5.39). At maximum K, it is possible to advance a working face 1.7 m/day (2.5 m/day at minimum K). The existing standards for calculating acceptable speed of advance of a working face do not consider minimum and maximum methane diffusion factors that vary for every coal seam, sometimes by a factor of 2 [146]. If a calculated load from an economic view appears insufficient, it is necessary to apply measures to decrease methane diffusion, for example, by injecting aqueous SASs as cited in Section 5.7.1.

5.8 Conclusions

It can be concluded that the safe load on working face and the preliminary arrangements to increase it if required by the mine economics can be established using the in-seam methane content and pressure measuring device and technique based on up-to-date research of molecular coal structure and phase state of occluded fluids.

References

1. Semenovich V.V. et al. 1987. *Fundamentals of Geology of Combustible Fossils.* Moscow: Nedra.

2. Park R. and Dunning H.N. 1961. Stable carbon isotope studies of crude oils and their porphyrin aggregates. *Geochim. cosmochim. acta* 22, 99.
3. Bakaldina A.P. 1980. Effect of material constitution of coals on their methane capacity and natural gas content. In *Gas Bearing Capacity of Coal Basins and Deposits in USSR.* Vol. 3. *Genesis and Distribution of Natural Gases in Coal Basins and Deposits of the USSR,* Krasvtsov A.I (Ed.). Moscow: Nedra, pp. 102–123.
4. Prasolov E.M. and Lobkov V.A. 1977. On conditions of methane formation and migration (by isotopic composition of carbon). *Geokhimiya* 1, 122–135.
5. Emets O.V., Lugova I.P., Kanin V.O. et al. 2008. Genesis of coal gases from carbon deposits on the territory Zasadko mine (Donets basin). *Dopov. NAN Ukr.* 4. 120–124.
6. Prasolov E.M. 1990. *Isotopic Geochemistry and Genesis of Natural Gases.* Leningrad: Nedra.
7. Nier A.O. and Gulbransen E.A. 1939. Variations in the relative abundance of the carbon isotopes. *J. Am. Chem. Soc.* 61, 679–698.
8. Trofimov A.V. 1950. Carbon isotopes in natural processes. *Priroda* 7, 663.
9. Landergren S. 1954. On the relative abundance of the stable carbon isotopes in marine sediments. *Deep Sea Res.* 1, 98–102.
10. Lebedev V.S. 1964. Isotopic composition of carbon of oil and natural gas. *Geokhimiya* 11, 1128–1137.
11. Colombo U. et al. 1965. Isotopic measurements of C^{13}/C^{12} ratio on Italian natural gasses and their geochemical interpretation. *Nature* 205, 4978, 1303–1304.
12. Galimov E.M. 1968. *Geochemistry of Stable Carbon Isotopes.* Moscow: Nedra.
13. Nagao K., Nakaoka N., and Matsubayashi O. 1981. Rare gas isotopic composition in natural gases of Japan. *Earth Planet. Ski. Lett.* 53, 175–188.
14. Medina J.C., Butala S.J., Bartholomew C.H. et al. 2000. Iron-catalyzed CO_2 hydrogenation as a mechanism for coal bed gas formation. *Fuel* 79, 89–93.
15. Stevens J.G. and Shenoy G.K. 1981. *Mössbauer Spectroscopy and Its Chemical Applications, Advances in Chemistry Series* 194. Washington: American Chemical Society, pp 135–137.
16. Shock E.L. 1994. Catalysing methane production. *Nature* 368, 499–501.
17. Mango F.D., Hightower J.W., and James A.T. 1994. Role of transition metal catalysis in the formation of natural gas. *Nature* 468. 536–538.
18. Price L.C. and Schoell M. 1995. Constraints on the origins of hydrocarbon gas from compositions of gases at their site of origin. *Nature* 378, 368–371.
19. Tannenbaum E., Kaplan I.R., and Low M. 1985. R-hydrocarbons generated during hydrous and dry pyrolysis of kerogen. *Nature* 317, 708–709.
20. Alexeev A.D., Ulyanova E.V., Starikov G.P. et al. 2004. Latent methane in fossil coals. *Fuel* 83, 1404–1411
21. Alexeev A.D., Vasilenko T.A., and Ulyanova E.V. 2005. Phase methane in fossil coals. *Solid State Commun.* 130, 669–673.
22. Alexeev A.D., Ulyanova E.V, and Vasilenko T.A. 2005. NMR potential for studying physical processes in fossil coals. *Uspekhi fiz. nauk.* 175, 1217–1232.
23. Rusyanova N.D. 2003. *Coal Fuel Chemistry.* Moscow: Nauka.
24. Malyshev Y.N., Trubetskoi K.N., and Airuni A.T. 2000. *Fundamental and Applied Methods of Solution of the Problem of Methane in Coal Beds.* Moscow: Academy of Science in Mining Engineering.

25. Suggate R.P. and Dickinson W.W. 2004. Carbon NMR of coals: the effects of coal type and rank. *Int. J. Coal Geol.* 57, 1–22.

26. Mogilnyi G.S., Skoblik A.P., Razumov O.N. et al. 2009. Effect of iron compositions on coal structure. *Metall. novey tekhn.* 31, 815–826.

27. Fischer F. and Tropsch H. 1923. Preparation of synthetic oil mixtures (synthol) from carbon monoxide and hydrogen. *I. Brenn. Chem.* 4, 276–285.

28. Falbe Yu. 1980. *Chemical Substances from Coal.* Moscow: Khimiya.

29. Curtis A. and Palmer C. 1990. Determination of 29 elements in 8 Argonne premium coal samples by instrumental neutron activation analysis. *Energy Fuels* 4, 436–439.

30. Huffman G.P. and Huggins F.E. 1978. Mossbauer studies of coal and coke: quantitative phase identification and direct determination of pyritic and iron sulphide sulphur content. *Fuel* 57, 437–442.

31. Taneja S.P. and Jones C.H.W. 1984. Mossbauer studies of iron-bearing minerals in coal and coal ash. *Fuel* 63, 695–702.

32. Herod A.J., Gibb T.C., Herod A.A. et al. 1996. Iron complexes by Mössbauer spectroscopy in extracts from Point of Ayr coal. *Fuel* 75, 437–442.

33. Lefelhocz J.F., Friedel R.A., and Kohman T.P. 1967. Mössbauer spectroscopy of iron in coal. *Geochim. Cosmochim. Acta* 31, 2261–2273.

34. Hodot V.V., Yanovskaya M.F., and Peremysler V.S. 1978. *Physicochemistry of Gas-Dynamic Phenomena in Mines.* Moscow: Nauka.

35. Hodot V.V. 1990. Energy sources of sudden coal and gas outbursts. *Vybrosoopasnost ugolnyh plastov* 186, 8–20.

36. Hodot V.V. 1961. *Sudden Outbursts of Coal and Gas.* Moscow: GNTIP.

37. Petuhov I.M. and Linkov A.M. 1983. *Mechanics of Rock Bumps and Outbursts.* Moscow: Nedra.

38. Airuni A.T. 1987. *Prediction and Prevention of Gas-Dynamic Phenomena in Mines.* Moscow: Nauka.

39. Zabigailo V.E. and Nikolin V.I. 1990. *Effect of Catagenesis of Rocks and Metamorphism of Coals on the Risk of Their Outburst.* Kiev: Naukova.

40. Standards of Ukrainian Organization 10.1.00174088.011. 2005. Mining works regulations for beds disposed to gas-dynamic phenomena. Minvugleprom Ukrainy. Kiev.

41. Agafonov A.V. 1999. *Methods and Means of Insuring the Safety of Developing Works on Outburst-Prone Beds.* Donetsk: Donbass.

42. Frolkov G.D., Fandeev M.I., Malova G.V. et al. 1997. Influence of natural mechanoactivation on outburst risk of coals. *Him. tverd. topl.* 5, 22–33.

43. Frolkov G.D. et al. 1992. On the effect of mode of deformation of the coal bed on the structure of organic mass and gas emission under sudden coal and gas outbursts. Preprint. Rostov on Don.

44. Gagarin S.G., Eremin I.V., and Lisurenko A.V. 1997. Structural–chemical aspects of defectiveness of black coals in outburst-prone beds. *Him. tverd. topl.* 3, 3–14.

45. Lisurenko A.V., Gagarin S.G., and Eremin I.V. 1997. On the nature of metastable state of coals in outburst-prone beds. *Him. tverd. topl.* 5, 34–43.

46. Alexeev A.D, Ulyanova E.V., and Starikov G.P. 1995. Methodological foundations of classification of gas-dynamic phenomena. *Fiz, tekhn. vysok. davl.* 1, 67–70.

47. Alexeev A.D, Starikov G.P., Ulyanova E.V. et al. 2000. Prediction of outburst hazard under development of steep and steeply inclined beds. Regulating normative document KD12.10.05.01-99.
48. Alexeev A.D, Zaidenvarg V.E., Sinolitskii V.V. et al. 1992. *Radiophysics in the Coal Industry.* Moscow: Nedra.
49. Ulyanova E.V. 2007. The effect of valence of iron inclusions on the outburst hazard and methane-bearing of coal beds. *Proc. Int. Conf.* Форум гірників. Dnepropetrovsk: National Mining University, pp.100–106.
50. Ulyanova E.V., Razumov O.N., and Skoblik A.P. 2006. Iron and its connection with methane accumulation in coals. *Fiz. tekhn. probl. gorn. proiz.* 9, 20–31.
51. Ulyanova E.V., Vasilenko T.A., Bitterlih V.O. et al. 1996. Structural transformations in fossil coals under mechanical effects. In *Proc. 6S Scientific School of CIS Nations,* Odessa, pp. 39–44.
52. Ulyanova E.V. 2009. On correlation between the risk of gas-dynamic phenomena and the presence of mineral inclusions in coals. *Fiz, tekhn. vysok. davl.*12, 15–24.
53. Polyakov P.I., Ulyanova E.V., Vasilenko T.A. 1998. *Changes in Fossil Coals Structure Occurring on High Pressure Treatment.* Donetsk: DFTI NANU, pp. 16–23.
54. Alexeev A.D., Serebrova N.N., and Ulyanova E.V. 1989. Investigation of methane sorption in fossil coals by means of high-resolution 1H and 13C NMR spectra. *Dokl. acad. nauk. USSR B* 9, 25–28.
55. Alexeev A.D., Ulyanova E.V., and Vasilenko T.A. 2005. Potential of NMR in the studies of physical phenomena in fossil coals. *Uspekhi fiz. nauk.*175, 1217–1232.
56. Hripunov S.V. and Ksenofontov V.G. 1988. Synthesis of metal complex compounds of coals in the reaction of ferrocene formation. *VINITI,* 147, 1–12..
57. Eremin I.V. and Bronevets T.M. 1983. On parameters of reduction of medium metamorphized humus coals of member states of Mutual Aid Council. *Him. tverd. topl.*4, 3–10.
58. Lifshits M.M. 1979. On determination of the reduction degree of black coals. *Him. tverd. topl.* 3, 3–11.
59. Matsenko G.P. and Kapra Z.S. 1988. Reflection indices of vitrinites of Donets high-metamorphized coals and anthracites of different reductions. In *The ways of Ukrainian Coal Processing.* Kiev: Naukova Dumka, pp. 94–97.
60. Rusianova N.D., Popov B.K., and Butakov V.I. 1985. Parameters characterizing the changes of organic coal bulk in the process of metamorphism. In *Structure and Properties of Coals in the Series of Metamorphism.* Kiev: Naukova Dumka, pp. 16–42.
61. Savchuk V.S. and Kuzmenko E.A. 2008. Effect of coals' reduction on gas-dynamic phenomena in Donets coal mines. *Geotech. Mech.* 80, pp. 84–90.
62. Mingju L. and Xueqiu H. 2001. Electromagnetic response of outburst-prone coal. *Int. J. Coal Geol.* 45, 155–162.
63. Bobin V.A. 2000. Estimation of the parameters of wane action on micro- and macrostructural formations in gas-saturated coal substance for the purpose of intensification of coal methane extraction. *Geotech. Mech.* 17, 56–60.
64. Mineev S.P., Prusova A.A., and Kornilov M.G. 2005. Estimation of wave action energy for methane molecules activation in microporous space of coal substance. *Geotech. Mech.* 54, 31–37.
65. Podoltsev A.D. and Kucheryavaya I.N. 1999. *Elements of the Theory and Numerical Calculation of Electromagnetic Processes in Conducting Media.* Kiev.

66. Ivanov V.V., Egorov P.V., Kolpakova L.A. et al. 1988. Dynamics of cracks and electromagnetic radiation of loaded rocks. *Fiz. tekhn. probl. razrab.* 5, 20–27.

67. Molchanov O., Kulchitsky A., and Hayakawa M. 2001. Inductive seismo-electromagnetic effect in relation to seismogenic ULF emission. *Nat. Haz. Earth Syst Sci.* 1, 61–67.

68. Spivak A.A. 1998. Relaxation processes and mechanical state of local Earth crust zones. *Dokl. akad. nauk.* 363, 246–249.

69. Kirillov A.K. and Slusarev V.V. 2006. Investigation of fossil coal consolidation by deformation of coal samples in high-pressure chamber. *Fiz, tekhn. vysok. davl.* 3, 137–143.

70. Alexeev A.D, Kirillov A.K., Shazhko Y.V. et al. 2007. Electromagnetic method of coal bed degassing. *Makeeva* 20, 26–37.

71. Alexeev A.D, Kirillov A.K., Mnuhin A.G. et al. 2006. Electromagnetic action on coal bed for activation of degassing process. *Fiz. tekhn. probl. gorn.* 9, 5–19.

72. Goldshtein L.D. and Zernov N.V. 1971. *Electromagnetic Fields and Waves.* Moscow: Soviet Radio.

73. Gabbilard K.D. 1972. Radiocommunication between underground and underwater stations. *Zarubez. radioelek.* 12, 16–34.

74. Landau L.D. and Livshits E.M. 1986. *Statistical Physics.* Moscow: Nauka.

75. Flory P. 1971. *Statistical Physics of Even Molecules.* Moscow: Mir.

76. Alexeev A.D, Kovriga N.N., Starikov G.P. et al. 2003. Bound methane in natural coals *Fiz. tekhn. Probl. gorn. proiz.* 6, 5–11.

77. Ettinger I.P. 1984. Methane solutions in coal beds. *Him. tverd. topl.* 4. 32–34.

78. Juntgen N. and Karweil J. 1966. Gasbildung und Gasspeicherung in Steinkohlen flozen. *Erdol u. Kohlr* 19, 251–258, 339–344.

79. Huck G. and Patteisky K. 1964. Inkohlungsreaktionen unter Druck. Fortschr. *Geol. Rheinld. Westf.* 12, 551–558.

80. P. de Gennes. 1982. *Scaling Concepts in Polymer Physics.* Moscow: Mir.

81. Kirillin V.A., Sychev V.V., and Sheidlin A.E. 1974. *Engineering Thermodynamics.* Moscow: Energiya.

82. Landau J.R. and Livshits E.M. 1986. *Hydrodynamics.* Moscow: Nauka.

83. Alexeev A.D., Ilyshenko V.G., and Kuzyara V.I. 1995. *Coal Mass State: NMR Analysis and Control.* Kiev: Naukova Dumka.

84. Galkin A.A. and Kichigin D.A. 1958. Investigation of paramagnetic resonance in black coals of Donets basin. *Him. tekhn. topl. masel.* 7. 8–14.

85. Stavrogin A.N. and Protosenya A.G. 1985. *Strength of Rocks and Stability of Openings at Big Depths.* Moscow: Nedra.

86. Landau L.D. and Livshits E.M. 1965. *Theory of Elasticity.* Moscow: Nauka.

87. Dmitriev A.P. and Goncharov S.A. 1983. *Thermostrengthening Processes in Rocks.* Moscow: Nedra.

88. Alexeev A.D, Starikov G.P., and Filippov A.E. 2003. Computer simulation of methane escape from coal with the account for unloading wave and opening of porosity under pressure changes. In *Problems Rock Pressure* 9, 120–151.

89. Alexeev A.D, Starikov G.P., Maluga M.G. et al. 1988. *Treatment of Outburst-Prone Seams by Aqueous Solutions (SAS).* Kiev: Tehhnika.

90. Ivanov B.M., Feit G.N., and Yanovskaya M.F. 1979. *Mechanical and Physico-Chemical Properties of Coals in Outburst-Prone Seams.* Moscow: Nauka.

91. Starikov G.P. 2001. Peculiarities of deformation and destruction of coals under volumetric compression. In *Geotechnologies at the Turn of XXI Century*. Donetsk: DonNTU, pp. 81–87.
92. Alexeev A.D., Ilyushenko V.G., and Starikov G.P. 1993. A method of decreasing of outburst intensity under developing workings. *Fiz, tekhn. vysok. davl.* 3, 38–44.
93. Alexeev A.D., Nedodaev N.V., and Starikov G.P. 1980. Destruction of volumetrically compressed gas-saturated coal at unloading. Preprint. Moscow: IPM SSSR.
94. Alexeev A.D. and Starikov G.P. 1981. Modeling of the processes of coal and gas outburst. *Dokl. akad. nauk. SSR* 259, 1072–1075.
95. Starikov G.P. 2004. Conditions of development of coal and gas outbursts and methods of prediction of outburst-proneness of coal beds. Coals. *Fiz. tekhn. probl. gorn. proiz.* 7, 214–223.
96. Feit G.N., Gaiko E.I., and Starikov G.P. 1992. Determination of outburst-proneness categories of coal beds by a complex of physical characteristics. *Sci. Rep. Skochinskii Min. Inst.* 2, 92–94.
97. Alexeev A.D., Feit G.N., Starikov G.N. et al. 1993. Method of determination of outburst-proneness of coal beds. Author Certificate 1788286.
98. Starikov G.P., Kuzyara V.I., and Kuzmenko I.V. 1994. Prediction and control of gas-dynamic state of rocks. *Fiz. tekhn. vysok. davl.* 4, 95–97.
99. Feit G.N. Investigation of the influence of rock pressure control method on outburst-proneness of the bottomhole zone of a coal bed. *Sci. Rep. Min. Inst. Skochinskii* 47, 12–18.
100. *Safety Instruction for Mining Works on Beds Prone to Coal, Rock and Gas Outbursts.* 1989. Moscow.
101. Starikov G.P., Alexeev A.D., Voloshina N.I. et al. 2003. Prediction of epicenters of coal and gas outbursts at the face of a stratal development opening. Branch Standard of Ukraine 101.24647077.001.
102. Alexeev A.D., Serebrova N.N., Ulyanova E.V. et al. 1982. C^{13} NMR high resolution spectra of fossil coals. *Dokl. akad. nauk USSR* 3, 3–5.
103. Vasilenko T.A., Starikov G.P., Zavrazhin V.V. et al. 2004. ^{1}H NMR study of methane sorption and desorption kinetics in fossil coals. In *Int. Symp. NMR in Condensed Matter: NMR in Heterogeneous Systems*, St. Petersburg, p. 130.
104. Starikov G.P., Vasilenko T.A., Voloshina N.I. et al. 2003. Activation energy of diffusion process as a characteristic of geomechanical state of coal seams. *Fiz, tekhn. vysok. davl.* 13, 104–113.
105. *Methodological guidelines on statistical analysis and processing of the results of investigation of rock pressure manifestations.* 1976. Leningrad: Ministry of Coal Industry of USSR. VNIMI.
106. Alexeev A.D. and Nedodaev N.V. 1982. *Limiting State of Rocks.* Kiev: Naukova Dumka, pp. 198–200.
107. *Temporary Manual of Safe Execution of Mining Works on Coal, Rock and Gas Outburst-Prone Beds.* 1983. Moscow.
108. Nadai A. 1954. *The Strength and Fracture of Solids.* Moscow: House of Foreign Literature.
109. Alexeev A.D., Revva V.N., and Ryazantsev N.A. 1989. *Fracture of Rocks in a Volumetric Field of Compressing Stresses.* Kiev: Naukova Dumka.
110. Alexeev A.D. and Nedodaev N.V. 1980. Method of determination of outburst proneness of coal beds. *Izobreteniya.* A.C. USSR 756034.

111. Zborschik M.P., Osokin V.V., and Sokolov N.M. 1984. *Prevention of Gas-Dynamic Phenomena in Coal Mines.* Kiev: Tekhnika.
112. Khodot V.V. 1979. Development of the theory of sudden outbursts and improvement of methods of their control. *Coal of Ukraine.* 4, 26–30.
113. Zorin A.N., Kolesnikov V.G., Sofiiskii K.K. et al. 1979. *Mechanics and Physics of Dynamic Phenomena in Mines.* Kiev: Naukova Dumka.
114. Nikolin V.I. 1978. A concept (hypothesis) on the nature of coal, rock and gas outbursts. In *Fundamentals of the Theory of Sudden Rock and Gas Outbursts.* Moscow: Nedra, pp. 122–140.
115. Bolshinskii M.I. 1978. Nature and mechanisms of gas-dynamic phenomena in mines and principles of their prevention. *Coal of Ukraine.* 9, 11–14.
116. Nikolin V.I., Lysikova B.A., and Tkach V.Y. 1972. *Prediction of Outburst Proneness of Coal and Rock Beds.* Donetsk: Donbass.
117. Rice, J. 1982. Mechanics of earthquake sources. In *Mechanics: News of Foreign Science.* Moscow: Mir.
118. Arshava V.G., Osipov S.I. Kucheba P.K. et al. 1979. *Elastic Properties of Rocks and Safety of Underground Operations.* Kiev: Tekhnika.
119. Glushko V.G. and Zorin A.N. 1972. *Rock Outbursts in Deep Mine Tunnels of Donbass.* Kiev: Naukova Dumka.
120. Zabigailo V.E. and Belyi I.S. 1981. *Geological Factors of Core Destruction in Drilling Stressed Rocks of Donbass.* Kiev: Naukova Dumka
121. Obert L. and Stechenson D.E. 1966. Stress condition under which ore dressing occurs. *Trans. Sos. Min. End. AIME* 235, 227–233.
122. Baiterekov A.B., Kolesnichenko E.V., and Leonov M.Y. 1981. Compression of a body with a semi-infinite plane. *Dokl. AN SSSR* 256, 1077–1081.
123. Barkovskii V.M. and Isaev A.V. 1980. On methodology of estimation of the stressed state of rocks by core disking. In *Diagnostics of the Stressed State of Rock Massifs.* Novosibirsk: IGD SO AN SSSR, pp. 56–67.
124. Brainer G. 1977. Determination of stresses in rocks of Ruhr's carbon. *Glukauf.* 9, 18–19.
125. Zorin A.N., Zabigailo V.E., and Mossur E.A. 1971. Prediction of outburst-proneness of sandstones by division of cores into disks. *Tekhn. bezop. okhrana truda gornos.* 7, 15–17.
126. Zabigailo V.E., Shirokov A.Z. Belyi I.S. et al. 1974. *Geological Factors of Outburst Proneness of Donbass Rocks.* Kiev: Naukova Dumka.
127. Rzhevskii V.V. and Novik G.Y. 1978. *Fundamentals of Rock Physics.* Moscow: Nedra.
128. Poturaev V.I., Zorin A.H., Zabigailo V.E. et al. 1986. *Prediction and Prevention of Rock and Gas Outbursts.* Kiev: Naukova Dumka.
129. Petrosyan A.E. and Ivanov B.M. 1978. *Causes of Coal and Gas Outbursts.* Moscow: Nedra, pp. 3–61.
130. Safe execution of mining works on beds prone to gas-dynamic phenomena. 2004. Official normative act on labor protection 1.1.30-1.XX.04. Kiev.
131. Khristianovich S.A. and Salganik R.L. Situations with dangerous of outburst. In *Rock Fragmentation: Wave of Outburst.* Preprint 152. Moscow: Institute of Mechanics.
132. Chernov O.I. and Rozantsev B.S. 1977. *Development of Mine Takes with Outburst-Prone Seams.* Moscow: Nedra.
134. Alexeev A.D. and Surgai N.S. 1985. Methods of magnetic resonance in coal industry. *Vest. AN Ukr.* 4, 33–43.

135. *Methodological guidelines on treatment of outburst-prone coal beds located in zones of geological deteriorations and elevated rock pressure with liquid solutions of SAS.* 1990. Donetsk.
136. Alexeev A.D., Melus A.N., Serebrova N.N. et al. 1986. A method of determination of residual gas content of black coals. Author Certificate SSSR 1213403.
137. Vasilev V.V. 1986. *Polymer Compounds in Mining Engineering.* Moscow: Nauka.
138. Grosberg A.Y. and Khokhlov R.V. 1989. *Statistical Physics of Macromolecules.* Moscow: Nauka.
139. Deryagin A.V., Churaev N.V., and Zorin Z.M. 1982. Structure and properties of boundary layers of water. *Izvest. AN SSSR* 1689–1710.
140. Orobchenko V.I., Tretinnik B.Y., and Kruglitskii N.N. 1981. Regulation of physicomechanical properties of outburst-prone coals for the purpose of relief of stressedly-deformed state of rock massive. In *Physicochemical Mechanics and Lyophilic Properties of Disperse Systems.* Kiev: Nauka Dumka, pp. 95–99.
141. Khristianovich S.A. and Salganik R.L. 1980. Situations with danger of outburst. In *Rock Fragmentation: Wave of Outburst.* Preprint 152. Moscow: Institute of Problems of Mechanics.
142. Alexeev A.D., Serebrova N.N., Sinolitskii V.V. et al. 1981. Magnetic resonance in fossil coals saturated by methane. *Fiz. tekhn. probl. razrab.* 5, 99–105.
143. Karkashadze G.G., Alexeev A.D., Starikov G.P. et al. 2009. Improvement of calculation procedure for mining face load with the account for methane pressure in a coal bed. *Gorn. Z.* 4, 47–50.
144. *Guidelines for Designing Coal Mine Ventilation.* 1994. Kiev: Osnova.
145. Puchcov L.A., Slastunov S.V., Karkashadze G.G. et al. Justification of optimum permitted exploration of working coal face related to coal seam gas content. *Symposium on Coal Methane,* Moscow State Mining University.
146. Karkashadze G.G. 2007. Procedure of calculation of methane debit from the zone of hydro-treatment on unloaded coal bed. In *Methane: Collection of Papers from Week of the Miner. Gorn. informats. analitic. bull.* 0B23, pp. 83–95.
147. Slastunov S.V., Karkashadze G.G., and Kolikov K.S. 2009. Calculation procedure of the allowable load on mining face by the gas factor. In *Collection of Papers from Week of the Miner.* Moscow: MGGU, pp. 83–95.

Index

Printed and bound by CPI Group (UK) Ltd, Croydon, CR0 4YY

18/10/2024

01776266-0008